Perspectives on the History of Chemistry

Series Editor

Seth C. Rasmussen, Department of Chemistry and Biochemistry, North Dakota State University, Fargo, USA

Commonly described as the “central science”, chemistry and the chemical arts have an extremely long history that is deeply intertwined with a wide variety of other historical subjects. Perspectives on the History of Chemistry is a book series that presents historical subjects covering all aspects of chemistry, alchemy, and chemical technology.

Potential topics might include:

- An updated account or review of an important historical topic of broad interest
- Biographies of prominent scientists, alchemists, or chemical practitioners
- Translations and/or analysis of foundational works in the development of chemical thought

The series aims to provide volumes that advance the historical knowledge of chemistry and its practice, while also remaining accessible to both scientists and formal historians of science. Volumes should thus be of broad interest to the greater chemical community, while still retaining a high level of historical scholarship. All titles should be presented with the aim of reaching a wide audience consisting of scientists, chemists, chemist-historians, and science historians.

All titles in the book series will be peer reviewed. Titles will be published as both printed books and as eBooks. Both solicited and unsolicited manuscripts are considered for publication in this series.

Anthony S. Travis

The Global Synthetic Ammonia Industry

History and Developments

Springer

Anthony S. Travis
Edelstein Center for the History
and Philosophy of Science, Technology
and Medicine
The Hebrew University of Jerusalem
Jerusalem, Israel

ISSN 2662-4591 ISSN 2662-4605 (electronic)
Perspectives on the History of Chemistry
ISBN 978-3-031-87977-7 ISBN 978-3-031-87978-4 (eBook)
https://doi.org/10.1007/978-3-031-87978-4

This Springer imprint is published by the registered company Springer Nature Switzerland AG
The registered company address is: Gewerbestrasse 11, 6330 Cham, Switzerland

Preface

Which chemical process has garnered three Nobel Prizes, was critical to the German war effort after 1915, and today is essential to the diet of over one-third of the global population? The answer is the synthetic ammonia process. The synthetic ammonia industry produces the fixed nitrogen that through fertilizer is the most important nutrient in the world's food supply chain. This book tells its story through the development of alternative technologies, all based on the same chemical principals, and global expansion. It also ties in the histories of the early participants—scientists, technologists, and entrepreneurs—and their trials and tribulations, during the startup phase, the first three decades of the twentieth century. Considerable prominence in this formative period is given to the studies of the German physical chemist Fritz Haber (Nobel laureate 1918), the scale up of his 1909 method at the BASF corporation by Carl Bosch (joint Nobel laureate 1931), and, especially, the synthetic ammonia process of the Italian chemist Luigi Casale. It should be explained at the outset why there is such great emphasis on Casale and his firm Ammonia Casale SA. There are two reasons. First, and unlike the case of Haber and certain other pioneers in the industry, there is no adequate biography of Luigi Casale. In addition, the history of the Casale enterprise during its first decade has received but scant attention, and what has been written contains errors of fact. And there are important omissions. And second, Casale's innovative strategy from 1921 for marketing his invention launched a trend that, through catalyzed high-pressure manufacture of important commodities, was the first contributor towards a transformation in the international chemical industry. Casale's business plan was simple: to globally patent, promote, and license his new high-pressure synthetic ammonia process. This worked because in the 1920s the BASF process was not available for licensing. The secrets of its construction and operation were safeguarded within the walls of BASF's two factories. Casale's headstart and his proven technology enabled the Casale ammonia process to become the most widely dispersed and utilized process in the world, mostly following its adoption in Japan during 1923 and shortly after in France and elsewhere. Casale's process and its wide availability for the first time ensured that the plentiful supply of nitrogen products as fertilizers ceased to be subject to geographical (mainly South American) constraints and industrial (BASF) monopolies.

By the mid-twenties, other synthetic ammonia processes were available, including those based on the inventions of Giacomo Fauser in Italy, and Georges Claude in France. There were also contributions from the UK, the US, Japan, and Germany. The several alternatives, and rivals, are included not only to create a more complete picture, and thereby show due reverence to the many actors in this story, but also because their inclusion broadens the traditional historical focus on Fritz Haber and the Haber-Bosch process, and the first decades of the synthetic ammonia industry. This enables us to reflect on the evolution of the sector as well as on its history. Thus the US would become a global force in the industry following the adoption in 1928 of the Casale process at Du Pont, and the scale up of Allied Chemical's process.

The developments outlined above constitute the first part of this hybrid monograph. The second part conveniently serves as a sequel to my *Nitrogen Capture*, published by Springer in 2018. Here the story transitions away from individual scientists and entrepreneurs, and small teams of co-workers, towards industrial corporations, new feedstocks, the needs of the 1930s and beyond, and the increasing role of large engineering contractors. This includes political and economic influences, especially preparations for war. The synthetic ammonia industry is also presented within the extraordinary context of the evolving events that surrounded the global commercialization of high-pressure, catalyzed chemical technologies. Of special significance is the industry's role in the production of pure hydrogen.

While this episodic journey is a series of stories that intersect with historical and technological events, it is also in parts a geopolitical travelogue. There are, for example, the roles of the UK's Imperial Chemical Industries (ICI) and large American engineering contractors, especially M.W. Kellogg, in the decade and a half following World War II; the enormous markets created by the Green Revolution from the mid-1960s as geographical boundaries of the ammonia industry expanded, especially towards Asia and the Middle East; America's changing trade relations with China and the Soviet Union in the 1970s; and the energy crises and environmental concerns, which are among our foremost concerns in the twenty-first century. The journey deals with the inventions and innovations which, one way or another, are relevant to an understanding of the industry's place in the world today. This includes brief descriptions of the mega nitrogen projects that have been incorporated into national infrastructure programs. The concluding chapter includes improvements in converter internals by a novel process change introduced by Ammonia Casale, the discovery of the mechanism of the ammonia catalyst by Gerhard Ertl (Nobel laureate 2007), and the importance of sustainable methods for producing hydrogen.

Apart from these developments, and in many ways because of them, it is important to emphasize that synthetic ammonia, through its derived products, contributes to the bumper crops that many regions of the globe have enjoyed since the 1960s. Synthetic ammonia is a story of vast proportions.

In sum, this book embodies the history of a technology that serves as a window into the rapid expansion of industrial chemistry on an international level. It is a shared history with a purpose and includes the technical detail required to enable an understanding of major challenges, and how they were overcome, in the development of the modern ammonia and nitrogen fertilizer industry. Readers interested in a more technical discussion are invited to refer to the monographs and articles on chemical technology found in the references. Most chapters are, as far as is possible, freestanding in an attempt to cater to a wide range of interests covering various epochs in modern history.

Sources

Archival Sources

Archive of Casale Société Anonyme

Du Pont archive, Manuscripts and Archives Department, Hagley Museum and Library, Wilmington, Delaware, Accession 500, series II, part 1.

Papers of Herbert Alfred Humphrey, Imperial College Archives and Corporate Records Unit, London.

Solvay Process Company archives (microfilm), the Sidney M. Edelstein Library, National Library of Israel, Jerusalem.

M. W. Kellogg documents located in Box 219, folder "Pullman-Kellogg (1)" of the U.S.-China Business Council Records, at the Gerald R. Ford Presidential Library. https://www.fordlibrarymuseum.gov/library/document/0029/7490136.

Luigi Casale documents, photographs, and papers (private)

René Leprestre documents (private)

Journals

Apart from the historical and technical literature, extensive use has been made of articles in trade journals as sources of information. They provide details of technical developments often several years before they entered into textbooks. Source material includes articles published in: *Chemical Engineering Progress*, *Chemical and Industrial Engineering*, *Chemical Industries*, *Chemical and Metallurgical Engineering* (from August 1946 *Chemical Engineering*), *Chemical and Process Engineering*, *Chemistry and Industry*, *Hydrocarbon Processing*, and its predecessor *Petroleum Refiner*, *Industrial and Engineering Chemistry*, and *Journal of the Society of Chemical Industry*.

Jerusalem, Israel — Anthony S. Travis

Acknowledgements

Among the many individuals and organizations whose unflagging and enthusiastic help enabled the writing of this story, I wish to especially thank Casale SA chemical engineer Luca Bianchi for tremendous assistance in providing material from the company's archive and elsewhere, frequent discussions concerning the history of the firm, and great efforts in undertaking translations from the Italian of old and sometimes fading handwritten documents and items of correspondence. His enthusiastic cooperation has made possible this first full account of Luigi Casale's venture into synthetic ammonia. This however would have been incomplete without the contributions of Lorenzo Francisci, who so generously shared with me his own studies into the early years of Casale's research. Critical to the more recent history of Casale's enterprise is the firsthand account of Giuseppe Guarino, former CEO at Casale SA, on the application of innovative axial-radial technology in the retrofitting of ammonia converters. Thanks also to: Francesco Baratto, Vittorio Cariati, Ermanno Filippi, Raffaele Ostuni, Sergio Panza, Fabio Sassi, and Elio Strepparola, all of Casale SA, for access to their writings; Ettore Comandini for sharing his reminiscences of Ammonia Casale going back to the 1950s; Elizabeth (Leprestre Cornell) Burrows for shedding light on René Leprestre, cofounder with Luigi Casale of Ammonia Casale; Gianni Bovini for access to his unpublished biography of Luigi Casale; Francesco Casale, grandson of Luigi Casale, for so generously sharing his collection of Luigi Casale's correspondence, documents, and family photographs, and for permission to reproduce items from his collection; Zehava Cohen, Rebecca R. Falkoff, Judith Goldstein, Kenneth J. Knoespel, Ernst Homburg, Michael Jewess, Erik Langlinay, Nicholas Lera, Michael Llopart, Chaya Meier-Herr, Peter J. T. Morris, Peter N. Reed, Jeffrey Seeman, Yona Siderer, Sara Tropper, Robin C. Travis, Yael Travis Lumer, the Sidney M. Edelstein Library at the National Library of Israel; the Harman Science Library, The Hebrew University of Jerusalem; Hagley Museum and Library, Wilmington, Delaware; the Othmer Library, Science History Institute, Philadelphia; the British Library, London; Imperial College Archives and Corporate Records Unit, London; and the Wellcome Collection, London. Thanks also to *Hydrocarbon Processing* for permission to reproduce figures that originally appeared in *Petroleum Refiner*. I thank the Asahi

Kasei Group for permission to reproduce a photograph of the Nobeoka ammonia factory under construction, and a group photograph, also at Nobeoka, both in 1923, and a photograph of the preserved Casale converter and Pignone compressor at that site; Barrett Industries: A Colas Company, for permission to reproduce a Barrett Division advertisement; BASF Corporate History for permission to reproduce images from its collection; Bechtel Corporation for permission to reproduce one of its advertisements; Wood Group plc for permission to reproduce Foster Wheeler advertisements; and Archiv der Max-Planck-Gesellschaft Berlin-Dahlem, The Museum of Modern and Contemporary Art of Trento and Rovereto, and Priska Pasquer Cologne for permission to publish images from their collections.

I am grateful to Casale SA for supporting the first part of this project. It started out as the story of how Luigi Casale made an entry into history with his synthetic ammonia process and his startup Ammonia Casale SA. Today rebranded as Casale SA, it is a diversified firm with a global presence. Moreover, it boasts a longer continuous existence than any other enterprise delivering the cutting-edge technologies that produce fixed nitrogen.

Finally, thanks to Seth C. Rasmussen, editor of the Springer *Perspectives on the History of Chemistry* book series, for suggestions for improvement; Springer chemistry editor Angeliki Athanasopoulou, for support and assistance; and Banu Dhayalan, also with Springer, for guiding the process through to final production.

Structure of the Book

Chapter 1 sketches the essential background on the early nitrogen, fertilizer, and related industries. It is important to step back in order to acquaint ourselves with the emergence of multifaceted strands of relevant industrial chemistry, as well as some key players, both individuals and firms, and communities of experts. This acquaintance includes recognition of the importance of the elements of life—nitrogen, phosphorus, and potassium—in agriculture, and their natural and other sources. Chapter 1 also includes the early uses of electricity in electrolytic separations, including for various industrial products, and in electrothermal processes for fixation of atmospheric nitrogen. It also briefly considers the global markets and supply chains, the main early source of electrical energy, and other advances relevant to chemical industry. Chapter 2 takes us through the early 1900s, to the 1909 invention of a method for producing synthetic ammonia by Fritz Haber. It enabled continuous production of ammonia from its elements at a high pressure in the presence of a catalyst. The method was taken up by BASF of Ludwigshafen, in Germany. Its application was not easy. This involved finding solutions to a number of engineering challenges. In 1913, BASF started the production of ammonia for refrigeration and the manufacture of the important fertilizer ammonium sulfate. This almost came to a stop after September 1914.

Following the outbreak of war in Europe, the previously global interdependence in nitrogen fertilizers and wheat and other food crops that had developed prior to 1914 collapsed. Supply chains became increasingly fragile. Embargoes, the war at sea with the resultant loss of merchant shipping, and land barriers created by military campaigns, contributed to major shortages. This included the supply of wheat in Italy, which had previously relied on Russia and Romania, of imported ammonium sulfate in Japan, and of South American Chilean nitrate fertilizer in Germany. Wartime measures in response to the shortage of nitrogen fertilizer included conservation and plowing of greater acreages of land for wheat crops. This was especially the case in the UK, which until 1914 had relied on inexpensive wheat imports, all free of duty. The UK increased production of ammonium sulfate fertilizer at coal tar and coke works in order to make up for shortages of nitrate fertilizer. Disruptions took place in the supply chains of all fertilizers. Moreover, the Chilean nitrate was also essential to the production of explosive materials. In Germany, now denied access to nitrate, BASF's Haber-Bosch process, as it became known, was consigned to munitions production, which included oxidation of ammonia to nitric acid.

These were the main factors that contributed to the urgent move after 1918 towards self-sufficiency in nitrogen products for both fertilizers and munitions and provided the stimulus for expanding the often secret wartime research conducted by Allied nations into technically viable synthetic ammonia processes. By all accounts, the synthetic ammonia and nitrogen products were critical to the defense of industrial nations. Some of the material here follows my *Nitrogen Capture: The Growth of an International Industry (1900–1940)* (Cham: Springer, 2018).

This is the prelude to Chap. 3. After 1918, the Haber-Bosch process, which BASF refused to license, stimulated a period of parallel technological developments, encouraged by considerations of national security in fertilizers and explosives, in alternative synthetic ammonia processes—using the same core high-pressure catalyzed reaction between nitrogen and hydrogen—especially in Europe. The front-runner was the remarkable process of Luigi Casale. He founded Ammonia Casale SA (Ammonia Casale Société Anonyme), in Lugano, Switzerland, jointly with René Leprestre, a man of many talents, but mainly of a remarkable business sense. Ammonia Casale was the first licensor of a rugged and reliable synthetic ammonia process. Luigi Casale's invention incorporated many advantages over the Haber-Bosch process, notably the use of a far higher pressure to give a greater yield and, directly, a liquid, anhydrous ammonia. This put Casale at a significant advantage to most other processes. His process was taken up in Japan and elsewhere. In mid-1923, plans for expansion included an American enterprise based on high-pressure chemistry and electricity. At the same time, there was a growing clash of personalities between Casale and Leprestre, in part involving family matters. This is exposed by correspondence between Luigi Casale and his wife, Maria, while Luigi was away on business. Here, the translations of correspondence in Italian tend to be free rather than literal, though keeping in mind the need to capture the spirit of the originals. Separately, through the French entrepreneur Hippolyte Bouchayer and his involvement with the Pechiney firm, the Casale process was licensed to users in Spain, France, Switzerland, and Belgium.

Chapter 4 opens with background on Noguchi Shitagau, the Japanese pioneer in industrial electrochemistry, including production of calcium cyanamide, and his interest in synthetic ammonia and especially Luigi Casale's process. Noguchi became Ammonia Casale's first major licensee. For his factory at Nobeoka the compressors and converters, as well as other machinery, were imported from Italy. However, it required Luigi Casale's skills to overcome the teething problems at what was now the first Japanese synthetic ammonia factory. The expansion of Noguchi Shitagau's enterprise, including in Japanese-occupied Korea, and alternative ammonia processes used in Japan, as described in *Nitrogen Capture*, are included here for sake of completeness. Then we move on to the French Claude process and rivalry with Casale Ammonia over contracts for the important French government works at Toulouse. Casale was the victor, but the gas production processes and the converters that were made in France created difficulties.

Chapter 5 is concerned with the falling out between Casale and Leprestre, and important business matters at Ammonia Casale, particularly the need to attract new investors in order to exploit the successes achieved in Japan and France. Hippolyte Bouchayer's role in introducing new business brought about his elevation to the presidency of Ammonia Casale. In 1925, Luigi Casale established Italy's first industrial research and development organization, dedicated to innovations in the area of high-pressure chemical processes. The growing reputation of Casale's enterprise attracted interest from the Soviet Union, and a British consortium. New, large converters were introduced. Luigi Casale drew up plans for future expansion. However, by late 1926, he was in poor health, the result of a debilitating accident that occurred while working on toxic substances during World War I. Luigi Casale died in February 1927. Ammonia Casale's future was ensured, but it would confront many challenges as the global ammonia industry continued to expand.

Chapter 6 provides a technical description of Casale's original ammonia process, and process improvements brought about by, for example, introducing an ejector into the synthesis loop.

In Chap. 7, the story of synthetic ammonia then moves on to summarize features that were common to all processes: pure gases, catalysts, compressors, and novel steels. Pure hydrogen was the main input, and the most costly item. The raw material for its production depended on available resources and feedstocks which in turn were determined by various geographical factors. As a result, hydrogen was generated by electrolytic, water gas, coking oven gas, and steam methane reforming processes. These developments took place in the 1920s and 1930s. The steam reforming and electrolytic processes are particularly relevant to the production of synthesis gas (syngas, carbon monoxide, and hydrogen) and hydrogen, respectively. The incorporation of novel steels into converters prevented the failures that had been experienced in the 1920s. Nevertheless, ammonia manufacture had to be conducted with care.

In Chap. 8, we review the fate of Ammonia Casale's American venture. While in the early 1920s, Japan and Europe embraced the Casale and other synthetic ammonia processes, this was not the case in the US, which showed little interest in large-scale high-pressure industrial chemistry and the use of synthetic ammonia in the manufacture of fertilizer. However, as already described in Chap. 3, in 1923, through his wheeling and dealing in New York, René Leprestre, Casale's partner, convinced the Electric Bond and Share Company to join with Ammonia Casale in establishing two businesses focused on high-pressure ammonia synthesis. From a commercial standpoint, the partnership fared badly. This enabled Du Pont to acquire US rights to Casale ammonia technology in order to overcome problems with its version of the rival Claude process. Du Pont integrated Casale technology into its Claude process and ordered new converters from Ammonia Casale. This was an important contribution to the growth of the US synthetic ammonia industry. Part of this account has been adapted from my article, "First Steps: Synthetic Ammonia in the United States," in *Substantia* 5, no. 2 (September 2021):55–77.

In Chap. 9, we move away from a narrative based largely on Ammonia Casale's microhistory, though without ignoring the firm's role in global nitrogen production, towards a wider perspective, with a focus on the US. It starts with preparations for war through the erection of ten state-sponsored synthetic ammonia factories geared towards munitions production. Significantly, six were based on natural gas (methane) for production of hydrogen using a steam reforming process developed by ICI in the UK. The post-war disposal of the government factories enabled growth of the ammonia industry and contributed to the US becoming the leading producer of ammonia. This growth was further encouraged by the perceived needs of the Korean War. By the 1960s, petrochemical firms had also taken up manufacture of synthetic ammonia.

Chapter 10 deals in the early part with application of the new ammonia technologies to the production of methanol and urea. Just as the synthesis of ammonia is hydrogenation of nitrogen, so the synthesis of methanol is the hydrogenation of carbon monoxide (using synthesis gas). Because the methods of manufacture were so similar, and so successful, with ready and growing markets for the methanol, it deserves a place in this account. There follows a brief account of war planning in Europe, in which high-pressure processes, as in the US, played vital roles. The doctrine of self-sufficiency as first taken up in the 1920s was intensified during the second half of the 1930s. Advances were made in production of synthesis gas from hydrocarbons by steam reforming and partial oxidation, in processes for purification of hydrogen-nitrogen mixtures, and in expanding the range of nitrogen fertilizers. After 1945, industrial reconstruction, the tremendous demand for fertilizers needed to restore wheat and other harvests and provide food security in devastated countries, and the growing international expansion of the ammonia industry, provided tremendous opportunities for European and US engineering contractors and chemical corporations.

From the 1950s, the synthetic ammonia industry would be transformed by multinational engineering contractors, often drawing on experiences and developments gained in wartime, including in styles of management. Chapter 11 describes how large, increasingly diversified, American contractors took over the design and construction of ammonia, urea, and nitrogen fertilizer plants. The story now unfolds in a very different era, attuned to new needs, political and economic. New markets appeared beyond the Western hemisphere, brought about by the necessity to confront growing famine and political changes, including the need to combat social unrest. Increasingly, self-sufficiency in synthetic ammonia took on a leading role in nation-state building. This was particularly the case in India. It was enabled from 1966 by innovations at the US engineering contractor M.W. Kellogg which led to a shift in the scale of production of individual ammonia units from 250 to 300 tons per day to 1000 tons per day. There were also changes in the developed economies. Older process were replaced, and in the US and Canada, in particular, the availability of natural gas as feedstock favored new sites of manufacture. American and Japanese engineering firms, often in collaboration with European contractors and chemical firms, in particular the UK's ICI, dominated the global industry, which had spread to the Middle East, Asia, and the Eastern Bloc.

Chapter 12 opens with the negative impact of the energy crises of the 1970s on Western firms. However, there were opportunities elsewhere, including in the Eastern Bloc, to where technology transfer had commenced in the 1960s mainly through British and Italian chemical corporations. During the 1970s ammonia became a leading diplomatic tool in American dealings with the People's Republic of China and the Soviet Union. Apart from the breaking down of political tensions, this provided important opportunities for American engineering firms during the years of the worst recession since 1929. Elsewhere self-sufficiency programs were encouraged, especially in countries with large resources of natural gas.

Chapter 13 describes process developments aimed at greater efficiency, including in gas purification and the synthesis step, and new ICI catalysts. Haldor Topsøe introduced a converter with a radial flow of gas (until the 1960s all converters operated with axial flow), and Braun a novel synthesis gas process. Despite the many improvements in the 1970s, certain of these innovations were not readily taken up. This was a consequence of the fourfold increase in the price of oil, followed a few years later by the collapse in the price. The resulting fluctuations in the price of energy, and as a consequence of ammonia, militated against new investments in the industry, especially in the US and Europe. ICI pulled out of ammonia manufacture. However, the industry continued to expand elsewhere, including in the Middle East.

The closing chapter, Chap. 14, deals with three topics that are relevant to the modern industry: the revamping of tired and inefficient converters; the study and improvement of catalysts; and sustainable production of hydrogen. Around 1980, many leading manufacturers of ammonia relied on aging and increasingly energy-inefficient converters to satisfy the demand for nitrogen fertilizers. This situation created an opportunity for the revival of Ammonia Casale, which, as a result of developments elsewhere, by the late 1970 had lost its innovative edge. It offered a service for revamping (retrofitting) the internals of existing converters. This included the modification of what had been almost exclusively axial flow ammonia converters to axial-radial flow. This not only saved the cost of total replacement but also enhanced yields. From 1986, in situ revamping, irrespective of the original process, was achieved on a large scale. Over the following two decades, continuous reinvestment propelled the Casale firm to a leading place in the field of ammonia synthesis. It also became involved in revamping of methanol converters, and diversified into related technologies, especially melamine and nitric acid. As for catalysts, notwithstanding great efforts to understand the action of the iron-based catalyst long used in the ammonia synthesis, little was achieved until the 1960s. Then academic work undertaken in surface physics provided the mechanism for take up of nitrogen at the surface of the iron catalyst. This work gained Gerhard Ertl the 2007 Nobel Prize in Chemistry. There have been other important developments, including in China, where academic workers discovered an important new catalyst which has successfully entered the market. M.W. Kellogg developed a ruthenium catalyst. Finally, there is production of hydrogen, the most costly feature of the manufacture of ammonia, which is so relevant to the rapidly developing hydrogen economy. It was the synthetic ammonia industry that first developed the means for large-scale production of pure hydrogen using a variety of feedstocks and techniques. Presently the most efficient and widely used process is steam reforming of natural gas, but it leads to release of carbon dioxide. Capture of the carbon dioxide is an effective way to slow climate change, and this has already been achieved, but is expensive. However, in the long-term methods for more sustainable production of what is known as green hydrogen, for green ammonia, will become necessary. Engineering contractors closely connected with the ammonia industry are at the forefront of research into viable methods that make use of solar, wind, and hydropower for advanced electrolyzers that generate pure hydrogen.

A Note on Units, Abbreviations, and Conventions

Weights

Note that many historical sources and much past trade literature used the term 'tons' without qualification. Thus US and British trade journals, especially in advertisements, often referred to capacities in US (short) tons. From the 1970s, capacities were increasingly given in metric tons. From Chap. 8 on all tonnages are in US, or short, tons, unless otherwise indicated.

The long, or British, ton is a unit of mass equal to 2240 pounds, or 1000 kilograms.

The short, or US, ton is a unit of mass equal to 2000 pounds, abbreviated as st.

One long ton is the equivalent of 1.12 short tons.

The British ton is almost equal to the metric ton (US usage), 1000 kilograms, abbreviated as mt. 900 metric tons (mt) are equivalent to 992 US tons.

Tonnages per day/year are shown as: tpd/tpy; stpd/stpy; mtpd/mtpy.

Pressures

The atmosphere, abbreviated as atm, was used as a measure of pressure until 1 January 1978. Because this is a historical text, and the technical literature often used atm well into the 1990s, its use is retained when given in the original source. The SI unit for pressure is the pascal, Pa. For sake of convenience, 100 atm is taken as equal to 100 megaPascals, or Mpa. However, the bar is often used in technical literature to express atmospheric pressure. 1atm = 1.01325 bar. Likewise, for convenience 100 atm is taken as equal to 1000 bar.

One pound-force per square inch gauge (1 psig) is the equivalent of 0.068 atm.

Plant Capacity

The United States Geological Survey (USGS) Mineral Handbook gives annual metric capacities in tons per year calculated as "engineering capacity adjusted for 340 days per year of effective production capacity," rounded to three significant digits.

The terms "converter" and "reactor" both refer to the high-pressure vessel in which the synthetic ammonia reaction takes place. A rated production capacity train, sometimes referred to as a loop, or stream, refers to the continuous rig for bringing about synthesis of ammonia. Here each train is taken to be a complete synthesis loop (or synloop), a continuous process stream including the ammonia reaction, separation of ammonia, and recirculation of unreacted gases. Until the mid-1960s, the rated production capacity of individual trains was generally not more than 300-tons-per-day.

Conventions

The spellings of sulphur, sulphuric, and sulphate as found in the earlier technical and scientific literature are given as sulfur, sulfuric, and sulfate.

As an aid to identification, birth dates and death dates are given in Chaps. 3–8 for a number of mentioned individuals. Britain, UK, and England are used interchangeably, according to context, including historical usage.

About This Book

This monograph is the first comprehensive historical overview of the global synthetic ammonia industry, highlighting its critical role in fixing nitrogen from the air in order to produce the fertilizers that support a significant portion of the world's population. While the first part includes the storied Haber-Bosch process, introduced in 1913 by the German firm BASF, its main focus is on the largely forgotten Italian chemist Luigi Casale, the inventor of an alternative synthetic ammonia process. Casale had the distinction of launching the global synthetic ammonia industry through, and unlike BASF, the licensing of his technology. The narrative here incorporates the first biography of Casale, with descriptions of his innovations, and the early years of his company, Ammonia Casale SA. Technical challenges, relationships with collaborators and investors, and patent issues are discussed. The second part, through a macroperspective, explores the industry's international expansion, the evolution of novel technologies, and contributions to the Green Revolution. Included are the roles of engineering contractors, high-capacity manufacturing units, and geopolitical contexts. The final chapter covers the recent past and briefly looks to the future with moves towards sustainability within the context of green hydrogen and ammonia production. This story of one of the world's most important industries is recommended for historians of modern science, technology, and business, chemical engineers, practitioners in the petroleum and chemical industries, and readers interested in technological advances that have shaped our world today.

Contents

About the Author

Anthony S. Travis is with the Edelstein Center for the History and Philosophy of Science, Technology and Medicine, at The Hebrew University of Jerusalem. His research into the history of technology mainly concerns the chemical industry in the 19th and 20th centuries. Publications include *The Rainbow Makers: The Origins of the Synthetic Dyestuffs Industry in Western Europe* (1993), and *Nitrogen Capture: The Growth of an International Industry (1900–1940)* (2018). He is recipient of four awards, including from the history of chemistry division of the American Chemical Society and the Royal Society of Chemistry.

Agriculture: Fertilizers for Feeding Growing Populations

1

Abstract

The nineteenth century was a period of unprecedented economic development during which the emerging modern science of chemistry as applied to agriculture contributed to the ability to feed the growing populations of industrial nations. This was based on an understanding of the role of fertilizers—containing nitrogen, phosphorus, and potassium—as essential elements of life, and their availability from natural sources. In the case of the most important fertilizers, those containing nitrogen, concerns that the main natural source, the nitrates imported to Europe and the US from South America, would be spent by the early twentieth century presented chemists with a great challenge: the fixing, with the aid of electricity, of the unreactive nitrogen from the atmosphere. This followed developments in academic and private laboratories and in certain seemingly unrelated sectors of the chemical industry.

1.1 Introducing Nitrogen Products

About 80% of the earth's atmosphere is made up of the unreactive gas nitrogen. This element was discovered by the Scottish botanist David Rutherford in 1772. Nitrogen exists as molecules, each molecule consisting of two atoms tightly bound together by a triple chemical bond. It is essential to all forms of life and is the most important inorganic plant nutrient. Nitrogen can be taken in, or fixed, by leguminous plants, with the aid of bacteria. Following the emergence of farming communities, these once wild plants became important sources of food. Since the time of the Romans, their ability to assimilate nitrogen (even though the chemistry was not understood) has been exploited in crop systems, by, for example, rotating soybeans, which fix atmospheric nitrogen, with corn. However, bacterial nitrogen fixation, by its very nature, takes place on a restricted scale. Lightning also fixes

A. S. Travis, *The Global Synthetic Ammonia Industry*, Perspectives on the History of Chemistry, https://doi.org/10.1007/978-3-031-87978-4_1

atmospheric nitrogen, “but,” as the Chemico corporation observed in its advertisements, “scatters its benefits without consideration for those who need them.” Most agricultural crops need nitrogen as and when required. For centuries, soil quality was enhanced by the addition of nitrogen-containing animal and human excrement (both dung and urine), plant remains, fish, and general manures. In the nineteenth century, scientists gained a chemical understanding of the roles of nitrogen-based and other fertilizers sufficient to transform agricultural practices [1, 2]. The Industrial Revolution stimulated rapid population growth, and with it a growing demand for food crops, especially wheat.

The German chemist Justus Liebig was a pioneer in the field of promoting agricultural chemistry (Fig. 1.1). His “law of the minimum,” based on a principal developed by Carl Sprengel in 1828, stated that not only is crop growth dependent on the availability of specific nutrients, but whichever nutrient is the least available will determine the final crop yield. That the yield is directly proportionate to the amount of the least available nutrient was supported through the results of numerous field trials [3–5]. Liebig’s book, *Die organische Chemie in ihrer Anwendung auf Agrikultur und Physiologie*, published in 1840, set out his program for placing agriculture on a scientific basis.

Liebig met with British landowners and agriculturalists and encouraged them to take a more scientific interest in methods that maximized crop yields. The British scientist Lyon Playfair translated Liebig’s 1840 book into English; it went through

Fig. 1.1 Justus Liebig’s laboratory at Giessen, around 1840. Liebig was ennobled in 1845. Lyon Playfair was a student of Liebig during 1840–1841. (From Hofmann JP (1842) Acht Tafeln zur Beschreibung des chemischen Laboratoriums zu Giessen. Heidelberg. Lithograph by Trautschold and Ritgen). (Edelstein Library, National Library of Israel. With permission)

several editions, including updated analyses of manures. The agricultural chemist John Bennett Lawes, jointly with Liebig's former student John Henry Gilbert, conducted field experiments with fertilizers on Lawes' estate at Rothamsted, Harpenden, 35 km north of London. This was the first modern agricultural research, or experiment, station.

It was mainly through Liebig that nitrogen was recognized as one of the three essential inorganic elements necessary for crop growth. The other two were phosphorus and potassium. As a result, in order to enhance the level of plant nutrients, and raise agricultural production, new types of fertilizers were introduced. First was the phosphorus-containing superphosphate of lime, as developed by the Irish physician Sir James Murray, who treated crushed bones with sulfuric acid. In 1846, Lawes acquired Murray's business, filed a patent for a superphosphate, and took up manufacture in London. Soon after, phosphate rock became the main raw material, notably in the UK, Denmark, and Germany. During the final two decades of the nineteenth century, superphosphate manufacture was commenced globally.

Because sulfuric acid was essential to the production of superphosphate and other major fertilizers, its manufacture, by the lead chamber process, was taken up in all industrial countries. In Japan, an English engineer built a sulfuric acid factory at Osaka in 1872, just four years into the Meiji Restoration. This enabled the production of superphosphate from 1888. In Italy, in the same year, the Montecatini Mining Company was founded to manufacture sulfuric acid for the superphosphate industry. In the years 1913 and 1920, the company absorbed two other leading Italian manufacturers of fertilizer; together, they were rebranded as the Montecatini Mining and Agriculture Company (Montecatini).

Chemical processing was also necessary for the production of potash, a potassium-containing fertilizer. Until the mid-nineteenth century, potash for fertilizer, and glass and soap making, was made from wood ash (predominantly potassium carbonate). A new source as fertilizer was discovered around 1860 by the German chemist Adolph Frank, who was employed at a beet sugar processing factory not far from Stassfurt, in north-central Saxony-Anhalt. After examining waste salts from the local salt (sodium chloride) industry, Frank devised a method for isolating potash (potassium chloride) from the waste and found that his product was an effective fertilizer. In October 1861, Frank opened the first potash factory based on his discovery, in Stassfurt [6]. Later he joined with other local manufacturers and almost singlehandedly introduced what was called "Stassfurt potash" to American farmers, who from the early 1870s became the product's main consumers. The success of the mineral potash eventually led to the collapse of the potash fertilizer trade based on wood ash [7].

The growing role of chemistry in agriculture encouraged the introduction of agricultural research stations, especially in Germany, where a station at Leipzig opened in 1851. Within four decades, there were over seventy research stations in the German Empire. The first permanent station in the US was the Connecticut Agricultural Experiment Station, established by legislative decree in 1875. According to the Yale-based agricultural chemist Samuel W. Johnson, its purpose was "to make a regular business of discovery for the use of farming." This station was followed by a similar one in California, opened in 1876. By the 1890s, there were also agricultural stations in Denmark, Holland, Italy, Japan, Norway, and Sweden.

At one stage, Liebig considered that the ammonia present in rainwater supplied sufficient nitrogen to enable commercial crops to thrive. Lawes and Gilbert, and other investigators, found this to be incorrect. They established that addition to the soil of nitrogen-containing fertilizers doubled the yield of wheat. However, unlike phosphorus and potassium, it was not possible to fix the atmospheric nitrogen as a fertilizer by the available chemical methods. Fortunately, the search for a useful and abundant nitrogen fertilizer was satisfied by a material found in the western part of South America, mainly on the three Chincha Islands off the coast of Peru. This was guano, the accumulation over thousands of years of vast amounts of bird droppings. It once supported the agriculture of the great Inca civilization. Guano was first exported to England in 1841. The fact that it also contained phosphate led to rapid uptake by farmers. By 1850, Europe was importing more than 200,000 tons of guano a year. Through a contract with the government of Peru, the British firm of Antony Gibbs & Sons monopolized much of the trade in guano. In 1854, the Irishman William Russell Grace arrived in Peru from where he exported guano. His enterprise was the predecessor of W.R. Grace & Co., a diversified company incorporated in Connecticut in 1899 [8, 9]. In August 1856, the US Congress passed the Guano Islands Act, empowering US citizens to take possession of unoccupied islands containing guano. Americans then set about shipping guano from islands in the Caribbean and Pacific to North America. In December 1863, a group of farmers in the Riverhead Region of New York formed the Riverhead Farm Agricultural Society, a cooperative that was incorporated in 1866, and through its agent shipped Peruvian guano from New York to a local wharf. This was the first of the large US cooperatives that were created to provide pooling, purchasing, and delivery services for farmers.

The growing economic importance of guano led to the Chincha Islands War (1864–1866), in which Spain attempted, but failed, to reassert influence over its former South American colonies. Guano production peaked in 1870, in which year 280,000 tons were sent on the three-month sea voyage to Europe.

Within a decade the supply of guano suitable for use as fertilizer became depleted. It was replaced as a source of nitrogen by the mineral caliche, also found in South America. Caliche is rich in sodium nitrate, which was also important for the manufacture of nitric acid. The acid was essential to the production of organic nitro compounds, required for both synthetic dyestuffs and modern explosives. The sodium nitrate was employed in the manufacture of blasting powder. The nitrate was also reacted with potassium chloride, available from the Stassfurt deposits, to form potassium nitrate, the main component of gunpowder. The use of caliche became particularly significant after Alfred Nobel's nitroglycerine and dynamite brought about dramatic changes in mining and civil engineering projects.

Caliche was extracted in the remote, rainless Atacama Desert, located west of the Andes Mountains and one of the world's most extreme environments. Until the 1870s, the extraction was carried out mainly in Bolivia, and to a lesser extent Peru. Chilean corporations invested heavily in the Bolivian nitrate deposits. The extracted nitrate was exported from the port city of Antofagasta. However, in February 1878, in order to gain greater income from the industry, Bolivia imposed tariffs on the export of nitrate, leading to tensions with Chile. In April 1879, the War of the Pacific broke out, with Bolivia allied with Peru, fighting against Chile. The struggle ended with Chile as victor. Under the Treaty of Ancon of October 1883, and a truce in 1884, Bolivia ceded its coastal territories, including Antofagasta, to Chile, which also acquired the nitrate deposits in the Atacama Desert and the southern province of Tarapacá. Chile became the principal world supplier of what was now called Chilean saltpeter (though the name Chili-salpetre was in use long before) [10]. An extensive narrow gauge railway network carried nitrate to the Pacific coast. By the early 1900s, Chilean saltpeter made up two-thirds of the total supply of fixed nitrogen consumed by industrial nations. Quality control was important. Nitrate was analyzed by qualified chemists, in Chile, and also in Europe, sometimes for trace amounts of contaminants in the material used in manufacture of explosives [11].

Another important source of nitrogen fertilizer was ammonium sulfate, made from the 1830s by reacting sulfuric acid with the ammonia generated during destructive distillation of coal at coal gas works, and later also at coking works (Fig. 1.2). In Europe alone, 140,000 tons of ammonium sulfate were produced annually. For the UK, the sulfate was a leading export commodity, and after 1900 sent to Japan, where it was preferred to Chilean nitrate, which was readily washed out of soil to form toxic nitrite. The British Ammonium Sulphate Federation, established in the 1890s, played an important role in coordination of the export market. In France, the importance of ammonium sulfate was such that its supply was controlled by the state. Until the 1940s, ammonium sulfate was the world's main nitrogen fertilizer.

The manufacturing equipment for most of the processes discussed so far was built in-house, or supplied by local firms. Typically, chemical factories used wooden barrels and iron mixing pans. Coal was the main source of heat and power, and reactions were carried out in batches. The coal gas supply industry provided piping and valves. In the medium-sized and larger factories, steam engines drove pulleys that turned batteries of stirrers. Changes began to arrive in the 1890s, with, for example, the multi-tube reactor designed by Rudolf Kneitsch at the German dye-making firm BASF (Badische Anilin & Soda-Fabrik), of Ludwigshafen, for use in the manufacture of sulfuric acid by the catalytic oxidation of sulfur dioxide. This important example of a heterogeneous catalytic reaction, in which the catalyst is in a different phase from the reactants, eventually replaced the lead chamber process for sulfuric acid. At the same time, gas and electrical power for driving machinery and electrically heated ovens made inroads into the chemical industry. Electrically heated ovens and furnaces not only led to changes in production methods but enabled the large-scale manufacture of certain chemicals by methods that were previously possible only on the laboratory bench. Within a short period, with the aid of electricity for both power and lighting, shop floors in factories around the world were transformed.

Fig. 1.2 Production of coal gas by destructive distillation of coal. (From Accum F (1819) Description of the process of manufacturing coal gas. For the lighting of streets, houses, and public buildings, with elevations, sections, and plans of the most improved sorts of apparatus. Now employed at a gas works in London. London). (Edelstein Library, National Library of Israel. With permission)

1.2 Electrochemical and Electrothermal Processes, and "Electric Nitrates"

The commercial development of early nitrogen fixation processes for agriculture followed a century of progress in the application of electricity to a variety of chemical reactions. This included methods, processes, and products that were not originally associated with nitrogen fixation. In 1789, Jan Rudolph Deiman and Adriaan Paets van Troostwijk in Holland brought about the splitting of water into hydrogen and oxygen by electrolysis using an electrostatic generator as the source of direct current. Interest in electrochemical methods was stimulated by the studies into electricity of the Italian scientist Alessandro Volta, notably his voltaic pile, the first electrical battery (1799); the creation of an electric arc by Humphry Davy; and the development of the laws of electrolysis by Michael Faraday. Batteries enabled electrolysis and the isolation of certain elements (Fig. 1.3). In 1807, Davy described how he had fixed dissolved atmospheric nitrogen as ammonia at the anode during electrolysis of water [12, 13].

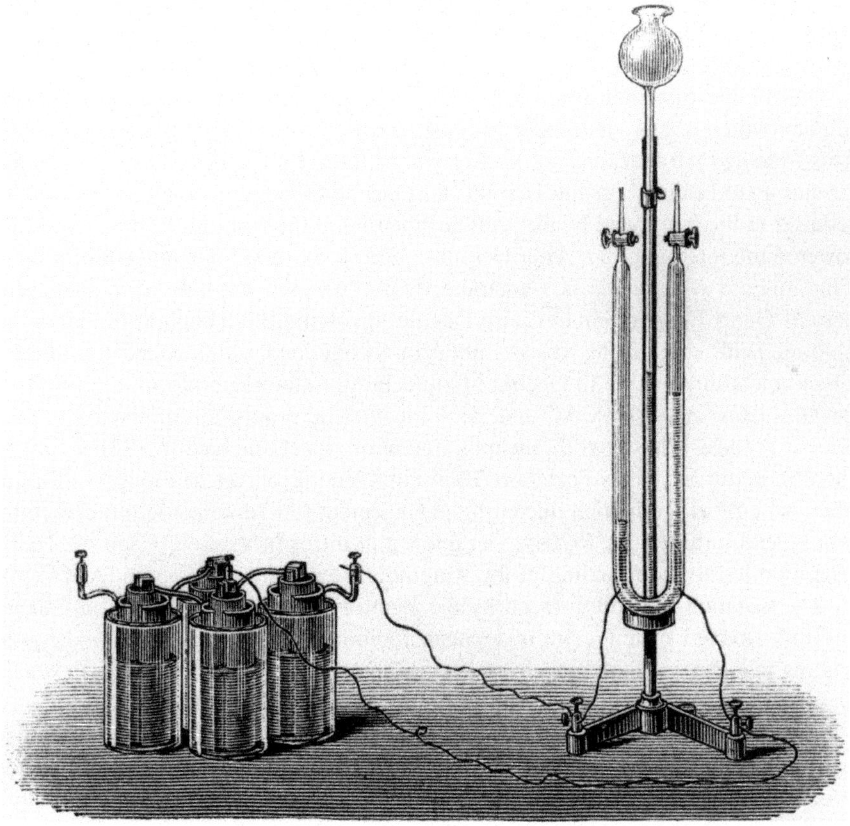

Fig. 1.3 Apparatus for demonstration of the electrolysis of water into hydrogen and oxygen, c. 1870. (Author's collection)

Chemical manufacture by electrolytic processes and with high-temperature electric ovens and furnaces required a continuous supply of low-cost electricity, which became available only late in the nineteenth century, following the construction of hydroelectric power stations. The natural energy available from fast-flowing rivers, waterfalls, lakes at high elevation and dams determined new geographical sites of chemical production, as was particularly the case in Italy, Scandinavia, Japan, and the Niagara Falls region in North America. For high heads, the water flow, carried down a cliff through almost vertical pressure pipes (penstocks), impinged on a series of vanes on the periphery of a large wheel, the rotor of a turbine, which, as employed at Niagara Falls, was coupled by revolving vertical shafts to an electrical generator. In later designs specially shaped buckets, arranged in a similar way on a water wheel (the Pelton wheel, or impulse turbine), and fixed on a horizontal rotating shaft, were connected to the rotor of the generator. Hydraulic flow was controlled by governors or flywheels fixed to the side of the generator. Because long-distance transmission of electrical current was difficult and wasteful until the early 1900s, chemical factories using this major source of energy were located close to the sites of power generation.

One of the first industrial electrolytic processes was the manufacture of aluminum, which was produced at a low cost, enabling its use in inexpensive products. This followed the independent discoveries, in 1886, by Charles Martin Hall in the US and Paul Louis-Toussaint Héroult in France, that the mineral alumina could be reduced to the free metal by electrolytic smelting in the presence of a material that lowered the melting point. After Héroult failed to interest Compagnie des Produits Chimiques d'Alais et de la Camargue (PCAC), which used the aluminum process of Henri Etienne Sainte-Claire Deville, involving the reduction of aluminum chloride with sodium, he raised capital in Switzerland and founded the Société Electrométallurgique de Française. Manufacture of aluminum began in 1887. Two decades later, in 1907, PCAC undertook electrolytic production of aluminum by a similar process. These two French aluminum producers merged in 1921, to create the firm better known as Pechiney. Their combined resources and long histories in electrochemical production, including management of hydroelectric infrastructure, would lead, in the early 1920s, to an interest in nitrogen fixation (Chap. 3). Hall's invention led to the founding of the Aluminum Company of America (ALCOA).

The isolation of sodium metal by the electrolytic process of Hamilton Castner in 1888 enabled manufacture of a cheap sodium cyanide, important for the gold mining industry in extractions from low-grade ores. The cyanide process, invented in 1887, was first applied in South Africa, and later in Australia and the US. At the turn of the twentieth century, cyanides were investigated as nitrogen fertilizers.

In 1897, the Castner-Kellner Alkali Company in Cheshire, England, began using its electrolytic process to produce, from salt, the important commercial products alkali (caustic soda, sodium hydroxide), and chlorine, the latter for use in the manufacture of bleaching powder (chloride of lime). The Castner-Kellner process was generally known as the chlor-alkali process.

From the mid-1890s, the electrochemical manufacture of chemicals was taken up at Niagara Falls and elsewhere in the US. In 1896, the Mathieson Alkali Works, of Saltville, Virginia, which was associated with the Castner-Kellner Alkali Company, began using electrolytic cells in production of bleaching powder and in the following year manufactured the alkali and chlorine at Niagara Falls. Around the same time, the Dow Chemical Company (established in 1897), at Midland, Michigan, manufactured bleaching powder on a large scale by a similar electrolytic process. Electrolytic production of bleaching powder was spurred on by an import tariff on the product. Among the newcomers in the early 1900s engaged in electrochemistry at Niagara Falls was the Development and Funding Company (1905), predecessor at Niagara of the Hooker Electrochemical Company, which became one of the main producers of bleaching powder and alkali. By the early 1900s, there were five main hydroelectric power stations at Niagara Falls.

Of somewhat greater interest here is the fact that the chlor-alkali process gave hydrogen as a by-product. It was used for fat hardening (hydrogenation) in the early 1900s, and also for the manufacture of anhydrous hydrogen chloride. As we shall see, the production of the reactive gas on a vast scale, and in high purity, including by electrolysis, was stimulated by the growth of the synthetic ammonia industry.

1.2.1 Sir William Crookes on a Modern Crisis

Electric arcs replicating nature's fixation of nitrogen through lightning are of special interest to the early history of industrial fixation of atmospheric nitrogen for fertilizer production. The processes were introduced not long after the newly knighted Sir William Crookes in 1898, in his inaugural address as president of the British Association for the Advancement of Science, predicted widespread starvation unless nitrogen fixation was achieved [14]. His approach, bordering on alarm, was unlike that of the usual presentations at prestigious gatherings of scientists [15]:

> My chief subject is of interest to the whole world—to every race, to every human being … I mean the question of food supply … England and all civilized nations stand in deadly peril of not having enough to eat. As mouths multiply, food resources dwindle. Land is a limited quantity, and the land that will grow wheat is absolutely dependent on difficult and capricious natural phenomena … I hope to point a way out of the colossal dilemma. It is the chemist who must come to the rescue of the threatened communities. It is through the laboratory that starvation may ultimately be turned to plenty … The fixation of atmospheric nitrogen, therefore, is one of the great discoveries awaiting the genius of chemists … The fixation of nitrogen is vital to the progress of civilized humanity … Nitrate can be produced artificially by the combustion of the atmosphere … The artificial production of nitrate is clearly within view …

Crookes believed that "electric nitrates," produced by processes using electric arcs would solve the world nitrogen fertilizer problem and liberate the world from hunger.

Even if they did not know it, chemists had been preparing for Crookes' speech. Technical problems abounded until 1903, when Norwegian physicist Kristian Birkeland, at the University of Kristiania (today Oslo), with the active support of the engineer and entrepreneur Samuel Eyde, fixed nitrogen as nitric oxide using an electric arc. The oxide was converted into nitric acid which, combined with limestone (calcium carbonate), produced the fertilizer calcium nitrate, also known as lime nitrogen. Eyde attempted to gain backing from BASF, which was also interested in nitrogen fixation. After this failed, the firm Norsk Hydro was founded in December 1905, with mainly French capital [16]. The first factory was opened in Notodden. It was followed in 1911 by the large Rjukan facility, which operated with electricity generated at the Vemork power station. Rjukan was initially a joint enterprise involving BASF, but BASF soon abandoned interest in the electric arc process, as a result of progress in a synthetic ammonia process (see Chap. 2).

Electrically heated industrial furnaces that operated at high temperatures led to the development of a thermochemical process suited to nitrogen fixation. This was the reaction between calcium carbide and nitrogen to give calcium cyanamide. It followed development of processes for preparing carbides. In 1891, the American Edward Goodrich Acheson discovered silicon carbide by strongly heating in a furnace of his own design a mixture of coke (carbon) and quartz (silicon dioxide). From 1894 he manufactured the carbide at Niagara Falls, New York State, and sold it under the name Carborundum. In 1892, the Canadian-born inventor Thomas Leopold Willson, at Spray (now Eden), North Carolina, heated lime (calcium oxide) with tar in an electrically heated furnace in an attempt to produce metallic calcium. Instead, he obtained calcium carbide. His partner was another North American entrepreneur, James Turner Moorhead. They commenced manufacture of carbide in 1894, and later in the year sold patent rights for the process to the Electro-Gas Company [17]. Moorhead, however, purchased a manufacturing license. His interests were later sold to the Union Carbide Company, founded in 1898, and built around Electro-Gas. The Union Carbide factory was at Niagara Falls, New York State. Willson retained all Canadian rights to the carbide process. In 1895, he returned to Canada and in the following year constructed a manufacturing plant in Ontario, and later built carbide factories in Ottowa and at Shawinigan Falls, southern Quebec. The Shawinigan Carbide Company started production in 1901, with an annual production of 2000 tons of carbide from twelve batch furnaces (Fig. 1.4).

The carbide was reacted with water to produce the highly reactive gas acetylene. For a time, acetylene was an important source of lighting, including for railway carriages, lighthouses, bicycles, and miners' lamps, until it was replaced by electric lamps. Of interest here is the fact that the ready availability of carbide, especially as its use in lighting declined, supported nitrogen fixation.[1] This followed experiments conducted in Germany from 1895 jointly by Adolph Frank, the potash pioneer, and chemist Nikodem Caro, who were initially interested in cyanides for use in extracting valuable metals in low-grade ores. Subsequently, they worked with the chemist Fritz Rothe, at Dynamit AG, and found that heating calcium carbide together with nitrogen in an electrical furnace at a high-temperature yielded a novel product, calcium cyanamide. In 1899, Frank, Caro, Siemens & Halske, and Degussa, with the backing of Deutsche Bank, founded Cyanid-Gesellschaft, to develop cyanide processes. In 1900, Frank's son Albert established that cyanamide released ammonia on contact with soil, and, following a number of trials, it was found to be a good fertilizer. Calcium cyanamide production was first undertaken, in 1905, by the Società Italiana per la fabbricazione di Proditti Azotati e di altri sostance per l'Agricoltura (SIPA), at Piano d'Orta, near Pescara, on the Adriatic coast, in the Abruzzo region of eastern central Italy. The plant drew on the abundant hydroelectric power available from the nearby power station of the Società Italiana di Elettrochimica that was fed by the waters of the Apennine range. However, there was a serious problem. The externally heated retorts could not withstand

[1] Later, carbide found use in the manufacture of chlorinated solvents, acetic acid, and acetone, and in oxyacetylene welding.

Fig. 1.4 Front cover of the April 1909 issue of the French journal *Revue d'Electrochimie et d'Electrométallurgie*. The main illustration shows what was probably a carbide factory (at left), supplied with electricity for the furnaces by a short overhead cable transmission system from a hydroelectric power plant (at right). Parallel water pipes (penstocks) are fed by a source high up the cliff at left. The water descends to the power plant at the base of the cliff. The small image at top left is a reminder that Ferdinand-Frédéric-Henri Moissan in Paris undertook early experiments on the thermochemical synthesis of calcium carbide. The image at bottom right credits Humphry Davy with experiments in connection with the application of electricity to chemistry. (Author's collection)

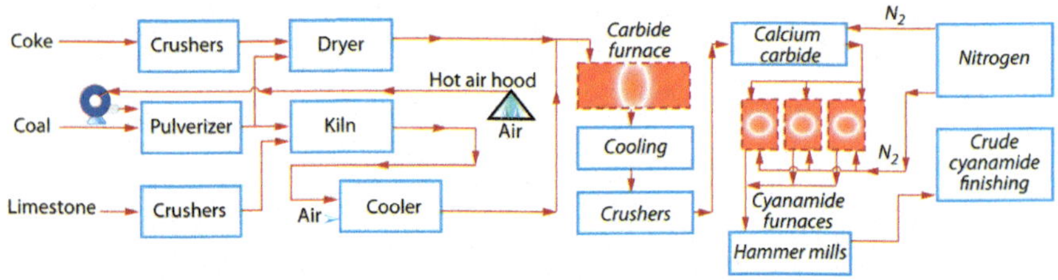

Fig. 1.5 Scheme for the manufacture of calcium carbide and its conversion into calcium cyanamide, both by electrothermal processes. Redrawn from Mantell CL (1940) Industrial electrochemistry, 2nd ed. McGraw Hill, New York, p. 521

the high operating temperature. This was overcome when Albert Frank found that by inserting an electrode into the charge to start the reaction, the heat released was sufficient to maintain the reaction in a sealed oven. The capacity of the factory was 4000 tons per year (Fig. 1.5). Significantly, the process required pure nitrogen. The Italian factory was the first user of nitrogen obtained by the process of Carl von Linde, developed for liquefaction and fractionation of air. Another process for obtaining nitrogen from the air by liquefaction was that of the French entrepreneur Georges Claude.

In July 1905, SIPA, the Società Italiana per il Carburo di Calcio, Acetilene e altri Gaz (Carburo, established in 1896), and Cyanid-Gesellschaft co-founded the Rome-based firm La Società Generale per la Cianamide (Cianamide), with a capital of three million lira, to license globally the cyanamide patents for the Frank-Caro process. Carburo's two carbide factories at Collestate and Papigno, both in the province of Terni, in Umbria, northeast of Rome, became major producers of calcium cyanamide (Fig. 1.6). Within a short time, Carburo, through its involvement in Cianamide, held financial interests in cyanamide factories that worked the Frank-Caro process in Switzerland, France, Norway, and Dalmatia. Cyanamide manufacture in Germany using the Frank-Caro process was taken up in 1908 at Trostberg, Upper Bavaria, after arrangements were made with the backers of the rival cyanamide process of Eduard Polzenius, which was worked at Knapsack [6, pp. 67–83, 86–89].

Frank Sherman Washburn co-founded the American Cyanamid Company in July 1907 to work the Frank-Caro process on the Canadian side of Niagara Falls. Calcium cyanamide production by this same process was taken up on the western Japanese island of Kyūshū by Noguchi Shitagau, who was engaged in electricity generation and production of calcium carbide. This was an important step in the emergence of the modern Japanese chemical industry [18]. Noguchi's firm was the forerunner of the industrial corporation Nitchitsu; he plays a significant role in this story (Chap. 4).

Research around 1900 into fixing atmospheric nitrogen for use in manufacture of fertilizer also included its direct reaction with hydrogen to form ammonia. In 1908, BASF assigned a consultant, the physical chemist Fritz Haber, at Karlsruhe

Fig. 1.6 Advertisement for calcium cyanamide fertilizer, manufactured at Papigno, near the city of Terni, and San Marcel, Aosta Valley (Val d'Orta), Società Italiana per la Fabbricazione di Prodotti Azotati ed altre sostance per l'Agricoltura, 1919. (Author's collection)

Technische Hochschule, to study nitrogen fixation. In July 1909, Haber demonstrated a small laboratory apparatus for the continuous production of synthetic ammonia under conditions of high pressure and high temperature in the presence of an osmium catalyst. Scaling up, under Carl Bosch at BASF, of what was a complex and highly problematic cutting-edge technology involved unprecedented engineering challenges. In the next chapter, we will learn how success was achieved.

1.2.2 Ammonia and Nitrogen

During the latter part of the eighteenth century, the study of gases was one of the most important topics that occupied the attentions of leading chemists. The English chemist Joseph Priestley was the first scientist to examine the chemical composition of the colorless, pungent alkaline ammonia (NH_3), which in 1774 he decomposed with an electric spark. Two years later, the French chemist Claude-Louis Berthollet showed that ammonia was a compound of nitrogen and hydrogen. Antoine-Laurent Lavoisier gave the name azote ("without life") to what we now call nitrogen. Lavoisier's interests included improvements in agriculture and gunpowder [19]. The natural nitrogen fixation pathways, particularly the roles of bacterial assimilation and lightning, and the nitrogen cycle, would be established in the nineteenth century [20].

References

1. Leigh GJ (2004) The world's greatest fix: a history of nitrogen and agriculture. Oxford University Press, New York
2. Smil V (2001) Enriching the earth: Fritz Haber, Carl Bosch, and the transformation of world food production. The MIT Press, Cambridge, Mass, pp 21–38
3. Brock WH (1997) Justus von Liebig, the chemical gatekeeper. Cambridge University Press, Cambridge
4. Schling-Brodersen U (1992) Liebig's role in the establishment of agricultural chemistry. Ambix 39:21–31
5. Munday P (1991) Liebig's metamorphosis: from organic chemistry to the chemistry of agriculture. Ambix 38:135–154
6. Travis AS (2018) Nitrogen capture: the growth of an international industry. Springer, Cham, pp 11–18
7. Lodge FS (1938) Potash in the fertilizer industry. Ind Eng Chem 30:878–882
8. de Secada CAG (1985) Arms, guano, and shipping: the WR Grace interests in Peru, 1865–1885. Bus Hist Rev 59:597–621
9. Skaggs JM (1994) The great guano rush: entrepreneurs and American overseas expansion. St. Martin's Press, New York
10. Melillo ED (2012) The first green revolution: debt peonage and the making of the nitrogen fertilizer trade, 1840–1930. AHR (Am Hist Rev) 117:1028–1060
11. Travis AS (2014) The emerging role of titrimetry in late nineteenth-century industrial problem solving: the example of trace analysis for perchlorate in Chile saltpeter. Ambix 61:67–94
12. Davy H (1807) On some chemical agencies of electricity. Phil Trans R Soc Lond 97:1–56

13. Jewess M (2017) Davy hits the headlines (again). Royal Society of Chemistry Historical Group Newsletter and Summary of Papers 72:28–30
14. Brock WH (2008) William Crookes (1832–1919) and the commercialization of science. Ashgate, Burlington VA
15. Crookes W (1898) Address of the president before the British Association for the Advancement of Science, Bristol, 1898. Sci New Ser 8(200):561–575
16. Sagatos OJ (2005) Progress of a different nature: Hydro 1905–2005. Pax Forlag, Oslo
17. American Chemical Society (1998) National historic chemical landmarks. Discovery of the commercial processes for making calcium carbide and acetylene. American Chemical Society, Division of the History of Chemistry, Washington DC. http://www.acs.org/content/acs/en/education/whatischemistry/landmarks/calciumcarbideacetylene.html. Accessed 8 Nov 2024
18. Molony B (1990) Technology and investment: the prewar Japanese chemical industry. Council on East Asian Studies/Harvard University Press, Cambridge, Mass
19. Donovan A (1993) Antoine Lavoisier: science, administration, and revolution. Cambridge University Press, Cambridge
20. Gorman HS (2013) The story of N: a social history of the nitrogen cycle and the challenges of sustainability. Rutgers University Press, Piscataway, NJ

The Great Nitrogen Challenge: Synthetic Ammonia

2

Abstract

The chemical equation for synthesis of ammonia from its elements is simple, but the method of synthesis is not. It requires a high pressure, an active catalyst, and a high temperature. Even then the yield is low. The potential for an industrial process was first demonstrated in 1909 by physical chemist Fritz Haber. He achieved around 5–8% yield per cycle in a closed-loop, steel apparatus. The process was scaled up by Carl Bosch of the BASF company, based at Ludwigshafen, Germany. There were many challenges, including the need for an inexpensive catalyst, the manufacture of powerful reciprocating compressors, the design of a converter that did not explode from attack by hydrogen, the assembly of control instruments, and the production of pure hydrogen and nitrogen. Manufacture commenced at Oppau, north of Ludwigshafen, in September 1913, with the intention of supplying ammonium sulfate fertilizer to farmers. World War I intervened, and what became known as the Haber–Bosch process was diverted to the manufacture of munitions. In 1917, a second factory was opened at Merseburg (Leuna), but wartime problems prevented it from reaching its intended capacity. Allied chemists and engineers were amazed at the industrial achievement in fixing atmospheric nitrogen. However, they were unable to replicate the process.

In the early twentieth century, industrial chemistry was transformed by a marvel of science-based technology: the high-pressure catalyzed fixation of atmospheric nitrogen as ammonia. It all began in 1909, when physical chemist Fritz Haber demonstrated a laboratory method to representatives of the German firm BASF. Four years later, in September 1913, heroic efforts on the part of Carl Bosch and colleagues at BASF led to industrial-scale production of synthetic ammonia. Force of circumstances, the outbreak of World War I, dictated that synthetic ammonia's first major contribution was to the production of munitions. Then, under peacetime

A. S. Travis, *The Global Synthetic Ammonia Industry*, Perspectives on the History of Chemistry, https://doi.org/10.1007/978-3-031-87978-4_2

conditions, what had become known as the Haber–Bosch process satisfied the demand for nitrogen fertilizers. Haber and Bosch, for their contributions to high-pressure methods, received the Nobel Prize in Chemistry, Haber in 1918 (awarded in 1919) and Bosch in 1931 (jointly with another high-pressure pioneer, Friedrich Bergius). This chapter recounts the scientific and technical achievements that laid the foundation for the synthetic ammonia industry, which in its first years was confined to Germany.

2.1 Nitrogen in Germany

Research into industrial nitrogen fixation was especially strong in Germany, a nation that was dependent on imported Chilean nitrate for its supply of nitrogen fertilizer and the manufacture of nitric acid. The perceived need for self-sufficiency in fixed and stable nitrogen products stimulated research within the technologically advanced synthetic dyestuffs industry, in which Germany was the world leader, mainly because such research offered a useful direction for diversification. The greatest interest in nitrogen fixation was shown at BASF, in Ludwigshafen. There, investigations at first involved the electric arc method and synthesis of various cyanides and nitrides. The firm also took an interest in the work of academic inventors. This included a method for combining nitrogen with hydrogen to produce ammonia as investigated by the physical chemist Wilhelm Ostwald at Leipzig. In March 1900, Ostwald informed BASF that he had prepared ammonia directly from its elements by applying a high temperature to a closed apparatus containing nitrogen and hydrogen in the presence of an iron wire catalyst. BASF investigated the method but rejected it after determining that ammonia was formed from the reaction of hydrogen with an iron–nitrogen compound in the catalyst and not from the reaction between nitrogen and hydrogen. Most chemists concluded that it was not possible to bring about the nitrogen–hydrogen reaction on an industrially viable scale.

2.2 Fritz Haber

Fritz Haber (1868–1934) was born in Breslau (today Wrocław), in Prussian-occupied Poland. He received his doctorate in organic chemistry at Berlin's Friedrich-Wilhelms-Universität in 1891 and in the following year joined the University of Jena as an assistant under Ludwig Knorr. Haber soon took up an interest in physical chemistry. In 1894, he was recommended to the provost of Karlsruhe Technische Hochschule, the chemist Carl Engler. Haber joined Engler's institute as research assistant under Hans Bunte [1].

Haber's research focused on the thermodynamics of reactions in the gaseous phase as applied to fuel technology. In 1896, Haber received his *Habilitation* working under Bunte on "Experimentaluntersuchungen über Zersetzung und Verbrennung von Kohlenwasserstoffen" (Experimental Studies of the Decomposition and Combustion of Hydrocarbons) and became a Privatdozent. In the same year, Haber turned towards the application of electricity to chemistry, and two years later, in 1898, he published *Grundriß der technischen Elektrochemie auf theoretischer Grundlage* ("Outline of Technical Electrochemistry on a Theoretical Basis") [2].

In 1904, Fritz Haber received an assignment from the Österreichische Chemische Werke in Vienna that would change history. The owners of the firm, the brothers Marguiles, had noticed the pungent odor of ammonia in one of their processes and asked Haber if this might be of value in devising a method to synthesize the gas. Haber started to consider the conditions for the combination of nitrogen with hydrogen, the ammonia reaction. This involved measuring the equilibrium between nitrogen and hydrogen and ammonia over a range of temperatures. Similar studies had been carried out by other researchers, including William Ramsay and Sidney Young at University College London in 1884. While the decomposition of ammonia could be followed, the reverse reaction, leading to formation of ammonia from its elements, had not been observed. Haber and a colleague, Gabriel van Oordt, decided to repeat the experiments using an elegant apparatus that enabled measurement of the formation and dissociation of ammonia at temperatures up to 1000 °C. The experiments were carried out in the presence of iron, manganese, nickel, and chromium as catalysts, out of which iron was found to work best. As the temperature was increased, ammonia was found to be present in smaller and smaller amounts. The yield of ammonia at 1000 °C in the presence of a finely divided iron catalyst at atmospheric pressure was, according to their calculations, 0.005–0.0125% [3]. Because ammonia decomposes upon being heated, any synthesis from its elements had to be carried out at a low temperature. Haber realized that under these conditions the forward reaction, that is, the formation of ammonia, required a highly active catalyst. No suitable catalysts were available, and the work was abandoned. The experimental results were included in Haber's *Thermodynamik technischer Gasreaktionen* (1905), in which the difficulties associated with fixing nitrogen were explicated [4].

In the summer of 1906, Haber was appointed full professor and head of the Karlsruhe Institute for Physical Chemistry and Electrochemistry (Fig. 2.1). His laboratory attracted researchers from Europe, the US, and Japan who were studying gas reactions and their practical applications. Also, during 1905–1906, the physical chemist Walther Nernst in Berlin formulated his heat theorem (later known as the third law of thermodynamics) based on data available from experiments published in the literature. Nernst adapted the heat theorem to the study of gas reactions. His findings suggested that there was a discrepancy between his theorem and Haber's work on ammonia. In the early autumn of 1906, Nernst advised Haber of the discrepancy. Haber immediately set about checking the results of his earlier work on the ammonia equilibrium. The experimental work was entrusted to a young English research assistant, Robert Le Rossignol, who had worked under William Ramsay at University College London and arrived at Karlsruhe in September 1906. As a result of their findings, Haber withdrew the higher figure for ammonia at 1000 °C that he had published with van Oordt. Haber and Le Rossignol's results now aligned

Fig. 2.1 Fritz Haber. (Bild-Nr VII/3 Haber, Archiv der Max-Planck-Gesellschaft, Berlin-Dahlem. With permission)

with Nernst's theorem. However, on 12 May 1907, at the meeting of the German Bunsen Society in Hamburg, Nernst made a point of publicly disparaging Haber's work, though apparently with little foundation, because Haber and Le Rossignol had already submitted their revised results for publication [5–7].

Nernst did, however, emphasize the role of pressure on the equilibrium between nitrogen and hydrogen and ammonia. His assistant Fritz Jost had designed an apparatus that withstood pressures of up to 75 atm. Nernst and Jost in their experiments employed a platinum catalyst. Haber and Le Rossignol were now stimulated to investigate the equilibrium under pressure, at around 30 atm, which was the pressure obtained from a pressurized gas cylinder of the type manufactured by the Mannesmann steelmaking firm. From these studies, Haber calculated that in the presence of a suitable catalyst an estimated 8% equilibrium yield of ammonia could be achieved at 600 °C and 200 atm.

The role of pressure in gas reactions had been investigated by the French chemist Henri-Louis Le Chatêlier at the end of the 1880s. He articulated the principle that when a reversible reaction at equilibrium is subjected to a physical force, the reaction takes place in the direction that reduces the force. This is observed in the reaction between nitrogen and hydrogen in the synthesis of ammonia.

$$N_2 + 3H_2(4\,\text{volumes}) \rightleftharpoons 2NH_3(2\,\text{volumes}) \qquad \Delta H = -91.8\,\text{kJ/mol}$$

As pressure is applied, the equilibrium moves in the direction that reduces its effect, in this case, to the ammonia side of the equation. The reaction is exothermic, meaning that lowering the temperature encourages formation of the product, to wit, the ammonia. However, this also slows down the rate of reaction. To activate the synthesis and bring about the rapid formation of ammonia, a high temperature was required, but care had to be taken to keep the temperature below the level that causes ammonia to decompose. Because high pressures and elevated temperatures work against each other in this case, the only way to bring about a viable reaction was by introducing a suitably active catalyst. Furthermore, to prevent decomposition of the ammonia, the gas mixture needed to be cooled sufficiently to bring about liquefaction of the ammonia, enabling its separation from the unreacted gases. When Le Chatêlier investigated the synthesis of ammonia in 1900, his apparatus blew up due to the presence of oxygen in the gas mixture, and with that, his ammonia research came to an end.

Carl Engler, as a member of the board of BASF, was well aware of the firm's investigations into nitrogen fixation, which included Otto Schönherr's long, slim electric arc furnace, as well as studies on the synthesis of metal cyanides and nitrides. Investigation into nitrides, focusing on barium nitride, was carried out in the Inorganic Department by Alwin Mittasch (1869–1953), who had studied under Ostwald, and joined BASF in 1904. Engler suggested to BASF that Haber be brought in to study nitrogen fixation, and the suggestion was taken up by August Bernthsen, director of the Central Research Laboratory. Two contracts were drawn up with Haber on 6 March 1908. One contract was between BASF and Haber and his colleague Adolf König for synthesis of nitric oxide by the electric arc method. The other was between BASF and Haber, allocating 6000 marks a year until 1911 towards investigations into the fixation of nitrogen by its reaction with hydrogen. All results and patents on inventions were to be assigned to BASF.

The status of nitrogen fixation as a leading research topic, with tremendous implications for practical application, was demonstrated at the Seventh International Congress of Applied Chemistry, which was held in London at the end of May 1909. Sir William Ramsay chaired the sessions that dealt with electrothermal methods for nitrogen fixation. Present in the audience were Kristian Birkeland, Samuel Eyde, Nikodem Caro, August Bernthsen, Otto N. Witt, a former consultant to Birkeland and Eyde, and the Russian high-pressure chemist Vladimir Nikolayevich Ipatieff. It was the largest-ever gathering of inventors and representatives of the electrothermal processes then used for fixing nitrogen. Birkeland reviewed progress in the application of his arc furnace, and Bernthsen described the Schönherr electric arc furnace of BASF. But it was elsewhere, on the laboratory benches at Karlsruhe, that the main progress—the high-pressure synthesis of ammonia from its elements—was unfolding.

2.3 High Pressures and Heterogeneous Catalysis

Haber knew that to advance the ammonia work, it would be necessary to fashion an apparatus for working at a much higher pressure than 30 atm. Robert Le Rossignol's new assignment was to devise a steel apparatus, of his own design, that would incorporate a novel compressor newly acquired by Haber that was capable of reaching a pressure of 200 atm. Le Rossignol and the laboratory technician Friedrich Kirchenbauer put together a steel reaction vessel capable of withstanding both high pressures and high temperatures. Le Rossignol, who had acquired skills in operating machine tools at University College London, was the inventor of a needle valve that enabled smooth control of the flow of hot, compressed gases. The gases circulated in a closed loop with continuous removal of the ammonia, thus favoring its formation. In this way, the mixture was not allowed to reach equilibrium. Unreacted nitrogen and hydrogen were recirculated. The heat of the exothermic reaction warmed up the incoming nitrogen–hydrogen gas mixture.

This experimental work became the basis of Haber's circulation patent, filed in Berlin on 13 October 1908 [8]. The patent stipulated the need for a heated catalyst and for application of pressure, but, apart from a description of gas circulation in a closed loop, it did not provide further details. The need for an active catalyst was critical. Iron, chromium, and nickel had been tried, but the results were poor.

In March 1909, finely divided osmium was found to be an effective catalyst, giving a yield in a single pass of around 5%. The more abundant uranium, as the free metal, or as its nitride or carbide, gave a yield of around 10%. Patents were filed for both osmium and uranium as catalysts, the initial filing being made on 31 March 1909. Tamaru Setsurō, a graduate of the Imperial University of Tokyo, and a member of Haber's group, undertook experiments on the synthesis of ammonia with the rare osmium as a catalyst in order to gather precise data on specific heats, used in the calculation of the dissociation equilibria for the reaction [9].

As described in the 1908 circulation patent, the compressed gas mixture (a 3:1 mixture of hydrogen and nitrogen) was introduced into the electrically heated converter where it came into contact with the catalyst. The ammonia was separated by liquefaction. Le Rossignol designed and assembled the double-acting recirculation pump that sent unreacted gases back to the converter (Fig. 2.2). Hydrogen was obtained by electrolysis, and nitrogen by air liquefaction.

Haber, after achieving a successful run with the 0.75-m high benchtop apparatus, advised Engler that liquid ammonia was dripping from the delivery tube: "Es tropft, Herr Geheimrat!"[10]. On 23 March 1909, Haber advised the BASF management of the results. Three days later, Carl Bosch and BASF directors Heinrich von Brunck and August von Bernthsen turned up at Karlsruhe to look over the laboratory apparatus. According to later accounts, Bosch was the only one of the visitors who showed much enthusiasm during the inspection.

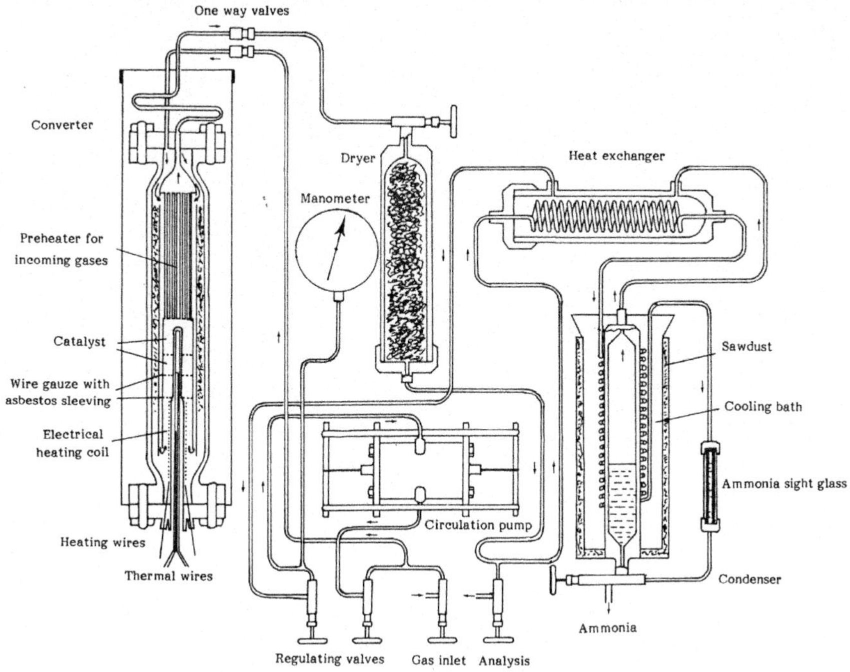

Fig. 2.2 Diagram of Haber–Le Rossignol laboratory apparatus for the continuous synthesis of ammonia from nitrogen and hydrogen using a complete loop. The mixture of hydrogen and nitrogen under pressure was heated over a platinum catalyst to remove traces of oxygen, dried, and then entered the electrically heated converter where the ammonia was formed over a catalyst. The mixed gases were cooled, and unreacted gas was separated from liquid ammonia. The double-acting recirculation pump directed unreacted gas back to the converter. This circulatory synthesis system became the standard in all commercial ammonia units, apart from the original Claude process. (after Fritz Haber and Robert Le Rossignol)

2.4 Carl Bosch and Fritz Haber

Carl Bosch (1874–1940) was the son of the owner of a metalworking business in Cologne that catered to the needs of coal gas suppliers and distributors (Fig. 2.3). He took up the study of chemistry and in April 1899 joined BASF. Not long after, assigned to the problem of nitrogen fixation, he demonstrated why Ostwald's ammonia process was flawed. In 1902, Bosch investigated methods for making cyanides and nitrides as possible sources of stable nitrogen products. The results did not suggest any commercial potential.

Despite Haber's achievement in 1909, and no doubt Carl Bosch's favorable report, August Bernthsen was not entirely convinced that the apparatus at Karlsruhe represented the basis of an industrial process, perhaps in part because of its seeming complexity. Haber was disappointed. On Haber's behalf, however, Engler asked BASF to investigate further. On 1 July 1909, Carl Bosch, Alwin Mittasch,

Fig. 2.3 Carl Bosch. (Image from *Erzeugnisse unserer Arbeit*. Frankfurt am Main: I. G. Farbenindustrie, 1938, p. 17) (Author's collection)

and BASF chief mechanic Julius Kranz arrived at Haber's laboratory to observe the apparatus in operation. Things did not go well. A bolt from one of the seals sprang under the great pressure. The apparatus had to be repaired. Bosch returned to Ludwigshafen in order to attend to other business. Mittasch and Kranz remained in Karlsruhe, and the next day they saw liquefied ammonia dripping into a glass collector. The apparatus was operated continuously for five hours.

Learning of this development, Heinrich von Brunck and his colleagues agreed to invest in further studies at Ludwigshafen. On 14 September, Haber filed another patent, this time specifying the pressure, about 100 atm, but adding that for the process to be truly useful, the pressure needed to be raised to 250 atm [11]. This high-pressure patent and the 1908 circulation patent defined the mode of operation of every ammonia converter constructed ever since. But many challenges remained, including finding a suitable catalyst, one that was less costly and more abundant than osmium.

Mittasch sought an inexpensive but active catalyst, using for test purposes a device that BASF chemist Georg Stern had put together: an air-cooled, internally heated laboratory apparatus well suited to the screening of metals and their mixtures. Uranium, though more readily available than osmium, was sensitive to oxygen and water and could not be produced in the particulate form required of a catalyst. Swedish magnetitite, a porous oxide of iron that had been examined on 6 November 1909 by Mittasch's colleague Hans Wolf, gave a yield of around 5%. Iron-containing compounds were soon activated as catalysts with other metal compounds, referred to as promoters. On 9 January 1910, Bosch, Mittasch, Stern, and Wolf filed a patent for mixed metal compounds as catalysts.

Carl Bosch worked on the design of a pilot plant with the help of the engineer Franz Lappe. Their goal was to devise a reaction tube, or converter, suited to continuous operation at the required high pressure (250 atm) and high temperature (around 400° C). Far more heat was produced than in the laboratory apparatus, and a way had to be found to remove this excessive heat. Removal by heat exchange required the installation, within the converter, of small-bore tubes (bundles) through which the process stream passed. This type of installation had already been used, for example, in the BASF catalytic sulfuric acid process of Rudolf Kneitsch. But the conditions accompanying that process were quite unlike those of the ammonia process. Experiments were carried out in 1910 with two Mannesmann hot rolled seemless carbon steel contact tubes, 2.5 m in length. They were filled with catalyst, and externally heated. For safe working at the high pressure, the tubes were embedded in concrete. After eighty hours of use, the tubes burst. Bosch established the cause: hydrogen diffused into the iron of the steel shell, causing it, by decarburization, to become brittle. The problem was overcome early in 1911, when Bosch, with the help of Lappe, designed a double-wall converter. The inner wall was made of soft iron, or low-carbon steel, which did not undergo decarburization and turn brittle. The hydrogen diffused from the interior through the inner wall, with a fall in pressure such that it had little to no deleterious effect on the outer wall. Small holes in the outer wall allowed a partial release of hydrogen to the surroundings.

Bosch and colleagues designed special leak-proof reciprocating gas compressors in which pistons, driven by crankshafts, delivered the mixture of gases to the converter.

On 11 September 1912, Bernthsen revealed the achievements of the new BASF synthetic ammonia process at the Eighth International Congress of Applied Chemistry in New York. The presentation included a demonstration apparatus. Present in the audience was Japanese engineer Jun Ōya, who sent back reports on the latest progress in nitrogen fixation to colleagues in Tokyo. Nitrogen fertilizers were in great demand in Japan, which was a major importer of ammonium sulfate and other nitrogenous products, and where calcium cyanamide was manufactured by the Frank-Caro process. In North America, by contrast, the application of Haber's method attracted far less attention, apart from the appeal of its technical brilliance.

For the pilot plant studies, as with Haber's experiments, nitrogen was obtained by Linde's process for liquefaction and distillation of air. Hydrogen came from the electrolysis of salt solution (brine). This was, however, too expensive for large-scale production. For a time the action of steam on iron, the Messerschmidt process, as used to provide hydrogen for aerial balloons, was investigated. Finally it was decided to use an existing coal-based process. Hydrogen was generated by the action of blowing steam on incandescent coke, which produced water gas, a mixture of carbon monoxide and hydrogen (today better known as synthesis gas, or syngas) (see also Chap. 7.3.2). Linde liquefaction of water gas allowed carbon monoxide, a catalyst poison, to be separated out.

$$C + H_2O \rightarrow CO + H_2 \qquad \Delta H = +188.7\,\mathrm{kJ/mol}$$

Removal of the final traces of monoxide was achieved by scrubbing with caustic soda (though this was soon found to cause problems of corrosion). The gas mixture was then made up and adjusted to 75% hydrogen and 25% nitrogen, the stoichiometric proportions required for the synthesis. Single pass conversion to 8% of ammonia through a single converter was achieved.

The success was such that in November 1911, BASF had decided to construct a new factory to produce synthetic ammonia, at Oppau, located three kilometers north of the Ludwigshafen facility. Construction started in the spring of 1912, when storage tanks, gasometers, piping, and piping bridges were put in place (Fig. 2.4). Novel instruments enabled monitoring and control of the pressure and temperature [12].

Mittasch was placed in charge of the new Ammonia Laboratory that was staffed with ten trained chemists. Production of synthetic ammonia began on 9 September 1913 at the rate of 20-tons per day (Fig. 2.5). The ammonia was sold in the anhydrous form, which required the installation of refrigeration equipment, for use as a coolant. The long-term goal was to convert it into ammonium sulfate for sale as fertilizer.

In 1914, BASF opened an agricultural experiment station at Limburger Hof, near Ludwigshafen, to engage in fertilizer and agrochemical studies. In the same year, Carl Duisberg, of the Bayer corporation, referred for the first time to the Haber–Bosch process. A few years earlier, in 1911, Haber, now a famous scientist, had been appointed head of the new Kaiser-Wilhelm-Institut für physikalische Chemie und Electrochemie in Berlin which opened in 1912.

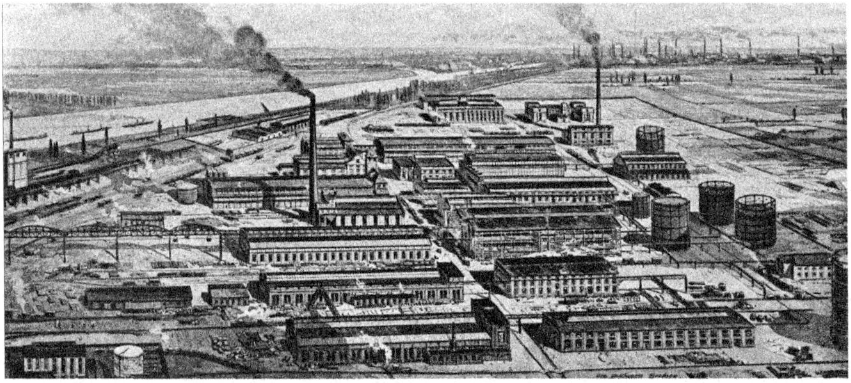

Fig. 2.4 BASF factory at Oppau c. 1914. From a painting by Otto Bollhagen. (BASF Corporate History. With permission)

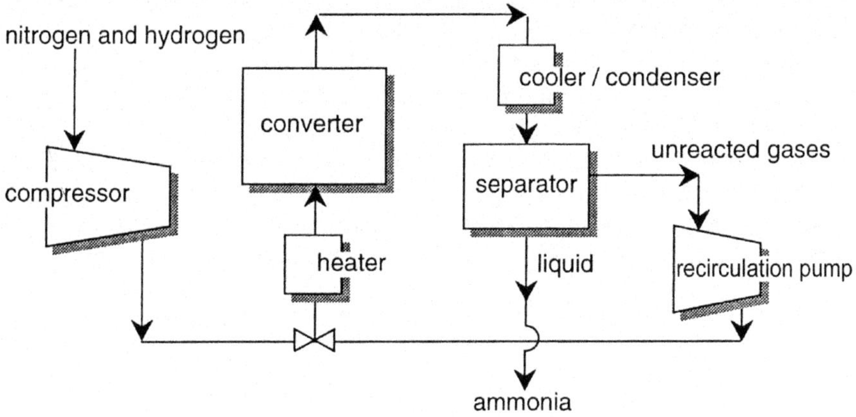

Fig. 2.5 Block diagram for synthesis of ammonia

BASF carefully guarded the secrets of its Haber–Bosch ammonia process. The company had no intention of offering licenses for its use outside of Oppau. Thus in the autumn of 1913, the American chemist Louis Cleveland Jones of the Solvay Process Company, while on business in Europe, wrote from Paris how his attempt to discuss the terms of acquiring a license from BASF met with a less than encouraging response: "Their determination to pursue independent ways, however, is indicated by their refusal to see me, on the ground that they would probably not favorably consider an American license" [13]. Jones discovered that Belgian and German Solvay works were purchasing anhydrous ammonia from BASF. Additionally, he learned that anhydrous ammonia, used in industrial refrigeration and for other purposes, was of more interest than BASF's ammonium sulfate fertilizer. Later, BASF acquired a firm engaged in the mixing and supply of fertilizers in order to engage in the marketing of its ammonium sulfate product to the agricultural sector.

2.5 Nitric Acid

A short time after the outbreak of war in Europe in August 1914, synthetic ammonia took on a new role as the starting point for the manufacture of nitric acid, which was essential to the production of explosive nitro compounds, such as TNT. This involved the oxidation of ammonia, achieved by a new process employing a platinum catalyst. In December 1914, Germany's supply of Chilean nitrate, from which the acid was made by reaction with sulfuric acid, was cut off by the British Royal Navy, and the process of catalytic oxidation, by necessity, had become top-priority.

The air oxidation of ammonia to nitric acid with a platinum catalyst had been used since 1908 by the Gewerkschaft des Steinkohlenbirgwerks Lothringen, starting with by-product ammonia from a coking oven plant at Gerthe, Westphalia (Gerthe gave its name to the facility), near Bochum in the Ruhr. The method had been developed by Wilhelm Ostwald and Eberhard Brauer, his future son-in-law, during 1900–1901. For large-scale production, a consortium of local coal producers was assembled.

The process took place in two steps. First, nitric oxide was formed. Second, the nitric oxide was combined with more oxygen to produce nitrogen dioxide.

$$4NH_3 + 5O_2 \rightleftharpoons 4NO + 6H_2O \qquad \Delta H = -909\,\mathrm{kJ/mol}$$

$$2NO + O_2 \rightleftharpoons 2NO_2 \qquad \Delta H = -115\,\mathrm{kJ/mol}$$

The dioxide, existing reversibly with its dimer, N_2O_4, was absorbed in water to produce dilute ("weak") nitric acid, not unlike the process in the electric arc factories. In 1906, Friedrich Uhde, of the Dr. C. Otto & Co. engineering consultancy, improved the process while working at Kohlenzeche Lothringen at Gerthe. However, in both cases the steel apparatus suffered from severe corrosion. This problem was overcome after World War I with the introduction of special steel alloys. BASF scientists and engineers introduced acid-resistant alloys containing silicon for the oxidation reactors.

Late in 1914, the Oppau facility ceased the manufacture of ammonium sulfate fertilizer and turned to the production of nitric acid and ammonium nitrate for use in the production of munitions. This new objective required highly concentrated nitric acid. At the time, it was made from ammonia in four steps: conversion by oxidation into dilute nitric acid, then to sodium nitrate (with sodium carbonate), evaporation to dryness, and finally reaction of nitrate with oil of vitriol (concentrated sulfuric acid). This lengthy procedure was necessary because it is not possible to achieve concentration by distillation because nitric acid forms an azeotropic mixture with water (a constant boiling mixture containing 68.5% acid by mass at 122 °C).

In May 1915, Mittasch and Christoph Beck introduced a two-step process to replace the four-step procedure. The new process involved the oxidation of ammonia into nitric acid followed by its concentration in a countercurrent operation with concentrated sulfuric acid. This took place in a tower packed with the newly invented Raschig rings. The resulting 99% nitric acid was ideal for use in the several nitration processes required in the manufacture of explosives. A cheap catalyst had become available based on iron. That was a welcome development because platinum had been imported from Russia, now an enemy nation. Thus, another process overcame the disruption of critical supply chains.

2.6 Hydrogen from Water Gas, and Leuna/Merseburg

The demands of war led to massive changes in the scale of production at Oppau. In 1913, the ammonia converters there were eight meters in length. In 1915, ten 12-m-long converters, each one weighing 75 tons, with an outside diameter of 108 cm, and assembled from two 6-m sections bolted together, were introduced. All of these were installed vertically, with axial flow of gases, a configuration that would be a characteristic feature of synthetic ammonia units until the late twentieth century. The maximum size of forgings from solid ingots dictated the length of each section. With the new, larger converters, pressures of 300 atm were introduced. In 1915, outputs of the large converters reached 85 tons per day. By this time, Bosch's team had developed reciprocating compressors of 3000 horsepower (2.2 MW) capable of working for six months with minimum maintenance. Early on, the external heating of converters was replaced by internal heating, in which air forced into the converter was ignited with hydrogen, creating a reverse flame. Internal heating helped reduce stress on the converter walls. With the introduction of larger converters, the countercurrent transfer of heat from the exothermic ammonia reaction to the incoming synthesis gas was sufficient to maintain the required reaction temperature. Another innovation, following the introduction of internal heating, involved flushing the gap between the inner and outer walls of the converter with nitrogen, thereby further reducing the pressure difference between the outer lining of the inner wall and the inner lining of the outer wall (the jacket). The restricted capacities of the reciprocating compressors meant that large-scale operations had to be conducted using a battery of identical closed trains or synthesis loops. Each synthesis loop was driven by a separate compressor.

A novel change in the production of hydrogen from water gas was based on the work of Wilhelm Wild. His method was developed jointly with Bosch in 1912. Water gas consists of almost equal volumes of hydrogen and carbon monoxide. The Bosch-Wild process involved a further steam treatment at 350–450 °C over an iron-chromium catalyst. The result was a mixture of carbon dioxide and more hydrogen. This is the "shift reaction."

$$CO + H_2O \rightarrow CO_2 + H_2 \qquad \Delta H = -41.2\,\mathrm{kJ/mol}$$

It was also called the Bamag process, after the engineering firm, Berlin-Anhaltische Maschinenbau-Akt.-Ges, which supplied the special equipment. Nitrogen was obtained by alternating the steam treatment with the blowing of air over the coke to form a mixture of nitrogen and carbon monoxide (producer gas). The mixed gases were then subjected to the shift reaction. Following removal of sulfur compounds over activated charcoal and carbon dioxide by water scrubbing, the nitrogen-hydrogen mixture was gradually compressed to 200 atm, or more. Karl Krauch found that removal of final traces of carbon monoxide could be achieved with ammoniacal copper liquor (cuprous) scrubbing, which did not corrode the apparatus.

At the end of 1915, Carl Bosch lobbied the Prussian government for backing for a new, large synthetic ammonia facility. The site chosen was at Leuna, near Merseburg, on the Saale River, twenty kilometers southeast of Halle, well out of the range of Allied aerial bombing. (The French bombed Oppau on 27 May 1915, in the first wartime strategic air raid on an industrial facility.) In 1916, Bosch was appointed a director of BASF, in recognition of his contribution to nitrogen fixation work and expansion of ammonia production.

Thirteen large boiler houses for power and other utilities, each with a 120-m-tall chimney, stretched almost the whole length of the Leuna factory. It first produced ammonia on 27 April 1917, just three weeks after the US declared war on Germany and just under three months before the British offensive in Flanders began. Plans were drawn up for the production of up to 240,000-tpy. However, shortages of raw materials as well as the collapse of the German rail network wreaked havoc on operations. At the time of the Armistice in November 1918, Leuna was not even able to meet the originally planned target of 30,000-tpy [14]. By the early 1920s, however, Leuna had grown into the leading producer of ammonium sulfate worldwide (Fig. 2.6).

During the war, British, French, Italian, Japanese, and American investigators had attempted, without success, to develop their own high-pressure processes for synthetic ammonia. When, after the cessation of hostilities, Allied investigators, including chemists and engineers, made visits to Oppau, then in the French zone of occupation, to glean information about the Haber–Bosch process, they came away disappointed. BASF hid its ammonia technology from prying eyes, including by shutting down operations before visitors arrived. This did not prevent the British firm of Brunner, Mond & Company, relying on innovation and imitation, and aided by extensive industrial espionage, from opening a synthetic ammonia factory at Billingham in northeast England that became fully operational in 1924. Following the merger of Britain's four largest chemical companies, including Brunner, Mond,

Fig. 2.6 BASF Leuna-Merseburg factory, c. 1920. From a painting by Otto Bollhagen. Note the pipe and utility bridges, moving cranes, coal conveyors, extensive railway network, and gasometers. (BASF Corporate History. With permission)

in 1926, the factory came under the ownership of Imperial Chemical Industries (ICI) [15].

More significantly, from this nucleus of outside interest there sprang an enterprise so great that within a decade it would transform the global chemical industry. The enterprise is especially associated with the Italian inventor Luigi Casale, who designed and developed the first successful synthetic ammonia process to rival, and even improve upon, the Haber–Bosch process. Unequaled in novelty and characterized by elegant simplicity as well as outstanding reliability, Luigi Casale's high-pressure process produced anhydrous ammonia directly. Casale's marketing and licensing strategy forged the way for his process to spearhead the rapid spread of a new technology around the world.

References

1. Stoltzenberg D (2004) Fritz Haber: chemist, Nobel laureate, German, Jew. Chemical Heritage Press, Philadelphia
2. Szöllösi-Janze M (1998) Fritz Haber 1868–1934: Eine Biographie. VCH-Beck, Munich
3. Haber F, van Oordt G (1905) Uber Bildung von Ammoniak aus den Elementen. Z anorg allgem Chem 13:111–115; 44:341–378; 47:42–44
4. Lamb AB (1908) Thermodynamics of technical gas reactions. Longmans, Green, & Co., London. A translation, with revisions, of Haber F (1905) Thermodynamik technischer Gasreaktionen. R Oldenbourg, Munich/Berlin. In the 1908 volume the yield of ammonia at 1000 °C was corrected to 0.0048%, p 204
5. Stoltzenberg D (1994) Fritz Haber: Chemiker, Nobelpreisträger, Deutscher, Jude: Eine Biographie. VCH, Weinheim, pp 154–155
6. Johnson B (2022) Making ammonia: Fritz Haber, Walther Nernst, and the nature of scientific discovery. Springer, Cham. https://doi.org/10.1007/978-3-030-85532-1. Accessed 12 Dec 2024
7. Sheppard D (2020) Robert Le Rossignol: engineer of the Haber process. Springer, Cham, pp 109–207
8. Haber F (1908) Prussian patent 235421, filed 13 Oct 1908, granted 8 June 1911
9. Tamaru Oyama H (2015) Setsuro Tamaru and Fritz Haber: Links between Japan and Germany in science and technology. Chem Rec 15(April):535–549
10. Schuster C (1976) Wissenschaft und Technik. BASF, Ludwigshafen, p 126
11. Haber F (1911) Prussian patent 238450, 28 Sept
12. Timm B (1960) The synthesis of ammonia. The contribution to the development of modern chemical technology and its economic importance for the nitrogen supply of the world. Chem Ind, 11 Mar, pp 274–281
13. Jones LC (1913) Report dated 2 October 1913. Solvay Process Company archives, h.4.9, microfilm reel 17. Sidney M. Edelstein Library, Israel National Library
14. Szöllösi-Janze M (2000) Losing the war, but gaining ground: the German chemical industry during World War I. In: Lesch JE (ed) The German chemical industry in the twentieth century. Kluwer, Dordrecht, pp 91–121
15. Travis AS (2018) Nitrogen capture: the growth of an international industry. Springer, Cham, pp 199–223

3 Luigi Casale and the Founding of Ammonia Casale SA

Abstract

Around 1917, Italian chemist Luigi Casale took up the study of synthetic ammonia at two electrochemical factories engaged in serving the Italian war effort. He was backed by the French-American entrepreneur René Leprestre who was involved with one of the factories, located in an old ironworks in the city of Terni. There, Casale developed the first viable alternative to the Haber–Bosch process. Casale employed a much higher pressure, which brought about an increase in the yield, and gave directly a liquid, anhydrous ammonia. His company, Ammonia Casale SA (Ammonia Casale Société Anonyme), located in Lugano, Switzerland, was founded in 1921 jointly with Leprestre. Unlike BASF, Ammonia Casale licensed its process. As a result, Ammonia Casale takes pride of place as the first engineering contractor that engaged in the development of the global synthetic ammonia industry. By 1923, Casale converters were producing ammonia in Japan and Europe. In that year, plans were drawn up for a manufacturing facility in the United States. While Casale's technology was outstanding, the relationship between Casale and Leprestre was less so. Relations soured dramatically, as evidenced by the correspondence between Luigi Casale and his wife Maria Sacchi-Casale while he was traveling on business. This and the two following chapters serve as the first full biography of Luigi Casale's industrial career.

3.1 Introduction

The Haber–Bosch process of BASF was the envy of Allied governments. It received extensive media coverage, especially in the technical and commercial press. However, the secrets of its success were confined to the factories at Oppau and Leuna. As demonstrated by Germany's great reliance on synthetic ammonia during the war, the process was recognized to be vital to the maintenance

A. S. Travis, *The Global Synthetic Ammonia Industry*, Perspectives on the History of Chemistry, https://doi.org/10.1007/978-3-031-87978-4_3

of national security in both food and munitions. Following the Armistice on 11 November 1918, teams of Allied technical experts visited Oppau, anxious to acquire information on the workings of the Haber–Bosch process. They came away with very little useful information apart from a few notes and handfuls of catalyst. Some corporations tried, but failed, to acquire licenses for operating the high-pressure process. It was unavailable for licensing because BASF planned to use the process to capture the world market in the nitrogen fertilizer ammonium sulfate. It was the reticence of BASF to share technical information that brought about a number of parallel, and urgent, research efforts—all drawing on Fritz Haber's method—towards the realization of original and alternative industrial processes for the large-scale manufacture of ammonia. The first successful process was that of the Italian chemist Luigi Casale.

Casale's process for the first time ensured that the supply of nitrogen products as fertilizers ceased to be subject to geographical (mainly South American) and industrial (BASF) monopolies. For reasons that will become apparent, the second Italian process, that of Giacomo Fauser, is also introduced in the following sections.

Significantly, the Casale ammonia process is less well-known than most of the other processes, which in part arises from their connections with major corporations, such as the Haber–Bosch process (with BASF/IG Farben), the Fauser process (Montecatini), the Claude process (Air Liquide), and the processes named after American companies that acquired and developed ammonia technologies, mainly Du Pont, Allied Chemical, and Chemico. However, and for whatever the reason, Casale's firm, Ammonia Casale SA takes pride of place as the first engineering contractor that engaged in the development of the global synthetic ammonia industry.

3.2 Luigi Casale

Italy at the dawn of the twentieth century. The country was undergoing rapid industrialization, sparked by the introduction of hydroelectric power. The Valtellina Railway in the north was the first in the world to employ locomotives powered by high-voltage three-phase alternating current (1902). The manufacture of calcium carbide and the nitrogen fertilizer calcium cyanamide drew on electrical power; the offices of the world's leading calcium cyanamide company were located in Rome. Northeast of the capital, the industrial city of Terni in Umbria relied on locally available hydroelectricity and was noted for its steel production for the military. Hydroelectricity offered independence from coal, as a source of energy, with which Italy was poorly endowed. Among the emerging industries was the factory of Camillo Olivetti, who in 1908 established in Piedmont (near Turin) his eponymous typewriter company. In the same year, the Fiat Motor Company of Turin produced its first aircraft engine. Two years later, the predecessor of what was to become Alfa Romeo was founded in Milan, the city in which Pirelli manufactured tires for automobiles. Machines and electricity were inextricably linked to the final decades of Italian liberalism, to academic programs and to new narratives in art and literature. The excitement was palpable. Artists of the Futurist movement were preparing to publish their first manifesto, on modern life, with its glorification of speed, transport, and industrialization (1909). Into this milieu of technological vitality stepped Luigi Casale.

Luigi Casale (1882–1927) was born in Langosco, in the Lomellina region of southwestern Lombardy (Fig. 3.1). He was the third of eleven children of Santino Casale and Maddalena (née Balocco). Casale, Sr was a farmer and administrator of the properties belonging to the earls of Langosco [1]. In 1908, Luigi Casale was awarded a degree in chemistry from the Royal Polytechnic of Turin (Reale Politecnico di Torino), where he studied under the tutelage of physical chemist Arturo Miolati (1869–1956). Miolati was professor of electrochemistry and had a particular interest in electrothermal and other nitrogen fixation processes, which he passed on to his students. During 1909–1912, Casale worked under Michele Fileti (1851–1914) at the institute of chemistry, University of Turin (Università di Torino), where in 1910 he received his postgraduate degree. It was there that he met fellow chemistry student and his future wife, and future business partner, Maria Rosa Sacchi (1889–1950). She was born in Pieve del Cairo (Pavia) on 13 May 1889 to Francesco and Ernesta Biscardi Sacchi. In 1907, Maria had enrolled at the University of Turin to study chemistry, also as a pupil of Fileti. She graduated in July 1911. In the same year, Luigi and Maria were married. Apparently Maria sold her small home to finance their studies.

The announcement of the successful outcome of Fritz Haber's work at Karlsruhe on synthetic ammonia no doubt further stimulated Luigi Casale's already great interest in physical chemistry and gas reactions at high pressures. Like many other researchers, he migrated to Berlin, where during 1912–1913 he and Maria undertook research in physical chemistry under Walther Nernst. Luigi's arrival took place shortly after the litigation in Berlin over the validity of Haber's ammonia patents was resolved. Several parties had contested the patents, including Griesheim-Elektron, whose consultant was Nernst. However, when offered a substantial honorarium by BASF, Nernst changed sides and supported Haber. This was an impressive display of the power of physical chemistry, and of money. After returning to Italy, Casale undertook research as head of the laboratory of organic synthesis in the institute of general chemistry at the University of Turin, where Maria was also an investigator [2]. As a research leader Luigi would soon be called upon to serve his country in a new branch of applied chemistry.

World War I transformed the relationship between science and industry in the major belligerent countries. Italy was no exception, especially after it joined the Allies on 23 May 1915, with a declaration of war against Austria-Hungary. This was just one month after Fritz Haber supervised the first major German poison gas attack, at Ypres, on French and Algerian troops. Luigi Casale moved to Naples and joined the university (Università degli Studi di Napoli Federico II), where he collaborated with physical chemist Oscar Scarpa, who had a strong interest in nitrogen fixation. Scarpa was also director of the Electrochemical Laboratory of the Royal Polytechnic High School of Naples. In addition to their joint interests,

Fig. 3.1 Luigi Casale as a young man. (Collection of Francesco Casale. With permission)

Casale became engaged in research into asphyxiating gases and protective measures for the war effort at the university's Institute of Pharmaceutical Chemistry and Toxicology (Fig. 3.2). At the university, Maria gave a course in methods of chemical analysis and lectured to first-year students (Fig. 3.3).

The institute was directed by Arnaldo Piutti (1857–1928), a member of the Army Commission for the Production of Poisonous Gases. This commission was hastily assembled immediately after Italy joined the war. In 1917, during a laboratory experiment with a new and highly poisonous gas, carried out without appropriate protection in place, Luigi Casale was involved in a serious accident. Two colleagues were killed instantly. Luigi Casale narrowly escaped. His lungs were damaged, and he was hospitalized for two months in a critical condition. He was discharged with a pulmonary abscess (eventually the cause of his death in 1927) [3].

There was now another wartime venture in which Casale was involved, one more long lasting. After recovering in 1917, and probably as a result of prompting from Miolati, a member of the supply committees set up by what was soon to become the Italian Ministry for Arms and Munitions, Luigi turned to the study of

Fig. 3.2 Luigi Casale in the laboratory, probably at Naples, around 1915. (Collection of Francesco Casale. With permission)

synthetic ammonia. For Italy, nitrogen products were needed not only for munitions, but also for fertilizers, because supplies of grain from Russia and Romania were cut off due to the impact of the war on the Dardanelles. After a short time he then became involved at two of the new electrochemical firms that had been established with the encouragement of the Italian government to supply products essential to the national defense [4].

Casale first joined the IDROS company (Società IDROS; its name derived from idrogeno and ossigeno), located in the city of Terni. IDROS was founded on 19 February 1916, to supply hydrogen to the Italian navy for dirigible and other aerial balloons, as well as oxygen for oxyacetylene welding. It relied on hydroelectricity from two rivers, the Nera and Velino, to feed its electrolyzers. The city's industrial backdrop would play an important part in Casale's painstaking groundwork on the direct combination of hydrogen with nitrogen and his later industrial activities. Terni was also the location of the first modern steel works in Italy, the Società degli Alti Forni Fonderie e Acciaiere di Terni (SAFFAT), incorporated in 1884. SAFFAT drew on water and, later, hydroelectric power from the local rivers. Nearby was an abandoned ironworks, La Ferriera, which had closed in 1905. The steelworks

Fig. 3.3 Maria Rosa Sacchi-Casale in the laboratory, probably at Naples, around 1915. (Collection of Francesco Casale. With permission)

flourished, however, producing specialty steels for the military, including cladding and cannons, and rails for the still expanding railway network [5, 6].

Wartime needs led to a revival of industrial activity at the site of La Ferriera, which became the location of IDROS. The objects of establishment of IDROS included, in addition to production of hydrogen and oxygen, the extraction of nitrogen from the atmosphere and generation of electricity. Prominent members of the board of directors were the French-American entrepreneur René Leprestre, as president, the lawyer Mario Santangelo, as CEO, and inventor Carlo Andreucci, as technical expert [7].

At IDROS during 1917, Luigi Casale started his research into synthetic ammonia jointly with Andreucci, encouraged by Santangelo, who took a more than passing interest in technical innovations in chemistry. In a small workshop in a corner of the old iron-making factory, with the backing of Santangelo, Casale and Andreucci developed a method for the continuous production of ammonia in a closed apparatus, with the recirculation of unreacted gases, as in the Haber method. The reaction required a novel catalyst, and was, according to patent documents, carried out at 350–600 °C. Santangelo acted as the legal representative for both Casale and Andreucci. They filed patent applications for the method in Italy on 18 July 1917, and in France on 19 October 1917 (the Italian patent was not granted; the French patent, no. 495725, was granted in 1919). Though the pipework was stated to be capable of withstanding gases at high pressures, no actual pressure range was specified, perhaps as a measure aimed at preventing imitation. For Casale, the neighboring SAFFAT steel works at Terni provided a steel pressure vessel. It was, according to later accounts, the barrel of a decommissioned naval cannon from the battleship *Dante Alighieri.* IDROS, however, was busy producing much-needed gases for the military, and did not have available suitable facilities for pilot plant studies.

In July 1918, Casale signed an agreement as a consultant for Alfonso Vitale, of the Società Rumianca (Rumianca; Stabilimenti di Rumianca Ingegner A. Vitale). This firm was founded in 1915, with headquarters in Genoa (Genova), to manufacture alkali and chlorine by electrolysis, using the Kastner–Kellner process, and to produce war gases, including, from 1916, phosgene. The works were located in the commune of Rumianca (later Pieve Vergonte), in the province of Novara, Val d'Ossola, northern Piedmont. During the war, the factory was under the administration of the Italian Army. The engineer Lorenzo Allievi (1856–1941), was a consultant to Rumianca. During 1905 he pioneered Italian aluminum manufacture, at Bussi in the Abruzzo region, through the Società Italiana per la Fabbricazione dell'Alluminio. Electricity for the Bussi factory was provided by the nearby hydroelectric power station of the Società Italiana di Elettrochimica, which also served the first calcium cyanamide factory, at Piano d'Orta, also opened in 1905. Allievi had a considerable interest in the calcium cyanamide industry. He was on the boards of Società Italiana per la Fabbricazione di Proditti Azotati e di altri Sostance per l'Agricoltura (SIPA), La Società Generale per la Cianamide (Cianamide) (until it was disbanded in 1914), and the calcium carbide and calcium cyanamide manufacturer Società Italiana per il Carburo di Calcio, Acetilene e altri Gaz (Carburo), with manufacturing facilities at Collestatte and Papigno, near Terni. Carburo had been managed since 1916 by the engineer Enrico Cairo, who held an interest in Cianamide. Carburo and Cairo will play a part in this story.

At the end of 1918, Casale moved to Rumianca, which offered two significant benefits for his research: the availability of by-product hydrogen from Kastner–Kellner cells, and an idle pilot plant capable of withstanding high pressures. The most reliable account of this early period in Casale's ammonia research derives from a court document, an expert report, drawn up in 1922 in support of a claim by Casale against Rumianca for damages and costs arising as a result of their falling out in around August 1919 (see later). In this case, Casale's expert in the lawsuit was the chemist Professor Giuseppe Bruni (1873–1946), who held posts at the Royal Polytechnic of Milan (Reale Politecnico di Milano) and the Royal University of Milan (Reale Universita di Milano). From 1917, he had headed the chemistry laboratories of Pirelli.

Bruni's document informs us that by May 1919, Luigi Casale had succeeded in producing synthetic ammonia at the rate of two hundred kilograms per day. Casale and Rumianca signed a fifteen-year agreement on 17 May, which entitled Casale to receive a royalty of 0.1 lira for each kg of ammonia produced, to be reduced to 0.08 lira per kilogram if production exceeded 1500 kg per day. It was understood that Rumianca's output of hydrogen would match the needs of the ammonia unit, which was designed to produce ammonia at the rate of around one ton a day by no later than 30 September 1919. Casale's process was observed in action by a party of interested American investors, accompanied by Professor Bruni on 14 and 23 July 1919 [7, 8]. Bruni had been commissioned as an independent expert to provide an opinion, as an initial proof of feasibility.[1]

3.3 René Leprestre

Bruni's assessment of Casale's pilot plant at the Rumianca works was undertaken at the behest of the potential American investors. They had been brought together by René Leprestre (1876–1941) of IDROS (Fig. 3.4). Leprestre now enters the stage in a leading role, and he requires more than a passing introduction. Brief biographical information on his family is also necessary for reasons that will later become apparent. Leprestre was born to Gregoire Leprestre in Nice on 5 January 1876. He studied physics at the University of Paris (the Sorbonne) in the mid-1890s, at a time when Marie Curie was completing her studies there. He lived with his wife, Jeanne, and two daughters in Paris until just after his wife died in late December 1905. In May 1906, Leprestre immigrated to the US. Based on what is known of his occupations, at least until around 1920, he was probably engaged in various commercial activities.

Leprestre resided in New York City, where in 1907 his occupation was given as financier, and his address was the Waldorf Hotel. He appears to have been frequently on the move, carrying out a trade in exports and imports.[2] During World War I, he was a partner of a certain Arthur C. Miller in what was described as the export business Leprestre, Miller & Co., Inc. Leprestre was president of this company. It was associated with a firm which handled the breeding and procurement of horses and mules for export to European armies and later to the U.S. Army abroad [9].[3] In the summer of 1916, he traveled to France, England, and Italy in connection with this business (Fig. 3.5).

Leprestre's sojourn in Europe was extended beyond the end of the war. He had made good use of his time in Italy. There, he had carefully cultivated his

[1] Judgment in favor of Luigi Casale was announced on 30 September 1922, and confirmed by the Turin Court of Appeal in 1923. Following review by a panel of experts, in 1929, Rumianca was ordered to pay Casale's heirs almost three million lira for loss of profits and compensation for delays in bringing his process to market.

[2] Leprestre's residential address on arrival in May 1906 was 610 West 116th Street, New York City. His address in 1907, as recorded on a document accompanying his May 1911 naturalization papers, was the Waldorf Hotel, New York. In 1923, his residential address was 501 5th Avenue, which was his home for over a decade. In 1923, at the founding of the Hydro-Electric Chemical Company, his address was given as Hotel Weylin, at 527–531 Madison Avenue and 40–54 East 54th Street.

[3] Leprestre, Miller Stock Farm, Inc., and Leprestre, Miller & Co., Inc., both located at 66 Broadway, New York. The Leprestre, Miller Stock Farm was located at Goshen, New York. Charles J. Hardy was a director of both firms and in 1923 became involved in Luigi Casale's operations in the US (see Chap. 8).

Fig. 3.4 René Leprestre, in 1916. (Elizabeth Burrows [family] Collection. With permission)

political and military contacts. Through these networks Leprestre, by all accounts a somewhat charismatic individual who described himself as an industrialist, was in 1917 in close touch with Italian politicians, in particular Minister of Finance, Francesco Saverio Nitti (later prime minister of Italy, 1919–1920). Leprestre—apparently possessed of certain political skills—and Nitti touched on arrangements for expanding industrialization in Italy in order to enhance the war effort, through hydroelectric power, and the import of much-needed war materiel from the US [10, 11]. There was also a dire need for nitrogen products, both as fertilizers and for use in the manufacture of explosives [12]. From 1918, starvation in central Europe and a rice crisis in Asia brought home the dire need for production and distribution of nitrogen fertilizers.

In the summer of 1919, Leprestre took up a serious interest in Casale's ammonia process, perhaps after he was informed of progress at Rumianca. On 2 August 1919, he came to a secret agreement with Casale, signed before a notary in Rome, to work together to bring Casale's ammonia process to market [7, p. 63]. It included their joint ownership of the Casale patents, and compensation to Casale. According to the agreement Casale would permit Bruni, the technical expert appointed by Leprestre, to draw on data from the ammonia unit until September 1919, in order to determine the cost for each kilogram of synthetic ammonia produced. On the basis of a satisfactory report prepared by Bruni, Leprestre would be committed to pay Casale the sum of one hundred thousand lira as compensation, in addition to four hundred thousand lira as the initial payment for the acquisition of processes and related patents, for use both in Italy and abroad. It was then

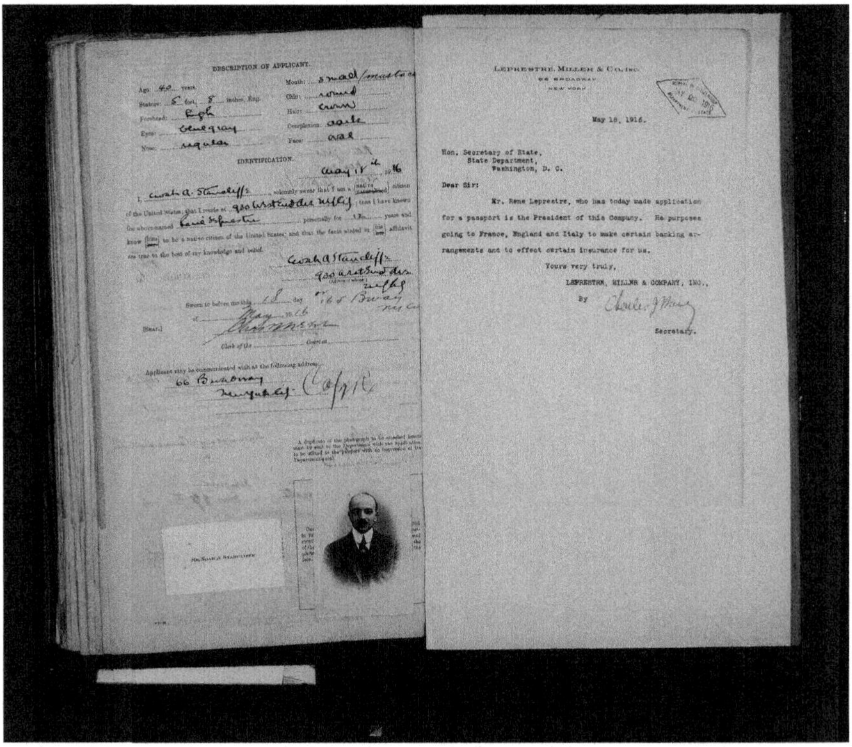

LEPRESTRE, MILLER & CO., INC.
66 BROADWAY
NEW YORK

May 18, 1916.

Hon. Secretary of State,
State Department,
Washington, D. C.

Dear Sir:

Mr. Rene Leprestre, who has today made application for a passport is the President of this Company. He purposes going to France, England and Italy to make certain banking arrangements and to effect certain insurance for us.

Yours very truly,
LEPRESTRE, MILLER & COMPANY, INC.,
By
Secretary.

Fig. 3.5 Letter from Leprestre, Miller & Company, Inc., to the Secretary of State, The State Department, Washington DC, in support of René Leprestre's application for a passport that would enable him to travel to France, England, and Italy in order to undertake business matters on behalf of the company, 18 May 1916. (Elizabeth Burrows [family] Collection. With permission)

reiterated that the agreement with Rumianca would remain in place subject to its meeting the conditions of its contract prior to 30 September.

However, in August there was a falling out between Rumianca and Casale, probably after Rumianca found out about the clandestine contractual arrangement that had been drawn up between Casale and Leprestre [7, p. 64]. The outcome was that Rumianca decided, without prior notice, to cancel its agreement with Casale and refused to provide Bruni and Casale with further access to the ammonia unit. There was also the fact that Rumianca's consultant, the engineer Lorenzo Allievi, held financial interests in both Carburo, the main producer of cyanamide fertilizer. For Carburo, synthetic ammonia was seen as a disruptive technology [13–15]. Whatever the reason, Bruni's 1922 expert report in support of Casale's lawsuit for damages against Rumianca was based on his observations made in 1919 [8].

According to the prominent American chemist Charles L. Parsons, Leprestre was around 1920 in the US promoting the Casale ammonia process with a view towards attracting investors [16]. After the falling out with Rumianca, Casale moved back to IDROS at Terni, which, following the cancelation of wartime contracts for hydrogen used in dirigible balloons, and other products, was now seeking new uses for its gases. Drawing on pure hydrogen available at the IDROS factory, Casale built an improved ammonia unit. In particular, he redesigned the converter in order to overcome the problem of fracturing of the outer wall of the converter, as had been experienced at Rumianca. Moreover, the ammonia unit was operated at around 800 atm.

Leprestre's decision to back Luigi Casale was to become one of the most important, and certainly most far-reaching, decisions of the entrepreneur's life. Luigi Casale soon discovered that he was entering into a partnership with a financial wizard.

3.4 Scale up of Casale's Ammonia Process

Leprestre's enthusiasm for Luigi Casale's process arose from its proven reliability through demonstrations and successful tests. IDROS ordered a novel multi-stage compressor, capable of working at 800 atmospheres, from Fonderia del Pignone, of Florence (Firenze), in Tuscany. Casale participated in the design of the final stage. By burning air in a stream of pure hydrogen, as obtained by electrolysis—the oxygen of the air reacted with hydrogen to form steam—it was possible to generate pure nitrogen. The mixture of hydrogen and nitrogen, made up in the correct (stoichiometric) proportions, was compressed and then fed to the converter from which, after drying and cooling, liquid, anhydrous ammonia was obtained. Unreacted gas was recycled to the converter.

As we shall see, the ammonia synthesis converter (reactor) was to become the defining feature of the synthetic ammonia industry. The various processes differed mainly according to the internal design of the reactor, the reaction conditions, and the nature of the catalyst. While the core fundamental synthetic ammonia technology had been developed by Carl Bosch and his team at BASF in Germany, there was certainly room for improvements, modifications, and the introduction of new features. Luigi Casale calculated that if the pressure was raised to around 800 atmospheres, the equilibrium concentration of ammonia as a function of pressure and temperature would be far in excess of that achieved in 1909 by Haber and his research assistant Robert Le Rossignol. At 350 °C (a somewhat lower temperature than the Haber–Bosch process) it might reach around 20%. This was the range that interested Casale. The technical challenge was how to achieve it on a large scale.

By 1920, Luigi Casale had come close to perfecting his high-pressure process, operating at 800–850 atm. Several challenges had been overcome, in particular the problem of hydrogen reacting with the carbon of the steel wall of the shell of the converter (the pressure vessel). Carl Bosch had established that at around 250 atm and even at 200 °C, ordinary steel became brittle due to the formation of iron hydride; he came up with a solution based on a double-wall converter design.

Casale also dealt with this difficulty brilliantly, still using ordinary steel, but at the far higher pressure. The interior of his converter was fitted with concentric steel tubes. The inner wall of the converter shell acted as the outer pressure tube, and was kept cool by incoming gas (Fig. 3.6).

It is important to emphasize a great advantage of working under the more extreme conditions of Casale's process, especially at that time. It produced liquid, anhydrous ammonia by simply drying and then cooling the gas mixture. The Haber–Bosch process required a refrigeration unit in order to produce liquid ammonia.

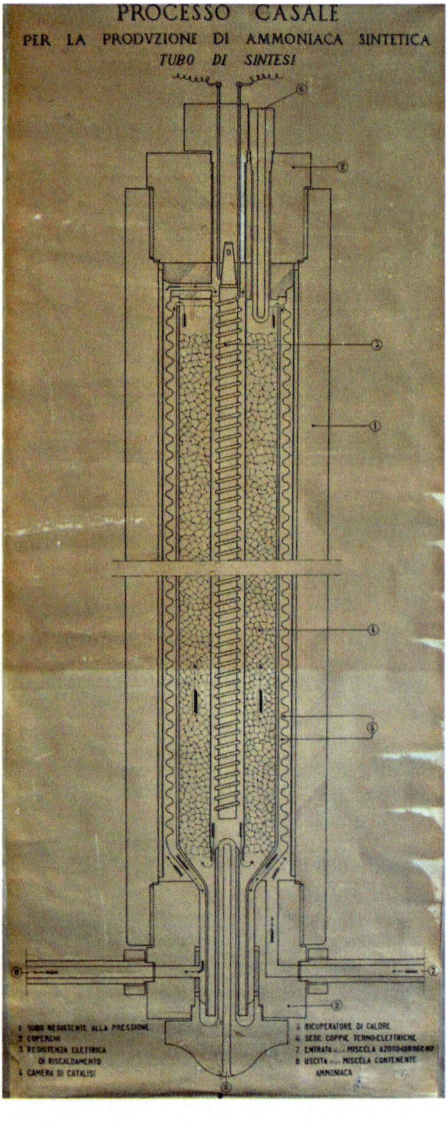

Fig. 3.6 Early Casale pressure vessel, or converter, for the synthesis of ammonia. (Casale SA. With permission)

Removal of the liquid ammonia shifted the equilibrium towards even more ammonia. As in the Haber–Bosch process, the unreacted mixture of nitrogen and hydrogen was recycled, with a new charge of the mixed gases. The catalyst at the very high pressure did not need to be as active as in the case of the Haber–Bosch process. It was based on iron filings with added promoters, or activators, generally oxides of metals. Casale had achieved the safe, successful combination of pure hydrogen and nitrogen at a high pressure and an elevated temperature in the presence of a catalyst on a pilot plant scale.

After Casale, the inventor-entrepreneur, and Leprestre, the early-stage venture capitalist, teamed up in 1919 to promote Casale's ammonia process, they spared no efforts in protecting Casale's inventions through patents in Italy and other countries. Some of Casale's first patents dealing with ammonia were assigned to both Casale and Leprestre. They included modifications to the electric heating coil, the separation of ammonia, and production of pure nitrogen. Casale filed patents in his own name, including for the production of a nitrogen–hydrogen mixture by burning air in hydrogen [17]; and for the recirculation of unreacted gases [18]. Casale also invented an electrolyzer for producing hydrogen [19].

Casale, Leprestre, and IDROS were, in 1920, now intent on capitalizing on Casale's invention and the growing interest in synthetic ammonia. At this stage, Enrico Cairo, as head of the Carburo firm, became involved. Cairo, who saw Casale's synthetic ammonia process as a threat to Carburo's calcium cyanamide business, had placed a major obstacle before IDROS. This happened after IDROS applied for a permit from the Terni municipality to draw additional water from the River Nera in order to generate sufficient hydroelectric power as required in the ammonia process. Carburo invoked its right of preemption in the use of local waterways, on the grounds of an agreement with the municipality signed in 1910, and blocked the application from IDROS.

However, by the close of 1920, following direct negotiations between Cairo and Leprestre and Casale, Cairo was persuaded to drop his opposition, and agreed to collaborate with IDROS. An important step was a provisional agreement signed on 3 January 1921 between Casale, Leprestre, and IDROS, on one side, and Cairo, of Carburo, on the other, in which Carburo, by agreeing to allow the supply of water required for generation of electricity to IDROS, received permission to send its technical experts to Terni in order to evaluate the Casale synthetic ammonia process. Carburo's intention was to bring about industrial exploitation in Italy, and in countries where it was well-established, those of the former Austro-Hungarian Empire, and Romania [20]. Cairo's change of mind reflected the increasing international interest in synthetic ammonia, and the postwar decline in demand for carbide for acetylene and cyanamide. Moreover, Carburo, through its involvement in carbide and cyanamide processes, had well-established contacts with foreign firms that worked the same electrothermal processes. One of them was the enterprise managed by Shitagau Noguchi in Japan. Because Noguchi was already keen to take up the manufacture of synthetic ammonia, Carburo saw an opportunity to introduce Noguchi to Casale. According to an agreement between Casale and partners, and Cairo, "Mr Enrico Cairo was able to secure the license for one plant by the Japanese group," and as compensation Casale and Leprestre signed the "well known letter of the 10%" [21]. This "well known letter," dated 31 March 1921, was an agreement between Casale and partners, and Cairo, in which Cairo was assigned 10% of "all present and future benefits of the new company" (Fig. 3.7).

Santangelo was called upon by Casale to assist in the creation of a new company to promote the ammonia process. The new company was Ammonia Casale SA.

3.5 Ammonia Casale Società Anonima

Casale and Leprestre decided on a course of action that involved founding a firm outside of Italy to license the Casale process worldwide rather than engaging in partnerships for undertaking large-scale ammonia manufacture in Italy. This independent firm, removed from the economic and political instability that was then afflicting Italy, would be well-placed to attract international investors from nations with stable currencies. For its headquarters, they chose Lugano, in Switzerland's southernmost canton, Ticino, the only canton where Italian was the sole official language. It was a shrewd move. In drawing up plans to provide licenses and technical assistance, Casale and Leprestre were following the strategy followed earlier by the Rome-based calcium cyanamide licensing company that represented Frank-Caro interests.

Casale and Leprestre established Ammonia Casale Società Anonima in Lugano on 27 April 1921 (Fig. 3.8 and Table 3.1). Of the two hundred issued shares in Ammonia Casale, Luigi Casale held seventy-eight and Leprestre, ninety. Thirty-two shares were assigned to minor shareholders, including ten to Cairo of Carburo, and one to Santangelo. Cairo had signed a further agreement with Casale on 23 April 1921, in which the 10% for all present and future benefits was replaced by ten shares in Ammonia Casale SA.

Vincent Handley, one of the group of Americans that had been brought together by Leprestre in 1919, was appointed the company secretary. Handley was the son of the accomplished American sculptor Francis Montague Handley (1841–1908), who lived in Rome and was a Privy Chamberlain of Sword and Cape to two popes (Pius X and Leo XII) as well as the first American to be made a Commander of the Order of St. Gregory. Early in the twentieth century, Vincent worked as a banker in the US, where he was married in 1911. After 1915, he became involved in enterprises connected to the Italian war effort, probably due to his status as a member of a prominent American family in Italy. He was appointed company secretary of the American Italian Commercial Corporation, established by Banca Commerciale Italiania in New York in 1917 to facilitate trade in steel and ships destined for Italy. It is likely that through such banking and finance activities that he would make the acquaintance in the US of René Leprestre, then engaged in shipping horses to Italy. Vincent Handley received his call-up papers to the U.S.

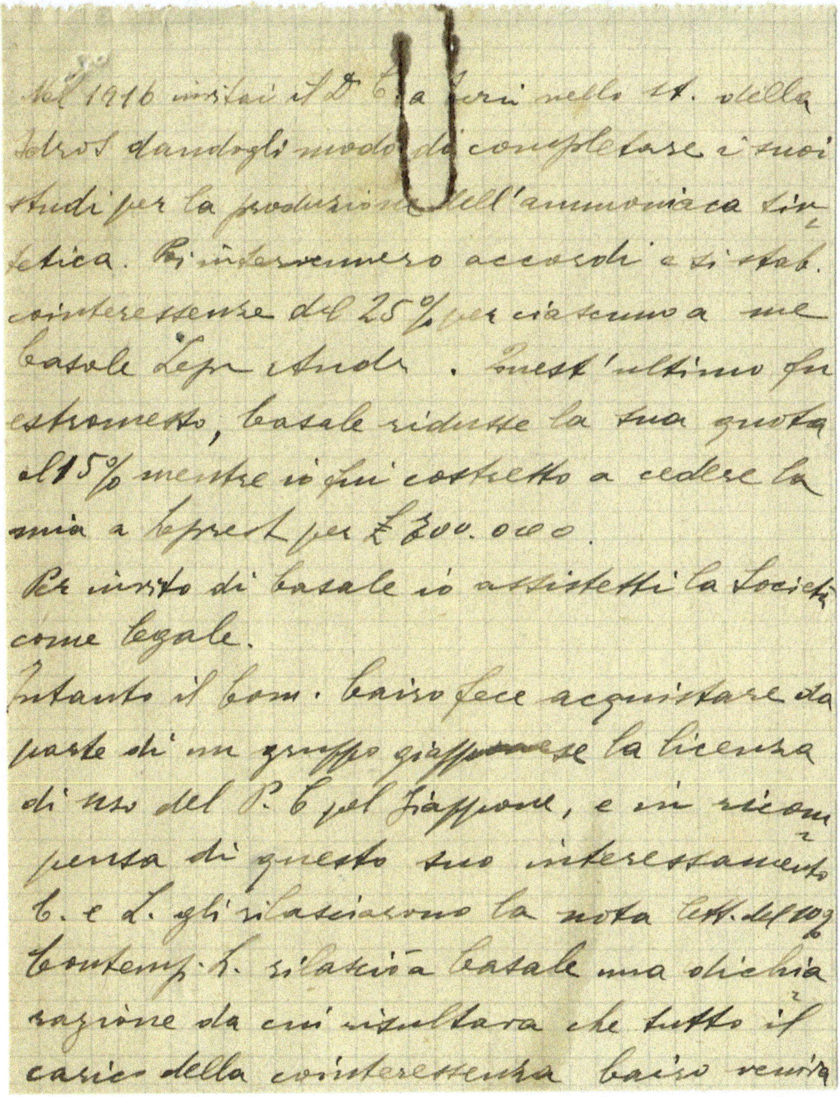
Nel 1916 invitai il D.r C. a Terni nello St. della
[illegible] dandogli modo di completare i suoi
studi per la produzione dell'ammoniaca sin-
tetica. Poi intervenimmo accordi e si stab.
cointeressenze del 25% per ciascuno a me
Casale Lepr. e Andr. Quest'ultimo fu
estromesso, Casale ridusse la sua quota
al 15% mentre io fui costretto a cedere la
mia a Lepres. per £ 200.000.
Per invito di Casale io assistetti la Società
come legale.
Intanto il Com. Cairo fece acquistare da
parte di un gruppo giapponese la licenza
di uso del P. C. pel Giappone, e in ricom-
pensa di questo suo interessamento
C. e L. gli rilasciarono la nota lett. del 10%
Contemp. L. rilasciò a Casale una dichia-
razione da cui risultava che tutto il
carico della cointeressenza Cairo veniva

Fig. 3.7 Memorandum or covering note in the hand of Mario Santangelo describing the change in the partnership arrangement between Carlo Andreucci, Enrico Cairo, Luigi Casale, René Leprestre, and Santangelo, immediately prior to the formation of Ammonia Casale SA in April 1921. This document also refers to Cairo's role in securing a contract for the erection of a factory based on Casale ammonia technology in Japan. (Casale SA. With permission)

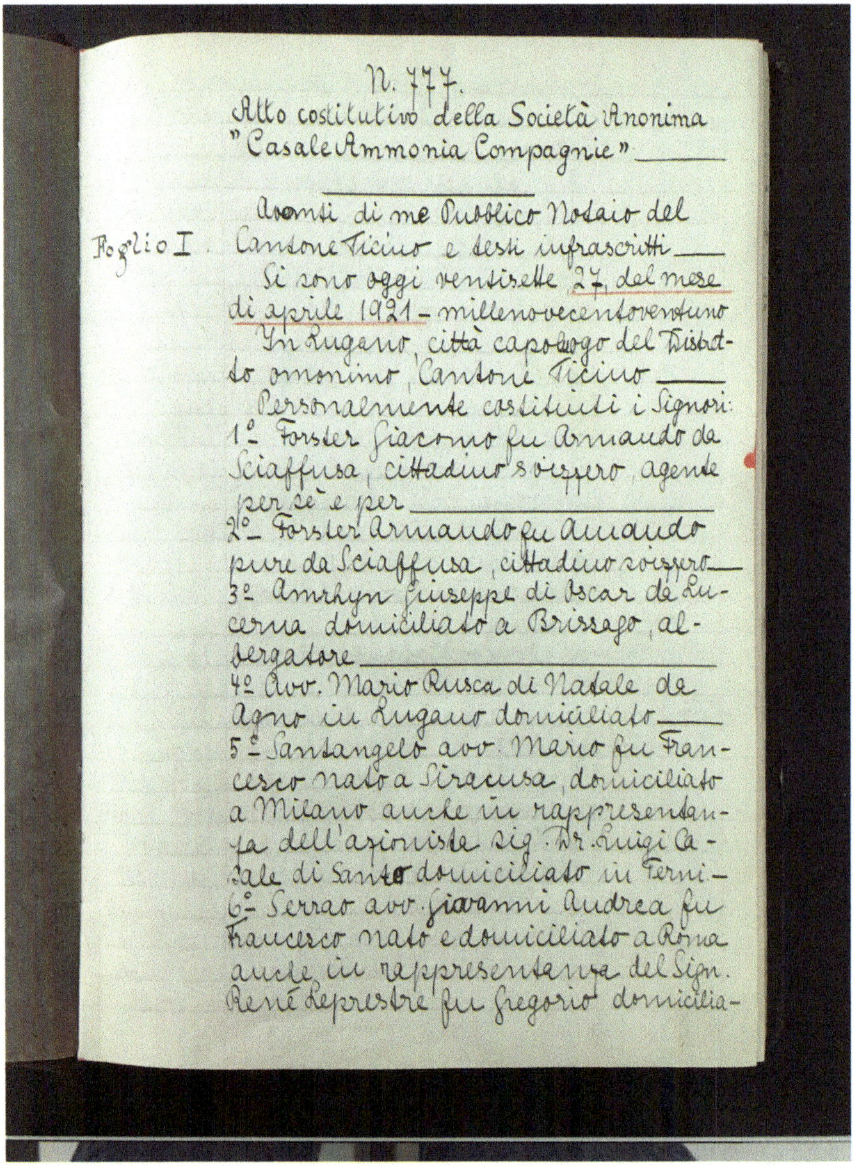

N. 777.

Atto costitutivo della Società Anonima
"Casale Ammonia Compagnie" ___

Avanti di me Pubblico Notaio del
Foglio I. Cantone Ticino e testi infrascritti ___
Si sono oggi ventisette, 27, del mese
di aprile 1921 - millenovecentoventuno
In Lugano, città capoluogo del Distretto omonimo, Cantone Ticino ___
Personalmente costituiti i Signori:
1° Forster Giacomo fu Armando da
Sciaffusa, cittadino svizzero, agente
per sè e per ___
2° Forster Armando fu Armando
pure da Sciaffusa, cittadino svizzero ___
3° Amrhyn Giuseppe di Oscar da Lucerna domiciliato a Brissago, albergatore ___
4° Avv. Mario Rusca di Natale da
Agno in Lugano domiciliato ___
5° Santangelo avv. Mario fu Francesco nato a Siracusa, domiciliato
a Milano anche in rappresentanza dell'azionista sig. Dr. Luigi Casale di Santo domiciliato in Terni -
6° Serrao avv. Giovanni Andrea fu
Francesco nato e domiciliato a Roma
anche in rappresentanza del Sign.
René Leprestre fu Gregorio domicilia-

Fig. 3.8 First page of the Act of Constitution of Ammonia Casale SA, Lugano, 27 April 1921. Those listed are: 1. Forster, Giacomo, of Sciaffusa, agent; 2. Forster, Armando, also of Sciaffusa; 3. Amrhyn, Giuseppe, son of Oscar, lawyer, of Lucerne, resident in Brissago; 4. Rusca, Mario, lawyer, son of Natale Agno, resident near Lugano; 5. Santangelo, Mario, lawyer, born at Siracusa and resident in Milan, also representing shareholder Dr Luigi Casale, resident in Terni; 6. Serrao, Andrea Giovanni, lawyer, born and resident in Rome, also representing René Leprestre. (Casale SA. With permission)

Table 3.1 Distribution of shares in Ammonia Casale SA at its foundation

Name	Shares
Luigi Casale	50 shares (represented by his patent holdings)
Luigi Casale	28 shares (corresponding to his capital investment)
René Leprestre	50 shares (represented by his patent holdings)
René Leprestre	40 shares (corresponding to his capital investment)
Giacomo Forster	5 shares
Armando Forster	5 shares
Giuseppe Amrhyn	5 shares
Mario Rusca	5 shares
Enrico Cairo	10 shares
Mario Santangelo	1 share
Andrea Giovanni Serrao	1 share
Total	200 shares, corresponding to a paid-up capital of 100,000 Swiss francs

Army in 1918. In 1919, he was back in Italy, where he joined his wife and son (who died in 1920).[4]

Promotion and licensing of Casale's ammonia process was the principal activity of the new company's Swiss office. Technical developments were confined to the Terni workshop. There was also an office in Rome. The business plan included supply of Italian-made equipment, mainly compressors and converters, to licensees of Casale's process. No doubt it was through Leprestre's networking that distinguished visitors were soon invited to Terni. They included, late in September 1921, the politicians Arlberto Beneduce and Luigi Sasparatto, both ministers under the government of Ivahoe Bonomi. Casale was praised for his work and vision in bringing about a viable process and Leprestre's important role as Casale's backer was acknowledged [7, p. 67]. Leprestre continued to focus on backing novel innovations in electricity and electrochemistry, which in November 1921 led to the establishment of another company in Lugano, the Pouchain Light Storage Company. It was founded to exploit the patents and processes of Alfredo Pouchain, of Turin, related to electric batteries and accumulators. The directors included Luigi Casale and two newcomers to the Casale business, Lieutenant Colonel Giuseppe Albisetti, a businessman, and Tomaso Quadri, an architect. However, this venture saw little if any success. In 1922, Miolati's interest in battery and accumulator chemistry led to the invention of an improvement on Pouchain's accumulator, the US patent for which was assigned to Leprestre.

[4] Vincent Handley was an accomplished artist, and provided the sketches for Piero Santini's book *Riding Reflections* (New York, 1932). Vincent's half-brother, Louis, was a well-known swimmer who participated for the US in the 1904 Olympic Games, as a member of the relay team, which achieved a gold medal, and who promoted the entry of women into swimming as a sport.

Luigi Casale, from his Rome office, continued to file patents for his ammonia and related processes in Italy, France, the UK, the US, Canada, and elsewhere. Casale's technical team of experts from 1922 included the engineers Renato Fabi (b. 1892) and Luigi Calissano. They were soon joined by another two engineers, Guido Santagostino and Iacopo Volpi (b. 1889). On 22 May 1921, Carburo absorbed IDROS and founded the Società Italiana Ammoniaca Sintetica-Processo Casale (SIAS), based at the IDROS works in Terni. The capital was 23 million lira: 5 million representing the Casale patents, some assigned jointly to Casale and Leprestre; 6 million IDROS; and 12 million Carburo. Carburo, apparently, had considerable control in the affairs of the consortium, including the final word in cases of disputes [22]. The first CEO of SIAS was Vincent Handley. The president of SIAS was Riccardo Bianchi (former general manager of Italian railways); and board members included: Arturo Bocciardio (of the Terni company), Enrico Cairo; Luigi Casale; and René Leprestre. Corporate changes elsewhere would dictate the future of this enterprise.

In September 1922, a 2-mtpd semi-industrial Casale ammonia converter was installed at Terni. John A. Harker (previously with the British Munitions Inventions Department's wartime Nitrogen Products Committee) traveled from London to examine the Casale process at work. The demonstration went well. In the report of his observations, Harker expressed surprise at the apparently casual attitude shown towards conducting the synthesis at an ultra-high pressure: "One of the things which struck me most about the working of the Terni plant, which had been only two days in operation, was that none of the individuals concerned appeared to regard the operations they were conducting as anything out of the common." Harker was not only impressed with the performance of the apparatus at Terni, but

Fig. 3.9 Manufacture of a Casale converter probably much later in the 1920s than indicated. (Casale SA. With permission)

also with its cost-saving features. "The process differs in details greatly from existing processes, is simpler and less costly to instal," he noted in a cable to colleagues in London [23, p. xvii]. It was no doubt on the basis of his report that Harker and civil engineer John F. Crowley, proprietor of the engineering consultancy J.F. Crowley in London, immediately agreed to act as representatives for Casale.[5] This was one year before the first full-scale plant for production of synthetic ammonia by the Casale process went on stream, in Japan (Fig. 3.9).

3.6 Casale in Japan and Europe

Just as with the promotors of the cyanamide process before 1910, Casale received great interest in synthetic ammonia from among certain entrepreneurs in Japan. One of them, as introduced by Enrico Cairo, was the aforementioned Shitagau Noguchi who, after visiting the Terni workshop in early 1921, paid a deposit for a license that covered use of Casale's process in Japan. Noguchi was backed by the Mitsubishi Bank, part of a major corporate group, or zaibatsu, involved in banking, chemicals, and other industrial activities. Before the war, Noguchi had studied the literature on synthetic ammonia processes. In 1921, he knew, mainly through Enrico Cairo, that there was only one viable process available for licensing: Luigi Casale's process. Shitagau Noguchi's license was granted in December 1921. His factory, at Nobeoka, in Miyazaki Prefecture, on the east coast of the island of Kyūshū was opened in September 1923. The daily rated capacity of the converters was of 7.5 tons. The success of the Japanese venture catapulted Luigi Casale and his ammonia process to fame. It is described more fully in the next chapter.

We now turn to growth in Europe. In 1920, SAFFAT, the Terni steelworks company, had come under the control of Banca Commerciale Italiana. From 1921, operations were managed by Arturo Bocciardo (1876–1959), who soon diversified into hydroelectric utilities. SAFFAT acquired concessions for the generation of hydroelectric power from the Nera and Velino rivers, and became an important supplier to local industries. Bocciardo also diversified into electrochemicals, including the fixation of atmospheric nitrogen, mainly by the cyanamide process. This followed the 1922 merger of SAFFAT, Carburo, and a branch of the engineering firm Vickers Ltd, of Sheffield, with the establishment of Terni-Società per l'Indùstria e l'Elettricità (the Terni company; during 1905–1906, Vickers had joined with Terni and shipbuilders Orlando and Ordero to form Vickers-Terni Società Italiana d'Artigliera di Armamenti). It was around this time that Carburo raised capital in order to acquire the shareholdings of Casale and Leprestre in the SIAS partnership.

[5] Dr John F. Crowley (circa 1884–1951) was born in Mourneabbey, County Cork and studied at the Royal College of Science, University College, Dublin, and University College, Cork. In the 1920s, he was involved with the Shannon hydropower scheme and was a member of the Water Power Legislation Committee and vice chairman of the Committee on the Water Power Resources of Ireland.

An outcome of these machinations was that the Terni company also assumed control of SIAS, and installed Casale ammonia technology at Nera Montoro, near the city of Terni, in buildings associated with the mothballed hydroelectric power station of the Società Idroelettrica di Villeneuve e Borgofranco (founded in Turin in 1913). A principal function of the revamped Nera Montoro site was the testing and development of ammonia units in collaboration with Ammonia Casale, work which was hampered at the Terni workshop by its limited facilities [24] (Table 3.2). The Terni company acquired a twenty-seven year lease on the Nera Montoro site, which had been used for production of chlorates prior to World War I by French and German firms, and from 1915 produced aluminum.

Table 3.2 Time line for early developments of Luigi Casale's Ammonia Process[a]

Year	Main events
1915	Establishment of Rumianca, at Rumianca (Pieve Vergonte)
1916	Establishment of IDROS at Terni, for production of hydrogen and oxygen for the military. The president is René Leprestre
1917	Luigi Casale joins IDROS, probably on the recommendation of Arturo Miolati. Casale collaborates with Carlo Andreucci on synthetic ammonia
1918	Casale moves from IDROS to Rumianca because Terni does not produce surplus hydrogen. For ammonia studies, Rumianca makes available redundant equipment, and by-product hydrogen from its alkali plant
1919	In May, Casale demonstrates synthesis of ammonia at the rate of 100 kg per day. Casale and Rumianca draw up a contract, signed on 17 May, specifying production by 30 September 1919 of one ton per day. Casale still owns his patents but agrees not to reveal confidential technical information for working the process developed at the Rumianca factory. Casale is allowed to deal with outside investors. In July, Leprestre visits Rumianca with Giuseppe Bruni, his technical expert, and representatives of a US group prepared to invest in Rumianca's ammonia, and promote the process outside Italy, subject to satisfaction of the conditions laid down in the Casale-Rumianca contract. Things fall apart because on 11 July Leprestre and Casale came to a separate agreement over control of the ammonia patents, and this was discovered by Rumianca. Rumianca in any case could not achieve the agreed output. The Casale-Rumianca agreement was declared void. Late in 1919, Casale returns to the now closed Terni works where he has sufficient working space, and a supply of hydrogen from the electrolyzers there
1920	In April, Casale's converter produces 200 kg of ammonia daily. In June, Casale completes the Terni ammonia synthesis unit and in July and September he files two patents for the ammonia process
1921	Ammonia Casale SA and SIAS are established. SIAS is created from IDROS and the calcium cyanamide manufacturer Carburo. Casale provides a license to the Japanese calcium cyanamide manufacturer Shitagau Noguchi
1922	A 2-metric-ton-per-day converter is installed at Terni (September). SIAS acquired access to larger premises at Nera Montoro for testing and development in collaboration with Ammonia Casale

[a]From reference [7]

The German interests in Nera Montoro were sequestered at the start of the war, and French firms, mainly Bouchayer et Viallet, of Grenoble, and Aluminium Français, in collaboration with the Italian government, completed the work. Bouchayer et Viallet requires an introduction. Founded in 1870 by Joseph Bouchayer, it was a prominent family business engaged in metalworking. From the early 1890s, it manufactured pressure pipes (penstocks) for several hydroelectric power stations. Over the following two decades the firm benefitted tremendously from the booming electrical industry, especially when applied to energy schemes for manufacture of aluminum and alkali. Aimé Bouchayer (1867–1928), eldest son of the founder, became head of the firm, which was turned into a joint stock company in 1912. Two of Aimé's brothers, both engineers, held important positions in the firm: Auguste Bouchayer (1874–1983), designer of hydraulic equipment, and in the 1920s considered the leading hydraulic engineer in France, and Hippolyte Bouchayer (1872–1957), whose network of business and political contacts was put to good use in garnering business from different sectors connected with the manufacture and uses of electricity. Following the end of the war, Bouchayer et Viallet brought out the other interests at Nera Montoro, and Hippolyte became head of Società Idroelettrica di Villeneuve e Borgofranco [25]. Looked at through a wider lens, this would have a significant impact on the growth of Luigi Casale's business.

Hippolyte Bouchayer was also prominent on the board of Tréfileries et Laminoirs du Havre (TLH), manufacturer of copper and other metal products for electrical transmission systems, and the third largest manufacturer of electrical equipment in France. TLH, a forward looking company with, through its leaders, important industrial and political contacts, during the war attempted to take over the largest French producer of aluminum, Société Electrométallurgique Française. When this failed, TLH in 1916 invested in another aluminum producer, Compagnie des Produits Chimiques d'Alais et de la Camargue (CPAC) as a result of which Bouchayer represented TLH on the board of CPAC. Though TLH did not gain as large a share of the aluminum market as it had wanted, it promoted interests in aluminum production in Italy, Norway, and elsewhere. In 1921, CPAC merged with Société Electrométallurgique de Française to create Compagnie des Produits Chimiques et Électrométallurgiques d'Alais, Froges et Camargue (Compagnie Péchiney), better known as Pechiney. Based in Lyon, Pechiney was diversifying its main business of electrochemical production of aluminum.[6] Around 1920, demand for aluminum had declined, as had the economy of Italy. The Nera Montoro complex was by then already shut down.

It was Bouchayer's involvement with hydropower and Nera Montoro that brought him into contact with Luigi Casale. Within a short time, they had developed a close working relationship. Bouchayer would now play a role in Pechiney's

[6] According to a late 1970s listing in the Casale archive, Pechiney ordered two ammonia units in 1922, with a total rated capacity of 5-mtpd.

adoption of the Casale ammonia process. He would have seen the 7.5-mtpd converters at Nera Montoro, manufactured by a subsidiary of the British engineering firm Sir W.G. Armstrong-Whitworth & Co., at Pozzuoli, in southern Italy. The associated machinery for the ammonia unit, including compressors, was installed at Nera Montoro during 1923–1924 [24] (Figs. 3.10 and 3.11). Ammonia was produced in 1925. In early 1927, the rated capacity of the ammonia facility was 31-mtpd.

In 1923, the stage was set for a special relationship between Hippolyte Bouchayer, representing Pechiney, and Ammonia Casale. As a result, by 1927, and through Pechiney as licensee, ten French factories had invested in the Casale process. Most of the licenses had been issued by 1924. They represented a far greater number of users of the Casale process than in any other country. Until 1927, the French factories generally operated with 5- or 7.5-mtpd converters, the standards until that time.

Ammonia Casale received several orders for ammonia units elsewhere in Europe. One of the first, in early 1923, was for the Franco-Spanish company Industrias Aragonesas SA (Sociedad Española de Industrias Aragonesas) in Sabiñánigo, Aragón, Spain. The Spanish company had been founded in 1918 by Hippolyte Bouchayer and Spanish capitalists to construct hydroelectric power stations in the Tena valley for the chemical industry. Once again, Hippolyte Bouchayer was instrumental in introducing the Casale process. Aimé Bouchayer and his nephew Henri Bouchayer were associated with Société des Forces Motrices et Usines

Fig. 3.10 Ammonia factory of the Terni company, at Nera Montoro, Italy. (Casale SA. With permission)

Fig. 3.11 Compressor for mixed gases, Nera Montoro. (Casale SA. With permission)

de l'Arve (FMUA), which in 1916 had been absorbed by the Pechiney forerunner CPAC. FMUA was involved in the procurement of electrical equipment for the Spanish venture. Pignone supplied compression equipment. Hydrogen was obtained from electrolyzers and nitrogen from both burning of air and a Linde plant. Ammonia was manufactured from 1924, making the site the second Casale factory to begin full production. Also in 1924, as a result of lobbying by Hippolyte Bouchayer and Pechiney, Ammonia Casale technology was adopted for a large state-sponsored ammonia factory at Toulouse (Chap. 4).

An important user of Casale's process was Etablissements Kuhlmann, which established joint subsidiaries with several leaders in the French coal and coking industry. They included Compagnie des Mines de Lens which in 1923 received a license for operating the Casale process in northern France. There, Mines de Lens and other mining firms had established jointly with Pechiney the Société Ammonia, which licensed the Casale process, including to Kuhlmann. In June 1924, Kuhlmann announced the formation of Anzin-Kuhlmann Chemical Products Co. (Société des Produits Chimiques Anzin-Kuhlmann) at Anzin in order to produce ammonia by the Casale process. Production began on 4 December 1926, using hydrogen from the coking ovens of SA Mines d'Anzin. Initially, three 7.5-mtpd ammonia units were installed.

The generation of hydrogen from byproduct coking ovens, as introduced in Belgium and France, was a significant development relevant to the expansion in

Fig. 3.12 Circulation pumps, at left, and compressors for mixed gases, Casale process, at Usines Electriques de la Lonza SA, Visp (Viège), Switzerland. (Casale SA. With permission)

the use of the Casale's process (see Chap. 7). This was through Société Ammonia. In December 1923, Belgian Solvay, which had earlier expressed an interest in investing in Ammonia Casale, acquired a 13.3% stake in Société Ammonia. This provided an opportunity for Solvay to become closely connected with the French chemical industry.

During 1925–1926, the Casale process was introduced at the French wartime Soulom facility (Hautes-Pyrénées) that hitherto had worked the Birkeland-Eyde electric arc process. The source of hydrogen was electrolytic, using the process of the UK inventor A. Edgar Knowles [26].[7] The six ammonia units at Soulom were each of 7.5-mtpd rated capacity. A Casale factory was under construction for Usines Electriques de la Lonza SA (Lonza), at the company's Valais works in Visp (Viège), Switzerland, of 2,500-mtpy capacity (opened in 1927) (Figs. 3.12 and 3.13).[8] Again, hydrogen was obtained by electrolysis.

[7] The Knowles process was controlled by the International Electrolytic Plant Company, of Sandycroft, Cheshire, England.

[8] Usines Electriques de la Lonza SA (Lonza) was founded in October 1897 in St Maurice, Switzerland. It originally engaged in carbide manufacture, and from 1915 also produced calcium cyanamide.

Fig. 3.13 Casale ammonia converters of rated capacity 7.5-mtpd, at Usines Electriques de la Lonza SA, Visp (Viège), 1930s. (Casale SA. With permission)

3.7 Giacomo Fauser's Process

By 1921, another successful Italian synthetic ammonia process was available. It was based on the studies of Giacomo Fauser (1892–1971), an engineer from Novara, west of Milan. Fauser's converter was worked at 250–300 atm, under conditions close to those of the Haber–Bosch process. Like Casale, Fauser was a semi-independent inventor.

As a teenager, Fauser apprenticed at his father's metalworking factory, where he assembled an electrolysis cell that produced oxygen for a welding machine. A version was later taken up, apparently, by the Electrochemical Company of Novara (Società Elettrochimica Novarese). Fauser studied mechanical engineering at Milan Polytechnic, where he became interested in ammonia synthesis, no doubt stimulated, as was Casale, by an awareness of what the Haber–Bosch process had achieved and the potential for its application to Italy's large agriculture sector. However, in common with Casale and Georges Claude in France there was no contact with BASF and Fauser worked alone. Fauser's early involvement, like his electrolysis cell, is partly anecdotal, and was probably on the following lines. For a source of pure nitrogen he investigated the burning in air of hydrogen, obtained from electrolysis cells. After devising the method of synthesis, and assembling all the necessary pipework, he then set to work in his garden using an old 250 mm howitzer as an experimental high-pressure converter to synthesize the ammonia. The shell was around 6–7-cm-thick, sufficient to withstand a pressure of up to 300 atmospheres, generated from a compressor perhaps available from his father's workshop. Here, he may well have benefited from the availability of surplus war equipment. The first catalyst consisted of trimmings from old iron horseshoes, which were low in carbon and sulfur. In what was probably a more advanced converter, an electrical heating element brought the gas mixture to the required temperature. The reaction was cooled by circulating water through tubes immersed in the catalyst. In 1920, he claimed production of ammonia at the rate of 100 kg daily, operating at around 500 °C and 250 atm [27]. Fauser's main ammonia patent was filed in Italy on 23 April 1921. On 14 May 1921, he filed a patent for an electrolyzer suited to the large-scale production of hydrogen.

Fauser's ammonia patent offered a significant barrier in keeping BASF out of Italy, provided of course that he could find a backer for industrial exploitation of his process. While Fauser, and for that matter Luigi Casale, were ambitious and highly motivated individualists, they could not bring about the commercialization of their processes alone. The necessary first step was to attract the attention of likely potential partners, licensees and investors. In Fauser's case, his prototype attracted the interest of Ettore Conti (1871–1972), president of the electrical energy firm Imprese Elettriche Conti. This firm, forerunner of the Italian Edison company, was founded in Milan in 1884, and was a major investor in hydroelectricity in Piedmont. Conti, wishing to exploit the Fauser process, arranged for Fauser to meet with Guido Donegani, head of Montecatini, who in 1921 took great interest in synthetic ammonia following the availability of the two promising Italian inventions. Apart from the Haber–Bosch process, they were in fact the only viable processes invented in Europe up to that time. Donegani considered the Casale process, but no doubt lost interest because of the ongoing dispute between Casale and Rumianca. As a result, Donegani began negotiations with Fauser to acquire his lower pressure process.

The outcome was that in May 1921, Conti and Donegani, with the Electrochemical Company of Novara, backed the development and scaling up of Fauser's process and Fauser joined the Montecatini enterprise. Within a short time, the Fauser ammonia unit had a capacity of one ton daily. However, as with the Casale process, there were many problems to overcome, including overheating of the catalyst. Great efforts were expended in finding a suitable and inexpensive catalyst.

The latter was hematite which had been reduced in a current of hydrogen with alumina and potash as promoters. The heat of reaction in the converter was recovered and then used to produce high-pressure steam.

3.7.1 Montecatini and Fauser

In 1924, Montecatini founded the Società Anonima Italiana Ammonia (SAIA; from 1929 Società Anonima Italiana Ammonia e Derivati), to exploit the Fauser process. SAIA established the Società Sarda Ammonia e Prodotti Nitrici, and the Società Alto Adige Ammonia, to manufacture synthetic ammonia and fertilizer in Coghinas, in northern Sardinia, and Merano, in the subAlpine South Tyrol (Venezia Tridentina; the present-day Trentino-Alto Adige region). Both facilities relied on hydroelectricity for generating hydrogen and for driving the synthesis loops, including the compressors and pumps. Fauser ammonia factories were also erected at Mas, Sedico Bribano, in the province of Belluno, Veneto (Venetia), in the north east; in the province of Crotone, Calabria; in the south between Taranto and Messina; at St Giuseppe di Cairo, in the province of Savona, Liguria; and elsewhere in Italy [28].

Though Fauser filed patents for his process outside Italy, Montecatini did not license or promote the Fauser process prior to 1926, when, with the Belgian firm Evence Coppée, it founded Société Anonyme Ammoniaque Synthétique et Dérivés (ASED) to license the manufacture of ammonia and its products and supply Fauser equipment (through Société Centrale pour la Fabrication d'Ammoniaque de Synthèse). Dutch State Mines (NV Nederlandse Staatsmijnen, DSM), awarded ASED a contract for gas preparation, and for the supply and installation of Fauser ammonia units and ammonium sulfate facilities at Geelen (province of Limburg) [29].[9] Other licensees included Dai Nippon Artificial Fertilizer in Japan. Licensing of the Fauser process declined in 1929, in part because of the Great Depression, but also because Montecatini had become deeply involved in Mussolini's autarky programs.

3.8 Reorganization at Ammonia Casale

By 1923, there had been a second round of structural changes in the ownership and management of Ammonia Casale. Now there were just four shareholders: Leprestre (138 shares), Casale (60 shares), Giuseppe Albisetti (1 share), and Tomaso Quadri (1 share). Casale and Leprestre acquired the shares of other investors, who had decided to, or been encouraged to, sell out. Previously Santangelo—as he later wrote, probably in support of litigation—Casale, Leprestre, and Andreucci,

[9] This was followed in 1929 with creation of the Compagnie Néerlandaise de l'Azote, located at the headquarters of Coppée in Brussels. In December 1930, through this initiative, an ammonia factory was opened in Dutch Flanders at Sluiskil, drawing on coke oven gas for hydrogen. By 1936, fourteen synthetic ammonia plants based on the Fauser process were in operation outside Italy.

"... came to an agreement which stipulated that all future benefits would be shared equally among us, i.e., myself, Casale, Leprestre and Andreucci." Leprestre's great leverage, as a result of his access to funds from American investors, as well as his wealth accumulated from the wartime business activities, subsequently forced Andreucci out of the partnership, reduced Casale's share of any profits to 15%, and enabled Leprestre to buy Santangelo out for 700,000 lira. After Enrico Cairo's death in 1923, his shares were purchased from his heirs by Santangelo, who appears to have passed them on to Leprestre.[10] Leprestre at that time was president, and owned 69% of Ammonia Casale. Luigi Casale, though a much smaller stakeholder than Leprestre, still held considerable control in the affairs of the company.

Following the granting of a license to Noguchi, a flurry of new orders for ammonia units in Europe, and the lack of competition from rival processes—the primary advantage of Casale's process at a time when others were held back by technical difficulties or confined to corporations such as BASF and Montecatini—the future appeared to lie in major collaborations with groups outside Italy, such as Pechiney and Solvay, that offered the possibility of financial and political support. These collaborations were built on the needs of chemical corporations. Leprestre, the propagandist for Ammonia Casale, was however dreaming of the creation of a diversified global behemoth in which Ammonia Casale would have a major interest. And the best place to start was in the US, the great postwar and economic power, where Leprestre's home and contacts—and the headquarters of most leading banks and industrial corporations—were in New York City. In July 1923, Casale set out to join Leprestre there, accompanied by the company secretary, Vincent Handley, on board the SS *Conte Verde*. Casale was then on his way to Japan.

[10] Cairo's heirs, Gustavo Scialoia and Ermidio Tesei, Cairo's son-in-law, attempted to sell Cairo's shareholding in Ammonia Casale to Leprestre, but without success. They succeeded to sell the shareholding to the lawyer Mario Santangelo. We must assume that the valuation was based on Leprestre's assessment, which Scialoia and Tesei subsequently contested. This precipitated litigation on account of a purported fraud. In short, the two men accused Leprestre of having cheated them into accepting a share price much lower than the actual value. Luigi Casale was held morally responsible, and was dragged into the matter. However, he died in 1927 while the suit was still unresolved. Maria Sacchi-Casale, in a memoir, probably drawn up in support of the defendants, claimed that Luigi was greatly aggrieved, and the dispute contributed to his death. This account, which has revealed important historical details on the early days of Ammonia Casale is drawn from four documents held at the Casale SA archive: carbon copies of what appear to be statements written by people involved in the litigation (but without signatures or means of identifying the writers); one letter signed by Luigi Casale; one handwritten memorandum by the lawyer Santangelo; and one long and detailed memoir written by, or at least on the request of, Maria Sacchi-Casale with the declared purpose of rehabilitating Luigi's reputation that had, apparently, been tarnished by the litigation. The writer may have been Santangelo, who Maria greatly appreciated. Santangelo, though he supported Maria in this litigation, probably resigned from the board of Ammonia Casale during 1923.

3.9 The Correspondence Between Luigi Casale and Maria During His Long Absence

The correspondence between Luigi Casale and Maria, as a result of what was to be their physical separation for several months from July 1923, provides a great deal of information about how Leprestre promoted the business, mainly by nurturing relationships with bankers and important businessmen, and also about family and personal relationships. As the sole record of a critical time in the affairs of Ammonia Casale, the exchanges, including the gossip and trivia, expose complexities that were emerging at the human level.

On 1 August, nearing the end of the transatlantic journey, Luigi Casale advised Maria [30]:

> In three hours, our ship will call at New York. The weather has been awful, but this morning I'm feeling much better than on the previous days because the sea is calm now. In the days before, I was not the only one who was seasick owing to the rough seas. The gale broke right after passing Gibraltar; it cleared for a day, but then the storm resumed stronger than before.
>
> There are several children on board, and though the small ones are not seasick, I saw a seven-year-old suffer throughout the entire trip. However, nobody complains, and we all try to hide our discomfort.
>
> I met on board a certain Maj. Falchi who told me he had been introduced to you by the widow of Mascagni [the Italian composer Pietro Mascagni]. He seems to me a decent person and quite well-to-do and shows great reverence for Mascagni. He says that he is the owner of some very important patents regarding aerial defence and we agreed to meet in Rome upon our returns, in relation to such patents.
>
> I had really hoped during the Atlantic crossing to both work and enjoy some rest at the same time, but I failed to accomplish either of them. I hope I shall be able to do so on the journey from Vancouver to Japan since I believe the present trip has sort of prepared me for the next one.
>
> Everyone tries to minimize the discomfort by reading, watching films, and listening to music. I personally read a lot by taking advantage of the library on board. By the way, I never saw such a nice bookcase as this one! It's a pity that I didn't see it before buying the one for my office.
>
> I just received two telegrams from Leprestre, who informs me that he'll be waiting at the quarantine area.

Two days later, Luigi Casale and Handley arrived in New York City's port. Leprestre assisted Casale with the customs and immigration process and then drove them in his brand-new car to the Astor Trust Company building, at 501 Fifth Avenue, where Leprestre lived in a large apartment, opposite Bryant Park. Leprestre had remarried in 1922. His wife Winifred and daughter Paule were in Canada which meant that there was room to accommodate both Handley and Casale. Casale, on his way to Japan with Leprestre to inaugurate the opening of the first factory that used Casale ammonia technology, was stopping over in North America for two weeks in order to attend to important business matters. In New York, he finalized negotiations that Leprestre had conducted in connection with the startup and promotion of synthetic ammonia manufacture based on the Casale process in the US. Judging by the dates of issue of certain stock certificates, 31 July, Leprestre had to act in some haste to conclude matters before the arrival of Casale [31]. He had, apparently, been nervous about whether or not he could pull off the deal before Casale's arrival in New York City.

3.10 The American Deal

Just prior to Casale's arrival in New York, and mainly through Leprestre's efforts, two American entities in which the "Casale Ammonia Company" (as the Swiss entity was called in the US) was the main shareholder, were created as holding corporations: The Ammonia Corporation and Chemical Investment Corporation. They were associated with, respectively, the Niagara Ammonia Company, and the Hydro-Electric Chemical Company, both formed at the same time.

The main investor in these concerns, after Ammonia Casale SA, was the large electrical utilities corporation Electric Bond and Share Company, founded in 1905. This subsidiary of the General Electric Company had a strong interest in the application of electrochemistry, and, for example, had opened the first magnesium plant in the US, in 1914, in Schenectady, New York. The president of Electric Bond and Share was Sidney Z. Mitchell, at one time "the richest man in America" (until he sustained large losses during the crash of 1929) [32]. Electric Bond and Share was the head of a financial empire based on a pyramid structure of holding companies that pursued domination of electric enterprises in North America and (from 1923) South America. Such a leader of industry was particularly attractive to Leprestre, who through gathering political and business patronage, and hobnobbing with the rich and powerful in New York, had transformed himself into a glory-hunting influencer and risk-taker. Even for the seasoned businessman Mitchell, the smooth-talking Leprestre appeared to have a lot to offer. Electric Bond and Share had enjoyed tremendous growth in revenue since 1921, and Mitchell was probably interested in promising new technologies and investments that drew on electricity [33]. The Casale technology was deceptively simple and proven on a semi-commercial scale. It did not seem to matter that in 1923 the US did not need a synthetic ammonia industry. The nitrogen market was saturated with nitrate imported from Chile, the by-product ammonia of coking ovens, which was converted into ammonium sulfate, and calcium cyanamide. Plus, there was a growing export trade in ammonium sulfate.

Mitchell was completely taken in by Leprestre's proposal. Expectations were high. Mitchell spoke enthusiastically about the creation of a chemical empire based on electricity and high-pressure reactions. By the end of July, Mitchell and Leprestre had completed negotiating the arrangements between Electric Bond and Share and Ammonia Casale in the US. The correspondence between Luigi Casale and Maria over the next few weeks provides a good sense of how things were progressing. On 3 August, Luigi Casale was optimistic: "The day when I left Italy,

Leprestre was... desperate. The day I arrived everything had been concluded. Now ... I hope in the quick rise of our company. I found Leprestre reasonable" [34]. Leprestre was certain about one thing. He needed Casale at his side in order to introduce the ammonia technology to industrial and financial leaders:

> [Leprestre] stated that he judges our separation a most dangerous and unreasonable thing. We continued discussing between the two of us, in front of Handley, business matters that arose yesterday evening after dinner with [Charles J.] Hardy [a director of Leprestre, Miller & Co.]. It seems to me that he has well understood the situation and we both agreed to entrust Hardy to frame an agreement which will constitute the basis of our cordial cooperation. Quite naturally, the finalization of such a rather complex agreement [with Electric Bond and Share] will not occur until the end of this year. Over the past two days, I have conferred with many big bankers and industrialists. Monday I'm going to have lunch with the president of General Electric. I will then take leave on the 9th to go and see the laboratories of General Electric [at Schenectady, New York]. On the 13th I will be in Niagara Falls and then we will leave, I think, on the 15th from Toronto bound for Vancouver. There will be Leprestre, Gerry [the moniker for Leprestre's wife, Winifred] and myself. Leprestre will stop over in Yokohama while I will continue towards Kobe and then Nobeoka.

The letters reveal the closeness that had developed between the Casale and Leprestre families, at least with Leprestre's daughters (Table 3.3). Winifred, however, appears to have been the cause of tension, and there were hints of growing disagreements over business matters between Luigi Casale and Leprestre. As for family matters, Casale put aside funds to enable Maria and their children to accompany him on a later visit to the US. Leprestre settled some unidentified debts of his own. His daughter Genevieve was then in Italy, perhaps staying with Maria at the Casale home in Rome.

> I was able to draw 9500 USD, 9000 of which I immediately deposited on account at the Bankers Trust. I decided to keep this amount here in order to facilitate your coming here next time with the children: if possible, in future I will never travel without my family.
>
> Leprestre was able to settle his debts and has got everything he needs for Japan. He's very happy that he has made contact with some of the most important people in the business world. This morning Hardy told me how powerful and important our partners are, as well as being the most decent ones in America. You can see yourself, how much the task is getting difficult for me! I hope that God will make me fit for this task. Handley is going to leave on the 9th [of August] on board of the "Conte Verde" and he's expected to be back in Rome on the 19th.
>
> Leprestre has asked me to convey his greetings to you as well as to convey his most affectionate ones to Gee [Genevieve] and the children. You may please also add my personal greetings to Gee.

In this (and his other letters), Luigi invariably expresses his great devotion to his wife and children in almost poetic terms. He tells Maria how much he misses her and the children (Fig. 3.14). He is never shy to make clear his love and affection, writing at one point that "...the greater our distance grows the more I feel homesick for that heart which beats at unison with mine, which has always beaten so for many years and will continue to beat."

Table 3.3 The Casale and Leprestre families

Name	Wife	Children	Moniker	Notes
Luigi Casale (1882–1927)	Maria Rosa Sacchi (1889–1950)	Lucia (1918–2008) Renato (1921–2009)		
René Leprestre (1876–1941)	Jeanne Dechappe (18?? -1905)	Génevieve (1900–1992)	Gee	Married Henry Watson Cornell Jr. (1899–1957) in 1932
		Paule (1903–1995)	Lee	
	Winifred Ashphant (1890–1938)		Gerry	Married René Leprestre in April 1922
	Margaret Feek Bouck (1892–1975)		Marjorie	Daughter of Alonzo B. Cornell (1832–1904) and Ellen Augusta Covert (1834–1893). Mother of Henry Watson Cornell Jr. (above)

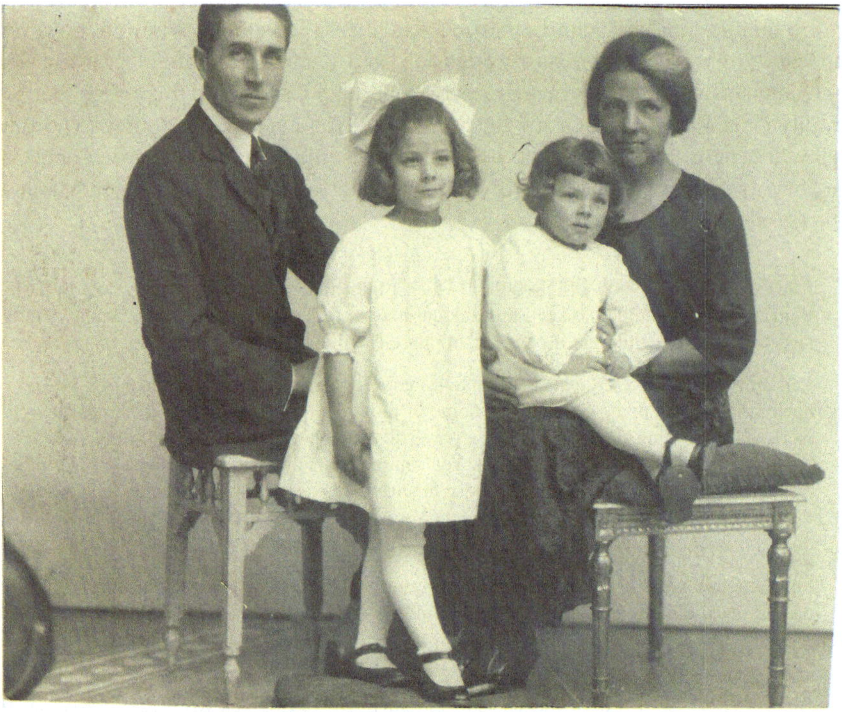

Fig. 3.14 Luigi and Maria Sacchi-Casale with their children, Lucia and Renato. (Collection of Francesco Casale. With permission)

Fig. 3.15 Stock certificate, The Ammonia Corporation, preferred stock, in the name of A. Iselin & Co., 30 July 1923. Iselin held the stock certificates of Ammonia Casale SA holdings in The Ammonia Corporation and the Hydro-Electric Chemical Company. (The Ammonia Corporation, A Preferred Stock. Records of E. I. Du Pont de Nemours & Co. Absorbed companies. Courtesy of Manuscripts and Archives Department, Hagley Museum and Library, Wilmington, Delaware, Accession 500, series II, part 1, box 67)

While making the rounds of venerable New York banks, Leprestre had managed to convince a few bankers and investors of the profits to be made from the new technologies of high-pressure chemistry and electrochemistry. Consumption of nitrogen products in the US was enormous, but, as noted earlier, was satisfied from various well-established sources. Synthetic ammonia could eclipse them all, Leprestre must have argued to his eager listeners, and subsequently investors.

Among them was A. E. Bonn, of A.E. Bonn & Company, Inc., who became the second largest investor in the Hydro-Electric Chemical Company. Bonn had previously been a director of the American Cyanamid Company, and secretary of its subsidiary, Air Nitrates Corporation, which had built a large wartime calcium cyanamide factory at Muscle Schoals, Alabama. Then there was the banker Ernest Iselin (1876–1954), of A. Iselin & Co.[11] He joined the board of directors of The Ammonia Corporation, Niagara Ammonia, and Hydro-Electric. Iselin & Co. held certificates representing securities in The Ammonia Corporation and in Hydro-Electric (Fig. 3.15).

[11] Bonn's business address was at 36 Wall Street, Iselin's at 501 Fifth Avenue.

The bankers would soon be invited to meet Luigi Casale, who made the most of his time on the East Coast in their company. They included Pierre Lagier, from the Canton of Vaud, Switzerland, soon to be a member of the board of Ammonia Casale, Bonn, and Iselin. Apart from meeting with investors there was work to be done on a new factory at Niagara Falls (Chap. 8), and talk of contracts from Sumatra and India, where the populations (240 million in India alone) were dependent on agriculture for their sustenance. Leprestre, according to Luigi Casale, was pleased with his achievements. His business relationship with Casale was, apparently, very satisfactory, and there was time to relax, to discuss personal matters, including a dispute within the Leprestre family, and reflect on life in the US. On 7 August, Casale described in a letter to Maria a visit to Bonn's home, in Great Neck, Long Island [35]:

> (.....) Last Sunday, Leprestre, Lagier, and myself travelled with Bonn to his country house, one hour from here. We needed the presence of Lagier because we planned to discuss certain financial arrangements.
>
> It was very hot and everyone was perspiring a lot except myself. I argued that, in my experience, the temperature was actually not so bad, but this was only to tease Leprestre, who just wanted me to admit the complete opposite and agree with him that I had never before experienced such a climate. He was mad at me until he realized I was just teasing him.
>
> At six we went for a swim nearby and it was very refreshing.
>
> Life here is certainly organized to foster business and you definitely feel the power of both fantastic organization and wealth, but I definitely do not envy the Americans. What would happen to them in case of a moral crisis such as that which befell us in the aftermath of the war [1919–1920]?
>
> Leprestre here enjoys a high reputation, far more so than in Italy, as I had the chance to understand from talking with his friends. He still has some difficulties to overcome. In difficult times he relies on me and I give him all the support he needs.
>
> Now that he has sealed the business everyone tries to profit from it, as it occurs in Italy.
>
> He told me he has taken out an insurance policy for the benefit of Gee [Genevieve], Gerry [Winifred] and Paule. He talked to me openly about the quarrel between Gee and Gerry and he believes that the rift between them cannot be mended even though he does not understand the true reasons behind it and how, in his opinion, the fault is on Gee's side.
>
> I had to do a lot of work for the plant at Niagara Falls since the buildings are already available and the engineers believe they are in a position to arrange the layout of the plant in the way that suits their need, which is different than ours.
>
> As for the catalyst, the patent has been rejected here and I had to discuss a lot with Gibbs and Norton on the basis for a new patent.
>
> I hope that upon my return here in October the contract for [Sumatra] will be signed and perhaps also that for India.

On the following day, just before Handley was scheduled to return to Rome, Casale wrote to Maria of a powerful partnership in the making [36]:

> Rene [Leprestre] is sending you through Handley a twenty-dollar piece in gold that represents his attendance fee at the first board meeting of the Hydro-Electric Chemical Co., i.e. the new company to produce nitrogenous fertilizers that we have set up under the auspices of General Electric Co. The president of this company is one of the most powerful men in the world and it seemed to me he intends to create with my process a colossal enterprise.

On 13 August, Luigi wrote to Maria from the Hilton Hotel at Niagara Falls, advising that Bonn's New York office had been turned over to Ammonia Casale (Casale Ammonia Co.), and construction work was continuing on the new factory [37]:

> Leprestre has moved into new offices at 501 5th Avenue, the Casale office in New York. This office still bears the name Bonn & Co., instead of Casale Ammonia Co. I spent all day at the [Niagara Falls] factory where our plant will be erected: the project director is Mr [G. A.] Hendrie who came to Terni and who I immediately liked very much.
>
> We leave tomorrow morning for Toronto [by train] and in the evening will travel onwards [to Vancouver] with Gerry.

Luigi also advises Maria that he has not been feeling well, and had lost three kilograms, though Leprestre has been doing his best to take care of him. There was more on the dispute within the Leprestre family, and now more than a hint of a serious difference, the cause of which is not altogether clear, but may have arisen from the extent of Leprestre's business expenses. However, Leprestre was going out of his way not to let this interfere with their well-refined opposites-attract dynamic when it came to the business relationship. Among issues arising were the patent rights in Japan for the recipe to Casale's iron catalyst, which, as with other catalysts at that time, were prepared by users on site (and only later purchased from specialists). As usual, business agreements were in the hands of Leprestre. Casale described these, interspersed with family matters, including animosities, to Maria:

> Leprestre tried his best to replace you in attending to me and nobody could have done it any better. For this I'll owe him a debt of gratitude.
>
> Leprestre received a letter last Friday from Gee [Genevieve] in which she informed him she was on my side in the dispute between the two of us [over family matters]. She added she was very sad about it since she was afraid that Maria was of the same opinion [as myself about Leprestre] and she insisted that I send a telegram to both of you making clear that everything was over now.
>
> As I have already written to you, Leprestre shows his best intentions in getting along with me and this is good even though his very special way of seeing things sometime drives him off the right path.
>
> Leprestre will come to Japan to negotiate an agreement for [the patent] recipe of the iron catalyst.

On 20 August 1923, shortly after arriving in Vancouver, from where he, Leprestre and Winifred would sail to Japan, Luigi Casale wrote, with no less enthusiasm, to Maria about the large venture, as described by Leprestre, with the American group (General Electric, and Electric Bond and Share) that apparently intended to create, using Casale technology, "a colossal enterprise much bigger than the Badische [BASF]"—the then most powerful science-based chemical corporation in the world—"by acquiring all companies around the world that use my patents: it is a fantastic programme that is being so outlined and it is the programme of Mitchell, General Electric's boss and the most powerful man in America, a man who has never failed in any of his business dealings."[12] To achieve this bold claim, Casale continued, "requires a process superior to the others from a technical-scientific standpoint, as I believe ours is, but also a financial genius or a mind-blowing fortune and this is the merit of Leprestre." Though Casale shared Leprestre's high optimism for the future of the venture, they were not in accord on many issues. Cracks in the personal relationship between the two men had begun to appear. However, for the sake of the business at hand they needed to work together, as Luigi Casale explained to Maria: "You may better understand now why Leprestre and myself are striving to bridge the abyss that fatally is widening between us, while it is urgent that we both stand united" [38]. Casale was writing from Vancouver's Canadian Pacific hotel, about to leave for Yokohama, in order to attend the opening of the first factory based on Casale technology at Nobeoka. Success in Japan was by no means guaranteed.

It is clear from this correspondence that Leprestre was not only the financial brains behind Ammonia Casale but that he also wanted to control the future direction of the firm. Moreover, Leprestre's grandiose plan, effectively to create one of the largest global chemical corporations, using the technology that would indeed define the chemical industry for at least the first half of the twentieth century, took shape at an awkward moment in the relationship between Leprestre and Casale. At the business level, it appears, Leprestre was cunningly manipulative, and his priorities probably lay elsewhere, but for this he needed Luigi Casale. The correspondence suggested serious disagreements. Thus Luigi, in the same letter to Maria, noted: "It is a pity that a gulf between the two of us is growing so large and that other gaps are trying to widen. When we both realize this, we strive to bridge it, but our efforts are becoming inadequate."

Two days later, still in Vancouver, and following the week-long journey across Canada by the Canadian Pacific Railway, with stopovers at its grand hotels in the company of Leprestre and his wife, Luigi could barely conceal his increasing frustration and growing disgust with the behavior of his partner [39]:

> Sometimes I wonder how a man of his age could possibly spend so much of his time travelling for the only purpose of amusing his young wife, leaving back at home his two daughters [Genevieve and Paule], both at an age when they must also enjoy traveling. And yet, day by day, he increasingly sticks to this wife and gets away from his daughters. If I ever win this battle, the battle of getting on stream with more new plants, I won't travel anymore without

---

[12] The president of General Electric was not Mitchell but Gerard Swope. Mitchell was president of Electric Bond and Share.

> the three of you. You will see yourself how, in a matter of two or three weeks, you will get used to traveling by sea.
>
> In your last letter you remind me, in reference to a letter by Gerry (Genevieve) to Lee (Paule), how phony Leprestre is and how wicked his wife is. I am totally convinced of this fact, and I strive, for fear of a mistake, to forget it. If you only knew how much it cost, this effort! Let's wait a few years more and then, maybe, we won't be needing anyone, and we can handle everything ourselves.

Luigi Casale, who once had such great admiration for Leprestre's rare business sense, now saw him as a person totally lacking in morals, and who posed a potential danger to their partnership. The correspondence between Luigi and Maria continued to hint that dealings between the two partners had been anything but easy. For the moment, though, while things seemed to be going well, it was preferable to accept an unspoken compromise rather than pay the price of an ugly separation.

As Luca Bianchi makes clear, "now we also understand the reason for the provisions in the Ammonia Casale bylaws, no doubt which Luigi Casale insisted upon, by which none of the partners were entitled to own more than 25 per cent of the voting shares: Casale had realized early on that he was dealing with a ruthless deal maker." Or what Maria Sacchi-Casale later called "a fishy international rascal."

In the end, the great expansion backed by Electric Bond and Share, the main stockholder in the American Casale firms, did not work out as Leprestre had envisaged. Had that happened, Ammonia Casale might have ended up like Thomas Edison's power company, which was transformed into General Electric by J. P. Morgan, with nothing left for Edison. Eventually, the Casale ventures in America did fall victim to a giant corporation, but it was Du Pont rather than Electric Bond and Share (Chap. 8).

3.11 Two Other European Processes

3.11.1 The Claude Process

In 1902, the French engineer Georges Claude (1870–1960) invented a system of air liquefaction that enabled, by fractional distillation, the separation of liquid oxygen, and, later, nitrogen and argon. The Claude process, based on compression and then decompression of air, competed with the earlier system of Carl von Linde. With Paul Delorme, Claude founded Société L'Air Liquide to promote the process. Claude, incidentally, also invented neon lighting. From 1917, as part of secret government wartime research, Claude undertook experiments on high-pressure ammonia synthesis at around 1000 atmospheres and 500 °C. Some success was achieved prior to the Armistice in November 1918. Pilot plant tests were carried out by Claude at "hyperpressures" in 1919 at the Montereau (Île-de-France) facility of the new Air Liquide subsidiary—formed jointly with Compagnie de Saint-Gobain—Société Chimique de la Grande Paroisse [40, 41].

For the first units, hydrogen was generated from water gas. The nitrogen–hydrogen mixture was passed through a series of converters without recirculation. When, by around 1930, coking oven gas became the source of hydrogen, the first converter served as the purifier, in which carbon monoxide was removed. This was followed by passage through the ammonia converters. After each stage the gas was cooled and liquid ammonia was collected. At 1000 atmospheres, the process gave 40% ammonia in the first synthesis tube, and 25% in the second. Apart from the Casale process, the Claude process was the only one that operated at such extreme pressures. Both processes involved handling smaller gas volumes than in the Haber–Bosch process, and use of less active catalysts. In the case of the Claude process, just 10% of unreacted nitrogen and hydrogen remained after passage through the converters. This gas mixture was not recycled.

In 1923, Claude, backed by two Swiss banks, established Società per la Fabbricazione dell'Ammoniaca Sintetica e Prodotti Derivati (Azogeno) in Milan to license and provide engineering services for his process. At the Bussi industrial site, in the Abruzzo region, Azogeno constructed a Claude ammonia unit of rated capacity 7-mtpd; the nearby works of Società Italiana di Elettrochimica supplied hydrogen. A second Italian facility was located at Vado Ligure, west of Genoa. Hydrogen was available from an adjacent coking oven plant. No doubt, the restricted capacities of these two factories did not pose a great threat to Montecatini which controlled ammonia production in Italy, particularly for use in fertilizers.

Operators of Claude's ammonia process confronted major technical problems, until at least the mid-1920s, as experienced in Japan, and in the US by Du Pont. Often performances were poor and the process did not produce the purported high yield. The enormous heat generated in the catalyst chamber was a major problem. Converters made of normal steel corroded under the reaction conditions and exploded, creating delays in introducing Claude's process. The difficulties were overcome with the introduction of special chrome-nickel steel converters first made in Sheffield, England, by the engineering firm Vickers Ltd (Vickers-Armstrongs Ltd from 1927). For smaller Claude converters, a chrome-nickel-tungsten alloy, trade name BTG, was manufactured by the French steel and mining firm Société de Commentry-Fourchambault et Decazeville, of Imphy. In the meantime Casale had a strong presence in France where Pechiney was the main licensee. There was also the government-backed factory at Toulouse, as described in Chap. 4.

3.11.2 The Uhde/Mont Cenis Process

In 1924, a second German synthetic ammonia process, often referred to as the Uhde/Mont Cenis process, based at first on coking oven hydrogen, was sponsored by a consortium of mining companies. It was named after the Gewerkschaft der Steinkohlenzeche Mont Cenis, of Herne-Solingen, Westphalia, which asked Friedrich Uhde (1880–1966), then a chemical engineering contractor based in Dortmund, to design a converter and ancillary equipment. The process operated at just 100 atm, which meant that it required an extremely potent catalyst, as provided by Ivar Walfrid Cederberg, who had undertaken catalyst investigations at the University of Stockholm (1916–1919, and from 1919 at Norsk Hydro). The new ammonia process was promoted from late 1925, at around the time when BASF became part of IG Farben. The patents were held by Gasverarbeitungs-Gesellschaft GmbH, also of Herne-Solingen, where the first facility was erected in 1925 and began operating the following year. Full-scale operation was achieved in 1928, with four converters, each capable of producing 25-mtpd. The synthesis loops included two-stage refrigeration for obtaining liquid, anhydrous ammonia. Subsequently Uhde engaged in the design of nitric acid plant, as well as a process for the manufacture of the fertilizer calcium ammonium nitrate.

In 1927, the Mont Cenis process was licensed for use in Holland by Royal Dutch-Shell. This happened after the board of Royal Dutch-Shell decided to counter IG Farben's monopoly on nitrogen fertilizers, for which the Netherlands was a good market—the second largest market for German ammonium sulfate after Japan—by entry into synthetic ammonia manufacture. The Mont Cenis and Casale processes were investigated and in September, Royal Dutch-Shell opted for the Mont Cenis process. The following month, contracts were signed. The Royal Dutch-Shell IJmuiden MEKOG facility was close to the source of hydrogen, the coking ovens of the Hoogovens steel works. The facility, including the synthesis gas process, designed by the Uhde engineering firm, began operating in September 1929. It was the first of three large Dutch ammonia factories to open by the end of 1930. Higher pressures were introduced into the Mont Cenis process following expiration of certain BASF ammonia patents. The two other Dutch ammonia facilities, those of DSM and the Compagnie Néerlandaise de l'Azote (a Montecatini subsidiary), used the Fauser process. The aggregate annual capacity, as fixed nitrogen, of Dutch nitrogen factories exceeded 90,000 metric tons, which included by-product nitrogen. The emergence of the Dutch ammonia industry would have far-reaching impacts on the growth of large engineering contractors in the Netherlands [42].

References

1. Bruni G (1927) Luigi Casale. Giornale di chimica industriale ed applicata 9(2):90
2. Crowley JF (1927) Luigi Casale: an appreciation. Chem Age, 5 Mar, p 232
3. Certificate giving account of the death of Luigi Casale, signed by Prof. Eugenio Morelli, of the hospital of the University of Rome, 24 Oct 1942 (Collection of Francisco Casale)
4. Petri R (1998) Technical change in the Italian chemical industry: markets, firms and state intervention. In: Travis AS, Schröter HG, Homburg E, Morris PJT (eds) Determinants in the evolution of the European chemical industry, 1900–1939. Kluwer, Dordrecht, pp 275–300
5. Bovini G, Covino R, Giorgini M, with Benvenuti E (1991) Archeologia industriale e territorio a Terni. Siri. Collestatte. Papigno. Catologo regionale dei beni culturali dell'Umbria. Electa Editori Umbri Associati, Perugia
6. Terenzi A et al (2009) Le frontiere dei materiali innovativi. Ricerca, servizi, tecnologie e produzioni industriali nel futuro della conca Ternana. Agencia Umbria Ricerche, Terni

7. Francisci L, Travis AS (2025) The early development of the Casale process for the production of synthetic ammonia (1917–1922): the protagonists, the technology, and a link between Italy and the United States. Substantia 9(1):57–72
8. Bruni G (1922) Perizia Stragiudiziale del prof. Dr. Giuseppe Bruni, Professore ordinario di Chimica nel Reale politecnico e nella Reale Universita di Milano, nella caus del Dr Luigi Casale contro la Società Anonima Per Azioni Stabilimenti di Rumianca. Expert opinion of Giuseppe Bruni on behalf of Luigi Casale (Casale SA)
9. Polk RL & Co.'s (1918–1919) Trow New York copartnership and corporation directory [1918], vol. 66. LR Polk & Co., Inc., New York, p 656
10. Falchero AM (1982) Banchieri e politici. Nitti e il gruppo Ansaldo-Banca di sconto. Italia contemporanea, nos 146–147(July):67–92
11. Falchero AM (1990) La Banca italiana di sconto, 1914–1921: Sette anni di Guerra. Franco Angeli, Milan
12. Pancaldi G (2006) Wartime chemistry in Italy: industry, the military, and the professors. In: Macleod R, Johnson JA (eds) Frontline and factory: comparative perspectives on the chemical industry at war, 1914–1924. Springer, Dordrecht, pp 61–74
13. Bovini G (2014) Terni tra siderurgia e chimica. In: L'industria chimica italiana a cinquant'anni dal Nobel di Giulio Natta. Arpa Umbria, Terni, pp 13–35, on 30–31
14. Travis AS (2018) Nitrogen capture: the growth of an international industry (1900–1940), Springer, Cham, pp 236–237
15. Bonifazi R (1986) Le attività chemiche dalla 'Terni'. In: Marianeschi E, Cipolla G, Papuli G et al (eds) La grande industria a Terni. CESTRES, Centro Studi e Ricerche Economiche e Sociali di Terni, Terni, pp 273–307
16. Muscle Schoals: hearings before the Committee on Agriculture and Forestry. United States Senate. Seventy-seventh Congress. Second session. S.3420 (1922). Government Printing Office, Washington DC, p 446
17. Casale L (1921) US patent 1384428, 12 July (application 7 July 1920); UK patent 148885-A, filed 30 June 1921
18. Casale L (1921) UK patent 194740-A, filed 19 Sept
19. Casale L (1922) US patent 1547362, filed 19 July, serial no. 576104
20. Gianni Bovini, "Luigi Casale," unpublished manuscript
21. Memorandum or covering note in the hand of Mario Santangelo describing the change in the partnership arrangement between Carlo Andreucci, Enrico Cairo, Luigi Casale, René Leprestre, and Santangelo (1921), April (Casale SA)
22. ICSIM (2003) Carburo, calciocianamide, ammoniaca sintetica polipropilene. Un secolo di industria chimica nella valle del Nera (2003). Catalogo della mostre Altrio di Palazzo Gazzoli, Terni, ICSIM-Osservatorio per il settore chimico, 10 Sept to 30 Sept, pp 29–32, on 30–31
23. John F Crowley, quoting Harker. In: Waeser B (1926) The atmospheric nitrogen industry: with special consideration of the production of ammonia and nitric acid. P Blakiston's Son & Co., Philadelphia, vol. I, p xx. Translation by Ernst Fyleman, a chemist associated with the London-based J.F. Crowley and Partners engineering consultancy, of Waeser, B (1922) Die Luftstickstoff-Industrie, mit besonderer Berücksichtigung der gewinnung von ammoniak und salpetersaure. Otto Spamer, Leipzig
24. Francisci L (2020) Il mancato contraltare alla Montecatini: Nera Montoro nelle politiche chimiche della Terni (1922–1964). Proposte e ricerche: economia e società nella storia dell'Italia centrale 84(1):121–132
25. Bertilorenzi M (2008) The Italian aluminium industry: cartels, multinationals and the autarkic phase, 1917–1943. Cahiers d'histoire de l'aluminium 1(41):42–71
26. Mantell CL (1940) Industrial electrochemistry, 2nd edn. McGraw-Hill, New York, pp 379–383
27. Maveri D (1981) Giacomo Fauser. Editrice Rapollo, Novara
28. Perugini M (2009) Grande impresa e Italia autarchica: Montecatini 1929–1943. PhD dissertation, Università Commerciale 'Luigi Bocconni', pp 56–165
29. Perugini M (2014) Il farsi di una grande impresa. La Montecatini fra le due guerre mondiali. Franco Angeli, Milan

30. Luigi Casale to Maria, 1 Aug 1923, on board the SS *Conte Verde*. (Collection of Francesco Casale) This and the following letters have been translated by Luca Bianchi
31. Stock certificate, The Ammonia Corporation, preferred stock, in the name of A. Iselin & Co., 30 July 1923. From: The Ammonia Corporation, A Preferred Stock. Accession 500, series II, part 1, box 67; and Hydro-Electric Chemical Company, Stock Certificate book, 1923–28. Accession 500, series II, part 1, box 848. Records of E. I. Du Pont de Nemours & Co. Absorbed companies. Manuscripts and Archives Department, Hagley Museum and Library, Wilmington, Delaware
32. Anon (1944) Once "Richest Man" dies at 81. Chicago Tribune, 19 Feb, p 12
33. Anon (1926) Electric Bond and Share is leader in its unique field. New York Evening Post, 10 April, p 19
34. Luigi Casale to Maria, from 501 5th Avenue, New York, 3 Aug 1923. (Collection of Francesco Casale)
35. Luigi Casale to Maria, on note paper of Bonn & Company, Inc., 7 Aug 1923. (Collection of Francesco Casale) The Norton mentioned in this letter may have been Frederick Harwood Norton (1896–1986), an expert in ceramic materials, who joined MIT in 1923
36. Luigi Casale to Maria, 8 Aug 1923. (Collection of Francesco Casale)
37. Luigi Casale to Maria, from Hilton Hotel, Niagara Falls, 13 Aug 1923. (Collection of Francesco Casale)
38. Luigi Casale to Maria, from Hotel Vancouver, Vancouver, 20 Aug 1923. (Collection of Francesco Casale)
39. Luigi Casale to Maria, from Hotel Vancouver, Vancouver, 22 Aug 1923. (Collection of Francesco Casale)
40. Emptoz G (2014) Les chemistes, leurs institutions et leurs sociétés savants l'entre les deux guerres. II. La nouvelle chemie de l'azote dans l'entre-deux-guerres. L'Actualité chimique, no. 381, Jan, pp 40–46
41. Anon (1921) The Claude process for ammonia synthesis. Nature 107:765
42. van Rooij A, Homburg E (2002) Building the plant: a history of engineering contracting in the Netherlands. Walburg Pers/Eindhoven, Foundation for the History of Technology, Zutphen

Ammonia Casale in Nobeoka, Japan, and Toulouse, France

4

Abstract

Luigi Casale's first major success was the sale in 1921 of an operating license for his synthetic ammonia process to a Japanese entrepreneur, Shitagau Noguchi, whose factory, at Nobeoka, was commissioned in late 1923. Hydrogen was produced by electrolysis and nitrogen by the Casale method of burning of air. During the first weeks, Luigi Casale was on hand to iron out major technical problems, much to the satisfaction of his client. This established Casale's reputation, and was followed with repeat orders. It also defined Ammonia Casale's marketing strategy, based on licensing and the supply of Italian-made machinery and know-how. In the late 1920s, the French government set up a large ammonia facility based on Casale technology at a former powder factory in Toulouse, mainly for production of munitions. However it was beset with several technical problems, including the French-designed gas purification process, until it was replaced by BASF gas purification equipment.

4.1 Japan: Shitagau Noguchi

The takeoff phase of Ammonia Casale SA in 1921 was stimulated by the Japanese entrepreneur Shitagau Noguchi, who introduced European technological innovations in nitrogen fixation to Japan. By 1910, Noguchi had undertaken the electrothermal production of calcium cyanamide, which represented an important early development in the twentieth-century Japanese chemical industry. In 1912, along with other Japanese entrepreneurs and inventors, he became interested in Fritz Haber's high-pressure method for synthesis of ammonia. The industrial ammonia process, like the calcium cyanamide process, relied on access to cheap and reliable sources of electricity, such as those readily available in Japan from hydroelectric power. In early 1921, while in Europe, Noguchi took note of

A. S. Travis, *The Global Synthetic Ammonia Industry*, Perspectives on the History of Chemistry, https://doi.org/10.1007/978-3-031-87978-4_4

Luigi Casale's synthetic ammonia process. The outcome was Casale's first major licensing agreement.

4.1.1 Western Technologies and Agriculture in Japan

In 1868, following the fall of the Tokugawa Shogunate, Japan returned to the rule of the emperor. Throughout the years of the Meiji Restoration (1868–1912), the country engaged in political, economic, and industrial modernization. Government-sponsored investigations into Western technologies were important during this period. In the 1880s, there was particular interest in products for and improvements in agriculture, which led to the development of a superphosphate industry based on treatment of phosphate rock with sulfuric acid. This was followed in 1893 with the establishment of the Agricultural Experiment Organization, based in Tokyo, with six regional stations. Its activities included investigation of fertilizers and chemical analysis of soils.

The importance of nitrogen products as fertilizer, as elsewhere, was recognized during the last decades of the nineteenth century; before 1900 the demand was satisfied by the import from China of beancake, the nitrogen-containing material remaining after oil was pressed from the seeds, and Chilean nitrate. After 1900, the demand for nitrogen fertilizers was increasingly met by imported ammonium sulfate. Ammonium sulfate fertilizer was well-suited to the wet-soil cultivation of rice and was preferred to the Chilean nitrate. Pure ammonium sulfate contains 21.2% nitrogen (compared with 16.5% nitrogen in Chilean nitrate). It was available as a white powder or in crystalline form, of low hygroscopicity. Ammonium sulfate was a source of both nitrogen and sulfur and its chemical stability enabled long-distance transportation and storage. Acid-forming when applied to the soil, it was thus useful for treating alkaline conditions.

Until the outbreak of World War I, Japan was the principal importer of ammonium sulfate fertilizer from the UK. In 1913 alone, 110,634 tons were imported [1]. Japanese production of the sulfate began on a small scale in 1901 at the Tokyo Gas Company. All ammonium sulfate and one-third of the Chilean nitrate imported was used for manurial purposes; two-thirds of the nitrate was employed in the manufacture of nitric acid for explosives. In 1914, Japan imported 120,000 tons of ammonium sulfate.

The nitrogen fertilizer supply situation began to change during the early 1900s, after young, technically qualified entrepreneurs developed hydroelectricity and exploited this resource to introduce Western electrochemical technologies, notably calcium carbide and calcium cyanamide manufacture. The first calcium cyanamide facility in Japan was opened by Shitagau Noguchi (1873–1944), whose interest in electrochemical enterprises led to the founding in 1908 of Japan Nitrogenous Fertilizer, Inc. (Nippon Chisso Hiryō Kabushiki Kaisha) on Kyūshū, Japan's third-largest and most southerly island. Under the management of Noguchi, the firm would become a leading industrial combine, or zaibatsu, known as Nitchitsu.

Noguchi had studied engineering at Tokyo Imperial University, and in 1903 joined the Japanese branch of Siemens & Halske (around the time that the power division was named Siemens-Schuckert). There, he became familiar with all aspects of modern electrical engineering. His technical expert was Tsuneichi Fujiyama, who had also studied engineering at Tokyo Imperial University and in 1901 undertook experiments into carbide manufacture. In 1906, using German electrical equipment, Noguchi constructed a hydroelectric power station at Kagoshima, Kagoshima Prefecture, on the west coast of Kyūshū, to serve local gold mines. In 1907, together with Fujiyama and Seiji Ichikawa, he cofounded the Japan Carbide Company, with a carbide factory nearby at Minamata, Kumamoto Prefecture, using surplus electrical power.

During 1909–1910, Noguchi commenced cyanamide production at the Minamata carbide factory. However, the cyanamide was generally not well-received as fertilizer. Noguchi converted it into ammonium sulfate that better suited the soil conditions and was a product already familiar to farmers as fertilizer.

Despite technical and financial difficulties—including the destruction in July 1912 of a plant for conversion of cyanamide to ammonium sulfate—and Fujiyama's departure earlier in the year, by 1914 Noguchi had opened another cyanamide facility, at Kagami, Kumamoto Prefecture, northern Kyūshū (present-day Kagami, part of Yatsushiro). Noguchi received considerable backing from the Mitsubishi corporation, a major zaibatsu.

The importation of ammonium sulfate effectively ceased at the start of World War I. In 1915, Fujiyama reorganized Hokkai Carbide as the Electro-Chemical Co. (Electrochemical Industries, Inc.), or Denka, an affiliate of the Mitsui corporation, and introduced a cyanamide process of his own design. The Denka and Noguchi concerns underwent rapid expansion. By 1918, Japanese domestic production of nitrogen fertilizer had tripled.

4.2 Interest in Synthetic Ammonia

The war encouraged scientific and technical research and development in Japanese government, private, and industrial laboratories, including into nitrogen fixation. As elsewhere, interest in high-pressure ammonia synthesis followed the announcement of Haber's method and the BASF presentation at the 1912 Eighth International Congress of Applied Chemistry in New York. Not long after attendee Jun Ōya had sent back his report on the German process, Noguchi published a series of review articles on the new synthetic ammonia technology in the *Electricity Review*, which, in 1914, were brought together in a monograph, *A Method for the Industrial Fixation of the Nitrogen of the Air*. This exercise demonstrated Noguchi's appreciation of the potential of the novel high-pressure process. Following the outbreak of war, the Japanese government sequestered the Haber–Bosch patents, but, as scientists and engineers in Japan and other Allied countries found, they were of little use for the development of an industrial process. The patent descriptions invariably provided insufficient information. Ammonia research undertaken during the war included the studies by Tatsuji Suzuki at the Yokohama Research Institute. In 1917, work began on construction of the Institute of Physical and Chemical Research (RIKEN), in Tokyo. Setsurō Tamaru, a former assistant of Haber, was head of physical chemistry, where the main focus of research was on nitrogen fixation. Four years earlier, in 1913, as a member of the Japanese Society for the Investigation of Ammonium Sulfate, he had lobbied for the adoption of a method similar to that of Haber. Tamaru was in Europe from 1918 until 1920 to procure apparatus for the new RIKEN laboratories. Two other institutes in Tokyo took up research into synthetic ammonia: the Special Nitrogen Research Laboratory, of the Department of Commerce and Agriculture (later Tokyo Industrial Experimental Laboratory, TIEL), and the Chemical Engineering Research Institute, headed by Fusajirō Kodera, who had also worked with Haber. Technical experts from four Japanese companies inspected the BASF Oppau facility in 1919, while it was still occupied by the French. In 1920, Japanese investigators in Europe included Kodera's colleague and high-pressure expert Buichi Yokoyama, who tried to gain access to Oppau, but was turned away. In Paris, he examined the Claude process, which was then still undergoing development.

A consortium of Japanese firms, including Sumitomo, approached BASF in an effort to acquire a license for the Haber–Bosch process. However, the price demanded by BASF was far too high to justify the risks of taking up this new technology. Independently, Masaya Suzuki, the director general of Sumitomo, failed in an attempt to negotiate a license for the Haber–Bosch process in December 1920. Earlier, Suzuki and the Sankyō Company had considered the Atmospheric Nitrogen Company's process of General Chemical (Chap. 8), but the development cost was considered prohibitive. In particular, Japanese engineering firms had no experience in the manufacture of high-pressure chemical equipment.

Noguchi, after some delay, in part due to illness, set off for Europe at the end of 1920 with the intention of investigating nitrogen fixation processes. It was fortunate that his visit to Europe had been delayed. His arrival, accompanied by two managers, Kannosuke Kasu and Kyoichi Taniguchi, coincided with Luigi Casale's perfection of his synthetic ammonia process at Terni. On learning of Noguchi's visit to Europe, Enrico Cairo the head of Società Italiana per il Carburo di Calcio, Acetilene e altri Gaz (Carburo), the leading Italian manufacturer of calcium cyanamide, arranged an introduction to Casale's workshop. Noguchi, who had dealings with Carburo from the time that he acquired a license for working the Frank-Caro process, was made aware by Cairo that the Casale process represented the only viable alternative to the Haber–Bosch process, which BASF was so reluctant to license. And, unlike the Oppau synthetic ammonia factory, the Terni workshop was available for inspection. By late 1920, both the Shōten Suzuki company and entrepreneur Fusanosuke Kuhara had sent their technical people to examine the Casale apparatus at Terni, perhaps also at the recommendation of Enrico Cairo, who was negotiating with Luigi Casale for a 10% commission on the introduction of new business.

4.2.1 Noguchi Visits Terni and Rome

Based on later accounts, events probably proceeded as follows. Early in 1921, Noguchi and colleagues traveled to Europe. Their itinerary included travel by train to Rome in order to meet with Cairo and discuss matters of mutual interest in connection with nitrogen fixation. However, at Cairo's suggestion, Noguchi decided to detour via Terni in order to investigate the Casale process.

Casale's disorderly workshop, set up in part of the abandoned ironworks, left a less-than-stellar impression on Noguchi and a colleague who had accompanied him to Terni. Even when Luigi Casale showed them a letter of interest from Suzuki Shōten's New York office, they remained uncertain of the commercial viability of Casale's process. But it was the only satisfactory process available, and Noguchi, long convinced of the future of synthetic ammonia, did not lose interest. He was no doubt fascinated by the application of far higher pressures (800–850 atm) than used in the Haber–Bosch process (250 atm), and by a demonstration of the direct production of anhydrous ammonia. No less remarkable was the fact that the converter, with its ingenious internal arrangements, was fabricated from ordinary steel and was designed to withstand the huge stresses. As with the Haber–Bosch process, unreacted gases were recirculated to the converter through a closed loop. The considerable energy requirements for producing hydrogen by electrolysis, and for driving compressors and pumps in the continuous production of ammonia, came from hydroelectric power, which in Noguchi's case was relevant because of his own involvement in hydroelectricity.

Noguchi, still somewhat hesitant, left the Terni workshop to catch a train to Rome, where he was scheduled to meet with Cairo and Carburo representatives. Then, acting on an impulse, he decided to remain in the city of Terni, maybe for a day or so. He needed to investigate further and again talk to Luigi Casale. Once his mind was made up, there was no further procrastination. He hurried to Casale's workshop, where Casale was not to be found, and then to Casale's home, only to be disappointed. Casale was working at the time in his Rome office. In the meantime, his wife Maria, alerted to Noguchi's interest, phoned Luigi and he set up a meeting with Noguchi in Rome. Noguchi was on the next train to Rome. His already great commitment to synthetic ammonia, along with his knowledge of the failures to master a high-pressure process in Japan and elsewhere, no doubt convinced him [2].

In taking on Casale's process, which had not yet been applied on a scale greater than perhaps 200 kg per day with the pilot plant at Terni, Noguchi was taking a tremendous risk, technical as well as financial. Casale demanded one million yen (around ten million lira) in exchange for a license. Noguchi agreed to this price—it was far less than BASF demanded of the other Japanese firms—but could not authorize the payment without approval from the board of Japan Nitrogenous Fertilizer (Nippon Chisso), which was dominated by the Mitsubishi corporation. He personally paid a 10% deposit and was given a two-week extension. The order and down payment must have impressed René Leprestre, Casale's backer, who probably assumed that funding for Noguchi was readily available from the powerful Mitsubishi. Moreover, Casale was impressed with Noguchi's knowledge of nitrogen fixation, to the extent that he willingly extended the grace period, giving Noguchi time to return to Japan to negotiate with his colleagues [3].

This confluence of circumstances, and timing, would now enable Casale to find the much-needed first licensee for his invention; the license, in turn, provided Noguchi with access to a technical process unprecedented in Japan. Negotiations were spread over twelve months, which gave Noguchi the time to convince his own backers of the need to invest in Casale technology. This became easier as Mitsubishi's control of Japan Nitrogenous Fertilizer declined during 1921. The formation in April of Ammonia Casale SA in Switzerland meant that further transactions would be in Swiss francs, or according to the gold standard. By this time, Enrico Cairo had exchanged his 10% share of new business for ten shares in Ammonia Casale.

Late in 1921, Noguchi and four Japanese colleagues, including Taniguchi, embarked on a lengthy international trip, including to London where arrangements were made for payment of the balance owing, now to Ammonia Casale SA, via a branch of the Mitsubishi Bank. The Japanese party traveled to Germany, and then on to Italy, accompanied by a researcher, Muneshiro Shiraishi, a former employee of Noguchi, who had received a scholarship from Mitsubishi to study in Germany. The group of six met with Casale in Rome [3, pp 125–129]. The contract, an exclusive license for Japan, was signed on 12 December 1921. The arrangements included provision for the supply of Italian-made machinery, including compressors manufactured by Pignone. Significantly, in 1921 the Hesegawa Iron Works assembled the first Japanese-made ammonia compressor, for use in the refrigeration industry.

4.3 The Nobeoka Ammonia Factory

Noguchi originally intended to build the synthetic ammonia factory adjacent to the Kagami cyanamide facility, but in March 1922, drainage problems, among other difficulties, led to the choice of Nobeoka, Miyazaki Prefecture, on the east coast, as the site of Japan's first synthetic ammonia factory.

Three months later, however, in June 1922, there was a threat to Luigi Casale's patent monopoly on synthetic ammonia in Japan. Noguchi, in a state of panic, advised Ammonia Casale that Fauser had filed a patent for his ammonia process with the Japanese Patent Office in Tokyo. Noguchi perceived this to be not only a threat to Casale but also to his own considerable investment in synthetic ammonia. He immediately requested permission from the Patent Office to invalidate the Fauser application. Around the same time, Leprestre, to calm down Noguchi, telegraphed him, confidant that the prior patent filing for the Casale process in the US effectively invalidated the Fauser patent in Japan because of the great rigor with which the US Patent Office examined applications. Moreover, at least according to Leprestre, Fauser's application would be voided, in the opinion of Luigi Casale, on the ground that it claimed rights to an earlier BASF patent. Not only did it infringe a BASF patent, but the Fauser process could not then be realized in practice because of the technical and economic difficulties of industrial application. For the time being, at least, the Fauser process was not a threat to Noguchi's enterprise.

In his dealings with Noguchi, Casale demonstrated that he had learned from Leprestre how to be an aggressive and capable negotiator. The arrangement between Casale and Noguchi included the services of Italian technical staff at Nobeoka.

4.4 Casale in Japan

Toward the end of August 1923, Casale, Leprestre, and his wife, Winifred, set sail from Vancouver on the Canadian Pacific's RMS *Empress of Canada* in order to attend the opening of the Nobeoka facility. The luxurious liner arrived at the port of Yokahama on 3 September but was forced to anchor in the outer harbor. A calamity had struck two days earlier: the Great Kantō earthquake had destroyed both Tokyo and Yokohama. It was Japan's worst natural disaster up to that time. The number of deaths was eventually estimated to be 140,000. The *Empress of Canada* took on board a number of Europeans, members of Yokahama's large expatriate community, and Chinese and Japanese survivors. In an emotional letter to his wife, Luigi Casale described the devastation he observed from the deck of the steamship [4]:

> My Dear Maria,
>
> We are anchored in front of what once was the city of Yokohama.
>
> A dense and large cloud of smoke and dust is lingering above the rubble, darkening the horizon. Since the very moment when we anchored, there has been a mass of refugees who were rescued by other steamships. They are in dire need of everything: they are ragged, dirty, shirtless and wounded. There are men, women, children. I was unable to offer help to our countrymen since there are none here. I have personally provided some clothing to the Spanish Consul and fed the Spanish ambassador, the head secretary, and a Swiss gentleman. I don't know the fate of our ambassador, De Martino. According to what was related to me by all these men and women rescued in Yokohama and Tokyo, this is the mightiest catastrophe ever experienced. The personal stories are horrible.
>
> Leprestre spares no efforts to provide help to the rescued dignitaries. Gerry [Winifred] looks after those wounded. I wanted to send you a telegram yesterday, but this was impossible since no Japanese telegraphic station was accepting messages and America was too far away. I hoped to do it today, but the situation is unchanged. I will try tomorrow from Kobe, if possible: I can imagine how much you are worried!

To allay Maria's concerns, Casale reported about his improved health and his changed travel plans for the return to Italy.

> I feel much better now. After sailing from Vancouver, I felt reasonably good but the day after I felt depressed and sad. I started to get better after seven days and now I am stronger.
>
> In the days before, I had planned to return to Italy via India but then someone warned me about the danger of typhoons in the China Sea and in the Indian Ocean and so I changed my mind, and I will return via America. However, I will avoid Vancouver passing instead via Honolulu and San Francisco. The Canadian Pacific line passes too far up north, and it was the bone-chilling cold, despite it being in August, which made me suffer a lot.
>
> After this disaster it makes no sense for Leprestre to continue his trip in Japan. I have no idea what they will do now. Maybe they will visit China.

On 5 September, the *Empress of Canada* docked in what was left of Yokohama's port. Noguchi was there to meet and assist his guests, as best he could, as Luigi explained to Maria [5]:

> I arrived four hours ago, and the director of Nippon Chisso [Shitagau Noguchi] immediately came to my cabin. An Italian speaking clerk was with him. They took my luggage so that I could immediately undergo the quarantine medical examination. The government has requisitioned all the hotels. However, Mr Noguchi was able to arrange two single rooms in Osaka for me and Leprestre, but only with great efforts. Noguchi was unaware that Leprestre's wife was with him, so she has decided to stay on board and then proceed to Shanghai where she will wait until a room in Kobe is available.
>
> Trains and steamboats here are loaded with refugees and wounded people, but everything is calm, and it seems that there is an efficient organization.
>
> I heard a great deal of news about the disaster and from many different sources: the disaster is beyond imagination, everything that will be published in the press will pale when compared with the reality.
>
> On our steamship we immediately organized a rescue committee and quickly collected 10,000 dollars. Leprestre contributed with 500 and I with 200. I will stay on board until I move on to Osaka with Noguchi.
>
> [Guido] Santagostino [Casale's engineer in Nobeoka] sent me a long letter from which it appears that everything is very good.
>
> …
>
> Greetings to all of you: I hold you very tight with our children.
>
> Oh, how many unfortunate children there are in this catastrophe!

Later in the day, Noguchi met with Leprestre, who was scheduled to attend meetings in Tokyo in connection with the filing of a patent for the Casale ammonia catalyst.

After resting in Osaka, Noguchi and Casale traveled west by train to Nobeoka. There, two ammonia units with converters—each of 7.5-mtpd rated capacity—compressors, pumps, and other machinery were in the process of being installed, under the supervision of Guido Santagostini, and with the assistance of Italian technicians (Figs. 4.1 and 4.2). Power for the facility came from hydroelectricity. Hydrogen was produced by electrolysis. The plant, managed by Taniguchi,

Fig. 4.1 Nobeoka ammonia factory, Japan, under construction. Taken from Mount Atagoyama. (Asahi Kasei. With permission)

Fig. 4.2 Pignone compressors, installed at Nobeoka, mid-1920s. (Casale SA. With permission)

was inaugurated by Luigi Casale, in the presence of Leprestre, and various dignitaries, on 20 September 1923. However, there were problems with converters, compressors, pumps, and other machinery [6].

In a letter to Maria from 9 October, Luigi wrote that he had sent telegrams to colleagues in Rome concealing an array of technical problems at Nobeoka that needed to be sorted out. He did not want to let his colleagues down. But to Maria he revealed the true, and difficult, situation [7]:

> I sent a telegram to Rome advising that we successfully started up the plant but, the truth is, instead, since the 5th of October we are facing many difficulties and we have failed to produce even 100 kilograms of ammonia. ...
>
> I send you these few lines because perhaps I will telegraph Rome again that all is proceeding very well, with the intention of encouraging the people in Terni and in Spain.

Luigi Casale's letters to Maria show that Casale and Leprestre, and Ammonia Casale, were at this time also facing a perilous financial situation. They had apparently placed all the funds at their disposal into Nobeoka, including payment for the equipment imported from Italy. Their pockets were empty. Perhaps there were additional reasons, including Leprestre's prior claims for expenses, and costs connected with the US venture. The correspondence leaves us with many unanswered questions. Whatever the cause of this situation, they now relied on the generosity of Noguchi to cover their expenses during these difficult weeks. Noguchi's support was also for his own benefit. Both sides had made massive financial investments and were taking enormous risks. The most pressing technical problem was the converter feed preheater, or heat exchanger. There was a mutual understanding that this and other problems could be solved only if there was an agreement to work together. Casale advised Maria that:

> Mr Noguchi promised Leprestre that he should expect the payment of 1,500,000 lira, because he [Leprestre] is broke and the company in Rome as well; all our hopes are now thwarted [unless we can solve the technical problems]. ... To overcome the difficulties with the inefficient [heat] exchanger we increased the power to the electrical heater [to bring the gas to the reaction strike temperature] but this wasn't nearly enough. Tomorrow, I'll give it another try with the modified reactor. Noguchi said I must not leave until the plant is on stream.

This was not the case for Leprestre and his wife Winifred, who were in Kobe. They left Kobe for Seattle on 19 October. Casale, much exasperated at their behavior, and much relieved at their departure, stayed on to deal with startup difficulties at Nobeoka, including a major failure with the preheater. Writing to Maria on 30 October, he explained the difficulties that were preventing proper operation. The tone is one of near-despair [8]:

> So far, we have failed to produce ammonia in the amount we normally readily make in Terni with the small reactor. The heat exchanger [the preheater] is insufficient and this is a big mistake by Santagostino. When I drew up the design [for this piece of equipment] I

> never had any bad surprises. [To give you an idea] just consider that this one [at Nobeoka] has a triple [heat exchange] surface—while Terni has one—but it is less efficient. You can imagine in what an embarrassing situation I am.

Luigi Casale was in a bit of a panic. He knew how high the stakes were as he made ready to confront the problems.

4.5 Breakthrough at Nobeoka

During the following days, laboring at all hours, Casale sweated his way through design modifications in order to tackle the shortcomings of the two ammonia units. The trial by pressure, heat, and catalyst lasted more than a week. By its end, Casale, near exhaustion, had overcome all the flaws in the machinery. For the first time, the commercial synthesis of anhydrous ammonia at the rate of around seven and a half tons a day from each of two units, at a pressure of 800 atmospheres, had been achieved. This was the real breakthrough. On 10 November 1923, Casale was euphoric. He had achieved the successful production of ammonia, and could barely disguise his joy at this triumph. Preparing for the future, he even found time to scribble down his thoughts on specifications for new, larger, converters. And he trumpeted his own capabilities. He could not wait to share his happiness, his dreams for the future, and news of his subsequent better health with Maria [9]:

> … based on the actual plant data I collected here, I have already completed the design of the [new] reactor of 15-mtpd capacity and I have begun to study the 32-mtpd one. I have already convinced Mr Noguchi how much more convenient it would be for him to buy the 15-mtpd reactor instead of the 7.5-mtpd one. And I'm sure I will likewise succeed with the French. We will go for the 32-mtpd reactor later on, for the American plant in Grand Rapids [Niagara Falls]. The continuous contact with the plant [Nobeoka] and my people has suggested to me an enormous range of improvements and modifications. Never before did I so fruitfully spend my time and, in addition, I'm leaving in better health then when I came!
>
> I realized while staying here that there were times when I was deeply distracted by so many things [concerning business matters], to the point that I almost lost touch with my co-workers and my process. I realize even more strongly now how no one can replace me, both for lack of intellectual ability and moral standing.

There was much cause for celebration. Casale, Noguchi, and colleagues gathered for a celebratory photograph (Fig. 4.3). Finally, Casale could relax (Fig. 4.4). On 14 November, with the Nobeoka factory fully operational, Luigi Casale left Japan, bound for San Francisco, and, eventually, Rome.

Later in the year, two more units of 7.5-mtpd rated capacity were ordered for Nobeoka. Noguchi soon cut back on the cyanamide process. He made further visits to Italy and built up an excellent working relationship with Luigi Casale, as did some of his senior managers (Fig. 4.5). News of the Nobeoka factory's success spread quickly abroad, and bolstered Casale's standing in Europe and the United States.

Fig. 4.3 Group photograph taken at Nobeoka in November 1923. *Front row (from left to right):* Mr Dutto; Seiji Ichikawa, junior managing director, Noguchi group (Nippon Chisso), associated with Shitagua Noguchi since around 1908; Luigi Casale; Shitagau Noguchi (Nippon Chisso); Mario Santagostino, engineer at Ammonia Casale, who supervised the installation of ammonia units at Nobeoka. *Back row (from left to right):* Mr Fujisaki; Gutō Koki (Nippon Chisso); Gizo Taki; Kyoichi Taniguchi, engineer, who visited Terni late in 1921 with Noguchi, and managed the Nobeoka plant; Yuhachi Nagasato, engineer, involved with electricity generation for Noguchi; Shikazo Shimada, engineer in charge of construction; Sojo Shiraishi, engineer with Noguchi; Tadayoshi Kitamura. (Asahi Kasai. With permission)

In February 1925, Noguchi opened a second Casale synthetic ammonia facility, at the Minamata cyanamide site. This was the first facility to operate 20-mtpd ammonia units. In 1927, it was noted of Casale converters that "smaller units have been given an output from ten to fifteen per cent higher than their nominal capacity. It was found that the same was true with larger units" [10]. Noguchi in fact achieved outputs of 26- to almost 32-mtpd from his 20-ton Casale units [11]. In 1930, Nobeoka was raised to 100-mtpd rated capacity, again with 20-mtpd units [12, 13].

In 1925, when imports of ammonium sulfate into Japan reached 203,500 tons, domestic production had already grown to 130,000 tons. In 1926, Noguchi's enterprise produced 85,000 tons of ammonium sulfate. In 1927, the Minamata ammonia facility was expanded to a rated capacity of 100-mtpd (Fig. 4.6). At Noguchi's invitation, an engineer from the Ministry of Agriculture and Commerce designed at Nobeoka a process that converted ammonia into nitric acid in three stages. The process enabled Noguchi from June 1928 to manufacture ammonium nitrate for explosives. The process was licensed abroad. By around 1930, Nitchitsu and Denka accounted for 40% of the total supply in Japan of ammonium sulfate, and

Fig. 4.4 Luigi Casale contemplates one of the two lion-dogs (*koma-inu*), guarding a Shintō shrine, during the time he was in Nobeoka to start up Shitagau Noguchi's synthetic ammonia factory in 1923. (Casale SA. With permission)

75% of domestic production of ammonia (Fig. 4.7). Nitchitsu emerged as a new zaibatsu.

In 1927, Noguchi planned to introduce the Casale process into the Korean peninsula, which in 1910 had been annexed by Imperial Japan. Noguchi's Korean enterprise at first met with strong resistance from the Mitsubishi zaibatsu, the conglomerate that until 1921 had backed him. Mitsubishi, in an attempt to put Noguchi out of business, acquired water rights in Korea that were essential to his enterprise.

Fig. 4.5 Representatives of Ammonia Casale and a guest from Japan, in Rome, c. 1924. *From left to right:* Renato Fabi; unknown; Vincent Handley; Luigi Casale (in light colored suit); Mr Dutto; a representative of Shitagau Noguchi; Luigi Calissano; and Iacopi Volpi. (Casale SA. With permission)

Noguchi put pressure on the Japanese military authorities in Korea to amend, in part, the apportioning of water rights. He gained access to a tributary of the Yalu River. Noguchi's chemical complex at Hûngnam, previously a small coastal fishing village in the northeast, would eventually extend over 320 acres. His technical people came from the Minamata factory. The original planned ammonia capacity of the Korean factory (Nippon Chisso Hûngnam; Chōsen Chisso Hiryō) was 120-mtpd. To provide the power requirements, Noguchi undertook construction of the Far East's largest hydroelectric power complex. Hydrogen, as on mainland Japan, was produced from electrolyzers. By the close of 1927, six 20-tpd ammonia units had been ordered from Ammonia Casale.

Ammonia and ammonium sulfate were produced in Korea towards the close of 1929. Full operation began in 1930, with, in all, 24 ammonia units, each of rated capacity 20-mtpd [15]. Other chemicals, including phosphoric acid, were manufactured at the site. By 1936 a second Korean factory, at Pon'gung, 4 km from Hûngnam, was producing both synthetic ammonia, by the Casale process, and calcium cyanamide [3, pp. 156–173]. By late 1938, the rated ammonia capacity for the Casale process at the two sites had increased four-fold, to 480-mtpd [14]. Hûngnam was claimed to be the world's third-largest producer of synthetic ammonia, with 45,000 workers at one stage (Table 4.1; Fig. 4.8).

Fig. 4.6 Battery of 20-mtpd ammonia converters, Casale process, at Minamata, prepared for an inspection by the Japanese Emperor, probably 1927. (Casale SA. With permission)

Fig. 4.7 Casale converter and Pignone compressor preserved at the site of the Nobeoka ammonia factory. (Asahi Kasei. With permission)

Table 4.1 Casale process in Japan

Place	Capacity	Date of contract	Opening date	Name of operating company
Nobeoka, Miyazaki Pref.	15-mtpd from two 7.5-mtpd units; two more 7.5-mtpd units ordered in 1923. With additional units total capacity raised to 100-mtpd in 1930	December 1921	1923	Nippon Chisso Hiryō
Minamata, Kumamoto Pref.	40-mtpd from two 20-mtpd units; with three more 20-mtpd units installed in 1927 capacity was raised to 100-mtpd		1925–26	Nippon Chisso Hiryō
Hûngnam/ Korea	120-mtpd from six 20-mtpd units; with additional units capacity was raised to 480-mtpd		1930–31	Nippon Chisso Hûngnam; Chōsen Chisso Hiryō
Pon'gung/ Korea	Included in Hûngnam		1936	Nippon Chisso Hûngnam; Chōsen Chisso Hiryō

4.6 The Followers of Noguchi in Japan

The technical and commercial success of Noguchi with the Casale process stimulated tremendous interest among Japanese corporations and entrepreneurs in the acquisition of synthetic ammonia technologies. While this interest already existed even before 1921, successful rival processes were not forthcoming, at least until after the mid-1920s. In the earliest case of the acquisition of a rival technology, by Shōten Suzuki, the venture turned out to be a technical and financial disaster.

In 1922, Shōten Suzuki , who in 1921 had lost interest in Casale's ammonia process, acquired a license for the French Claude process, which operated at 1000 atm. Pilot-plant trials were underway by 1924. A works was erected under the name of Suzuki's Claude Nitrogen Industries, on Hikoshima Island, Yamaguchi Prefecture. However, as at other early Claude factories, there were numerous technical difficulties. They were compounded by major explosions, in the hydrogen production unit, and from failure of a converter; very little ammonia left the converters. Suzuki also faced financial difficulties, in the form of a large unsecured debt owed to the Bank of Taiwan, whose failure in 1927 contributed to a major Japanese banking crisis. Suzuki closed down, along with its ammonia business. In 1929, Mitsui purchased the Suzuki ammonia factory. Its subsidiary, Mitsui Mining Company, failed to improve on the Claude process. In 1933, with government support, Mitsui started up ammonia production at Hikoshima using Du Pont technology, which had been developed from Casale's ammonia process (see Chap. 8)

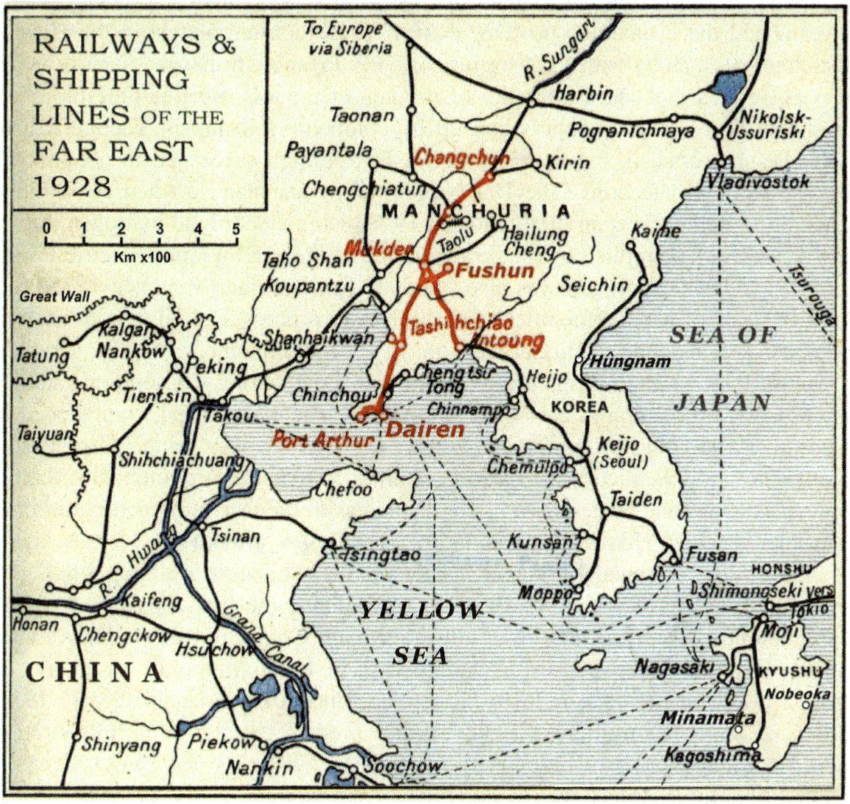

Fig. 4.8 South–East Asia, showing northern Republic of China (including Manchuria), Korea, and Kyūshū, Japan. Nitrogen products manufactured in this region were: cyanamide (Minamata, 1909–1910; Kagami, northern Kyūshū, 1914; Fushun, 1916–1920; near Hûngnam, mid-1930s); synthetic ammonia (Casale process, Nobeoka, 1923, Minamata, 1926, Hûngnam, 1930; Mont Cenis process, Dairen, 1935; Nitrogen Engineering Corporation process, Nanking (Nanjing), 1937; ammonium sulfate (Mond gas process, Fushun, 1916–1920; and distillation of shale oil, Fushun, 1927–1928). The railway connection to Hûngnam was constructed in the late 1920s to serve Noguchi's factory. (Adapted, with additions, by Nicholas Lera, from Kinney HW (1928) La Mandchourie Moderne et la Compagnie du Chemin de Fer Sud-Mandchourien. Paris) (Nicholas Lera. With permission)

[3, p. 293]. Mitsui Mining and other mining companies diversified into the manufacture of chemicals. For the production of synthetic ammonia, they relied on coking oven gas for hydrogen.

In 1926, the superphosphate manufacturer Dai Nippon Artificial Fertilizer acquired a license from Montecatini to operate the Fauser process. This was after any problems with the Fauser patent, insofar as there were similarities with the Haber–Bosch process, had been overcome. Dai Nippon was one of the first firms to gain access to Fauser technology outside Italy. With an interest in nitrogen compounds, especially ammonium sulfate, its management had in 1917 briefly investigated the cyanamide process, then in 1919 took an interest in the Haber–Bosch process, along with a consortium of other Japanese firms, and in early 1925, was made aware of the availability of the Fauser process. By this time the other ammonia processes had been taken up by competitors in Japan. Though Mitsui was also interested in the Fauser process, Dai Nippon succeeded in obtaining a license from Montecatini. The Dai Nippon ammonia plant, erected at Toyoma, Toyoma Prefecture, began production in 1928, using electrolytic hydrogen, drawing on electricity supplied by the power station of the nearby Japan Electric Power Co. (From 1938, hydrogen was also obtained from coking ovens.) In 1927, perhaps because of early difficulties with the Fauser process, Dai Nippon purchased Nitchitsu's cyanamide facility at Kagami. In 1928, Sumitomo, of Osaka, acquired a license to work the Chemico NEC (Nitrogen Engineering Corporation) ammonia process. Production of ammonia, at a rated capacity of 15-tpd, commenced in December 1930, and of ammonium sulfate shortly thereafter. In the early 1930s, Sumitomo and Chemico set up a joint venture to manufacture ammonia. Sumitomo, as a coal mine owner, used coking oven gas to produce hydrogen. Sumitomo was later renamed Nissin Chemical Co.

At the TIEL nitrogen laboratory, catalyst expert Shibata Katsutarō, collaborating with Buichi Yokoyama, had by 1922 developed the first Japanese-designed ammonia process. It was used by Shōwa Fertilizer Company (Shōwa Hiryō; later Shōwa Denkō), established by Nobuteru Mori in 1928. Ammonia was produced from 1931. Ube Nitrogen Industry, Ltd (Ube Yuka Kogya), established in 1933, produced ammonia from July 1934, using a process developed in collaboration with Heinrich Koppers of Essen [15].

In 1929, Julius Bueb, who had served as the German wartime commissioner for nitrogen products, and now represented mainly the behemoth IG Farben, was on a second visit to Japan and China, promoting increased imports of ammonium sulfate from Germany. At first, the Japanese showed no interest. They also declined requests to join the international nitrogen cartel. From 1930, they introduced among themselves various agreements over control of prices and production. Then in April 1931, Noguchi reached an arrangement with IG Farben over reduction of nitrogen imports from IG Farben and ICI [3, pp. 174–185]. IG Farben, however, would not then agree to license its Haber–Bosch synthetic ammonia process to Japanese companies. Within a few years, however, overproduction at Oppau and Leuna, and competition from other ammonia processes, convinced IG Farben to review its licensing and exchange of know-how policies. Subsequently there were several arrangements between IG Farben and Japanese firms in the mid-1930s over the licensing of synthetic ammonia technologies [16]. They were in part enabled by the Japanese agreement to join the international nitrogen cartel. Early in 1937, Mitsubishi Mining's subsidiary Japan Tar started the manufacture of ammonia according to the Haber–Bosch process. Hydrogen was made from coking oven gas.

None of Nitchitsu's competitors in Japan ever matched its output of nitrogen products. In 1937 alone, Nitchitsu produced around 50% of Japanese-made ammonium sulfate, some 585,000 tons [3, p. 283]. Hûngnam, together with Pon'gung, remained one of the world's largest producers of synthetic ammonia and ammonium sulfate. Today, the Nobeoka facility belongs to the Asahi Kasei Group.[1]

4.7 France: The Toulouse Ammonia Factory

Outside of Japan, the largest early Casale synthetic ammonia factory was at Toulouse in southwest France. It resulted from somewhat complex negotiations and deliberations of commissions of enquiry that began with an arrangement concluded with BASF. Following the Armistice in November 1918, the French, as occupiers of the BASF Oppau factory, attempted, including with site visits, to gain access to its Haber–Bosch technology (Fig. 4.9). On 11 November 1919, representatives of the French state, headed by Minister of Industrial Reconstruction Louis Loucheur (1872–1931), and the management of BASF signed an agreement over certain conditions dictated by the French. In return for technical assistance and a license for the Haber–Bosch process, the French agreed not to permit destruction of the all-important Oppau ammonia works [17].[2]

At the time, France faced severe shortages of fertilizers, mainly as a result of land neglect during the war. Loucheur, himself a leading engineer, and co-founder in 1899 of the construction company Giros et Loucheur, brought together a consortium of industrialists, headed by Joseph Frossard and Emile Lambert of Société Lambert, Rivière & Cie, called the Société Techniques d'Etudes et d'Enterprise pour l'Industrie, with the intention of operating the Haber–Bosch process in France for fertilizer production, and developing related technologies. Loucheur declared that even if BASF refused to cooperate, the French had the necessary technical means to go ahead with the production of ammonia. This alarmed supporters of the Claude process, who claimed that its higher operating pressure made the French process more economical. Moreover, rather than favoring the process of the French patriot Georges Claude they could not comprehend "how the government could afford to promote a process covered in French blood" [18].

In April 1920, the French government decided to promote a large synthetic ammonia facility at Toulouse. The choice of Toulouse—the site of a former powder factory—for the ammonia facility was made with national security in mind. For this reason alone, its remoteness from Germany—800 km from the border with Germany—was warranted. The same considerations applied when during the recent war the French built hydroelectric power stations in the central and western Pyrenees. It was the remoteness of Toulouse from French coalfields that deterred

[1] In 1950, the Nitchitsu concern, then no longer involved in northeastern Korea, was renamed the Shin Nippon Nitrogenous Fertilizer Co., Ltd.

[2] Loucheur was economic advisor to Georges Clemenceau at the 1919 Paris Peace Conference, which led to the Treaty of Versailles. He was Minister of Industrial Reconstruction until 20 January 1920.

Fig. 4.9 A French commission, some members in military uniforms, visit the occupied BASF works at Ludwigshafen-Oppau, 1919. Note the armed guard at right (BASF Corporate History. With permission)

investors from entering into what was intended to be either a private or a state-private partnership. In February 1921, the Académie des Sciences was directed by the government to form a commission to investigate the merits of the rival Claude and Haber–Bosch synthetic ammonia processes.

Attention was soon drawn to the Casale process. Promotion of the Casale process in France came about through the business and social networking of the Bouchayers of Grenoble. The network included Louis Loucheur, whose firm had been involved with the Bouchayers and Aluminium Français in the construction of the wartime Nera Montoro hydroelectric power station and aluminium factory near Terni. During the war, Loucheur served the French government as undersecretary of state for artillery and munitions, and from September 1917 to November 1918 as minister of armaments. Through Loucheur's close friendship with Aimé Bouchayer and family, Bouchayer et Viallet received large orders for military equipment [19]. No less relevant was the fact that, among his other wartime posts, Loucheur was until September 1917 administrator of the large firm Tréfileries et Laminoirs du Havre (TLH). Moreover, Loucheur had played a prominent role in TLH even before the war and introduced modern management methods similar to those practiced in the US. As noted in Chap. 3, Hippolyte Bouchayer was a member of the board of TLH, which during the war had invested in Compagnie des Produits Chimiques d'Alais et de la Camargue (CPAC), and as a result of which Bouchayer represented TLH on the board of CPAC, the forerunner of Pechiney [20]. There were also close links with Italy, in part through the wartime transfer of

German interests in aluminium production in both France and Italy to Bouchayer et Viallet and Aluminium Français [21]. Hippolyte Bouchayer and other members of the board of TLH were on the board of its subsidiary Le Trafilerie e Laminatoi di Metalli, which served the Italian market, including the military. TLH board members were also active in various other industrial enterprises, mainly those engaged in electrical technologies.

In early 1924, soon after the French commission for investigating ammonia processes was formed, Hippolyte Bouchayer and Pechiney's Emile Bouchard were drawing on their extensive political connections to lobby for the Casale process. As a result, in addition to investigating the Claude and Haber–Bosch processes, state officials showed a strong interest in the Casale process, for which a preference was soon shown, including by the military engineers Andre-Paul Marot (1890–1956), and Georges Patart (1869–1944). Patart also dabbled in high-pressure chemistry with work on a methanol process.

Apart from the networking efforts of Bouchayer and Pechiney executives in order to promote the Casale process, there was another consideration: the technical difficulties then being experienced with the Claude converters. The Casale process, with its efficient converter and recirculation of unreacted gases, appeared to offer distinct advantages over the Claude process. Claude interests contested, without success, the validity of the Casale patents [22].

On 11 April 1924, the French state, unable to attract private funding for the ammonia project, established the National Industrial Nitrogen Office (Office National Industriel de l'Azote, ONIA), administered by the Ministry of Public Works. Patart, president of the ONIA decided, after having made comparative estimates between the Haber, Casale, and Claude processes, to adopt the Casale ammonia process. Some members of the group responsible for drawing up plans for the factory were surprised by this choice. They included René-Paul Duchemin (1875–1963) of Kuhlmann (and CEO of Kuhlmann from 1929). He considered the choice of Casale risky, because in 1924 the Italian process was still worked with small converters. Duchemin recommended the setting up of a synthetic ammonia unit based on BASF technology, at least as a first step. But the deal was done. In December 1924, no doubt for services rendered, through Pechiney, Hippolyte Bouchayer was appointed president of Ammonia Casale (Chap. 5).

Construction of the new synthetic ammonia factory of the Poudrerie Nationale de Toulouse—based on the Casale process—was intended mainly for the manufacture of explosives, with up to 40% of the ammonia intended for agriculture. In April 1927, the factory was started up with six Casale ammonia units, each of rated capacity 20-mtpd. There were nine units altogether, three kept as spare. Unusually, the converters were not manufactured in Italy but in France, by Compagnie des forges et aciéries de la Marine et d'Homécourt. Toulouse was now one of the largest factories, not only in France, but also in the world, to be equipped with Casale ammonia units [18, pp. 23–25].

Hydrogen was manufactured by the BASF water gas process as developed in collaboration with Bamag (Bamag-Meguin from 1924). The purification, however, involving the removal of carbon oxides, was according to the method devised by

Patart for Société Techniques d'Etudes, and eliminated the BASF copper washing step. Despite early warnings from Ammonia Casale and Pechiney engineers that Patart's process was not viable, the Toulouse management, under pressure to commence production, went ahead with full operation. Things did not go according to plan. Between October 1927 and April 1928, the factory was plagued with damage to machinery and accidents. Then, in April 1928, ruptures were discovered in the walls of the converters. Ammonia Casale engineers also found major faults with the gas generation and purification equipment, and recommended closedown of ammonia production. On the advice of both Ammonia Casale and Pechiney, production was brought to a complete halt in May 1928. Half the workforce, some 1200 employees, was laid off. Inspection and analyses of the steel of the converters showed that the French-made machinery did not withstand the conditions of continuous operation. Replacement converters were ordered from the Italian factory of Vickers-Armstrongs (formerly Armstrong Whitworth), which had recently produced 20-mtpd Casale converters for Japan and was manufacturing similar converters for use at Ruhrchemie AG.

The mismatch of processes, the difficulties experienced with gas purification, and the failure of the converters, led the ONIA to resort to the earlier agreements between the French state and BASF. On 28 September 1928, ONIA ordered from BASF its total hydrogen generation and purification process, and at the same time placed an order for six Haber–Bosch converters. All the BASF equipment was manufactured in Germany. Satisfactory production was achieved after the BASF water gas technology was installed. The Haber–Bosch units, operating at the usual lower pressure than the Casale units, were installed in a separate production area (three more Haber–Bosch units followed later). In order to protect the proprietary information of both BASF and Ammonia Casale, BASF engineers were strictly forbidden from entering into the Casale production areas and workshops, and Casale engineers had to keep out of the BASF areas. Industrial secrecy was thus preserved. A group photograph dated November 1928 and filed away in the Casale archive appears to show a commission made up of Italian, French, and German experts, accompanied by one American, on an inspection of the Toulouse factory (Fig. 4.10).

In 1933, Toulouse was operating with five Casale 20-mtpd converters and six Haber–Bosch 15-mtpd converters, a total rated output of 215-mtpd [26]. The arrangement for using both processes continued until at least the late 1950s, with nine Casale and nine BASF ammonia units installed [27]. The Casale process had the advantage of producing anhydrous ammonia without the need for refrigeration (Fig. 4.11). A nitric acid facility, installed in 1931, enabled manufacture of ammonium nitrate, which, following an IG Farben process, was mixed with limestone to give a granulated fertilizer.

Of the fourteen French synthetic ammonia factories operating by around 1930, ten were concentrated in or near the Pas-de-Calais and Nord departments, with two located at Henin-Lietard (Pas-de Calais). Five of these employed the Casale process, three the Claude process, and two the NEC (Nitrogen Engineering Corporation) process (see Chap. 8). Of the other four factories, three utilized the Casale

Fig. 4.10 Group photograph, dated November 1928, possibly in connection with an inspection of the Toulouse synthetic ammonia factory. Until then, Toulouse used only Ammonia Casale technology in the synthesis of ammonia. Present are (*from left to right*): Constantin Mallinckrodt, Dr Deschamps, Professor Arturo Miolati, Dr Schmidt, Dr Maller, Vincent Handley (Ammonia Casale), Hippolyte Bouchayer (Ammonia Casale), and Dr Wollinnweber. Miolati was on the board of SIRI and a consultant to Ammonia Casale. Mallinkrodt represented the Mallinkrodt company in the US, a major supplier of anhydrous ammonia since 1889. (Casale SA. With permission)

Fig. 4.11 Storage tanks for liquid, anhydrous ammonia, manufactured by the Casale process, at Toulouse. (Casale SA. With permission)

process: Toulouse, which in 1928 also ordered Haber–Bosch converters; Firminy, with coking oven gas hydrogen; and Soulom (Pyrenees), with electrolytic hydrogen. The fourth factory used the Claude process, at Grand-Quevilly (near Rouen, Normandy). French Casale facilities installed mainly 7.5- and 20-mtpd ammonia converters.

The Casale process was also adopted in the 1920s in Belgium, at Ostend and Selzeate, and in the 1930s at Le Louviere and Tertre, in connection with coking oven processes for hydrogen.

The examples examined in this chapter have demonstrated how during the 1920s Ammonia Casale emerged as a specialized design and engineering contractor through licensing of patents and provision of specialized know-how. This was a policy that was followed elsewhere in the synthetic ammonia industry, and, as we shall see, in the chemical industry in general mainly after World War II through international engineering contractors.

References and Notes

1. Anon (1918) Japan: the nitrogen industry. J Soc Chem Ind 37(15 Nov):416R
2. Walton RD (1977) The rising companies and the establishment 1905–1931: the forerunners of Japanese expansion and their struggle for emergence. PhD dissertation, University of British Columbia
3. Molony B (1990) Technology and investment: the prewar Japanese chemical industry. Council on East Asian Studies/Harvard University Press, Cambridge, Mass
4. Luigi Casale to Maria, 3 Sept 1923, on board the RMS *Empress of Canada.* (Collection of Francesco Casale)
5. Luigi Casale to Maria, 5 Sept 1923, on board the RMS *Empress of Canada* in note 4. (Collection of Francesco Casale)
6. Asahi Kasai centennial homepage. https://bit.ly/3rgmnUm. Accessed 12 Nov 2024
7. Luigi Casale to Maria, from Nobeoka, 9 Oct 1923. Spain was the site at Sabiñánigo, where Casale converters and machinery were being installed. (Collection of Francesco Casale)
8. Luigi Casale to Maria, from Nobeoka, 30 Oct 1923. (Collection of Francesco Casale)
9. Luigi Casale to Maria, from Nobeoka, 10 Nov 1923. (Collection of Francesco Casale)
10. Anon (1927) News from Washington: Casale process operations. Chem Metall Eng 34(Mar):187
11. Miolati A (2009) Synthetic ammonia and the Casale process. Amplified edition of a lecture delivered the 27th February 1927 at the Institute of Chemistry of the Polytechnic School of Prague. Lugano, Ammonia Casale SA, p 34. Translation by G. Impallomeni of Miolati A (1927) L'ammoniaca sintetica ed il processo Casale. Ammonia Casale SA/L'Universale, Tipografia Poliglotta, Rome, p 37
12. Minutes of the meeting of shareholders, Ammonia Casale, 12 June 1930. (Casale SA)
13. Minutes of the meeting of shareholders, Ammonia Casale, 24 June 1931. (Casale SA)
14. Capacité des usines à ammoniaque synthétique procéedé Casale. List of factories in operation and under construction, Dec 1938. (Casale SA)
15. Travis AS (2018) Nitrogen capture: the growth of an international industry (1900–1940). Springer, Cham, pp 329–330, 334–338
16. Kudo A (2000) Dominance through cooperation: I.G. Farben's Japan strategy. In: Lesch JE (ed) The German chemical industry in the twentieth century. Kluwer, Dordrecht, pp 242–283
17. Meinzer L (1998) "Productive collateral", or "economic sense?": BASF under French occupation, 1919–1923. In: Travis AS, Schröter HG, Homburg E, Morris PJT (eds) Determinants in the evolution of the European chemical industry, 1900–1939. Kluwer, Dordrecht, pp 51–63
18. Gannon F (2021) La refondation de l'industrie chimique française de l'azote au lendemain du Traité de Versailles à travers le parcours de l'un de ses protagonistes: Georges Patart (X 1889). Revue de l'OFCE 171(1):199–238
19. Smith RJ (2001) The Bouchayers of Grenoble and French industrial enterprise, 1850–1970. Johns Hopkins University Press, Baltimore
20. Lagana M (1990) Le parti colonial Français: éléments d'histoire. Presses de l'Universite du Québec, Quebec, esp. pp 104–110
21. Bertilorenzi M (2008) The Italian aluminium industry: cartels, multinationals and the autarkic phase, 1917–1945. Caheirs de l'histoire de l'aluminium, 41(Dec):43–71
22. In 1925, Ammonia Casale representatives met with officers of the Claude concern, perhaps to discuss mutual interests. Ammonia Casale S.A. Massagne-Lugano. Conseil d'Administration, Réunion du 7 Nov 1925, p 15. (Casale SA)
23. Llopart M (2021) The French nitrogen industry during the interwar period: the ambiguous relationship between the state and manufacturers. Jahrbuch für Wirtschaftsgeschichte/Economic History Yearbook 62(1):191–212

24. Langlinay E (2018) La création de l'office national industriel de l'azote (ONIA) et la construction d'une filière chimique nationale (1919–1931). In: Emptoz G, Fauque D, Breysse J (eds) Entre reconstructions et mutations, les industries de la chimie entre les deux guerres. EDP Sciences, Les Ulis, pp 263–292
25. Friedlander HE (1943) The story of the nitrogen process in France: an experiment in post-war planning. J Bus Univ Chicago 16(4)(Oct):247–252
26. Information kindly supplied by Michael Llopart
27. Gadoin J, Litaise A, Pellenic M, Mlle Rapuzzi (1957–1958). Rapport d'Information. Conseil de la République, No. 306, Session ordinaire de 1957–1958, pp 7–11

Reorganization, Research, and Development at Ammonia Casale. The Final Years of Luigi Casale's Remarkable Life Story

5

Abstract

At the end of 1923, Ammonia Casale faced a financial crisis. Notwithstanding the great success achieved in Japan, there was a dire need of fresh capital, particularly for sustaining the expensive development work at Terni. There were also concerns of competition in ammonia production and marketing. No less problematic was the already difficult personal relationship between Luigi Casale and René Leprestre. Luigi had come to despise Leprestre as a person. The outcome was a falling out, followed by a major reorganization at Ammonia Casale, including new investors. The French entrepreneur Hippolyte Bouchayer, who had introduced much new licensing business for Casale technology through his connection with the French aluminum company Pechiney, was appointed president. New markets were sought, and new technical innovations were introduced. In 1925, Luigi Casale inaugurated Italy's first industrial research and development organization, dedicated to innovations in the area of high-pressure chemical processes. Luigi Casale was in an expansive mood. However, he had sustained a serious accident in 1917 while working on asphyxiating gases, resulting in his untimely death a decade later, while in his forties.

5.1 Introduction

Luigi Casale's first major success was the sale in 1921 of an operating license for his ammonia process to a Japanese entrepreneur who was deeply involved in the nitrogen industry. The Japanese factory was commissioned in September 1923. Hydrogen was produced by electrolysis and nitrogen by the Casale method of burning of air. Luigi Casale was on hand to iron out major technical problems, much to the satisfaction of his client. This established Casale's reputation, and was followed with repeat orders, and great interest from several European countries. It also defined Ammonia Casale's marketing strategy, based mainly on supply of

A. S. Travis, *The Global Synthetic Ammonia Industry*, Perspectives on the History of Chemistry, https://doi.org/10.1007/978-3-031-87978-4_5

Italian-made machinery and know-how. While the technology was outstanding, the financial status of Ammonia Casale, and the relationship between the partners Casale and Leprestre, was less so.

5.2 Casale and Leprestre: The Falling Out

At the end of 1923, Ammonia Casale faced a financial crisis. Notwithstanding the great success achieved in Japan, there was a dire need of fresh capital, particularly for sustaining the expensive development work at Terni. There were also concerns of competition from BASF, and "French companies," no doubt Air Liquide/Société Chimique de la Grande Paroisse, which owned the patents of the Claude synthetic ammonia process. No less problematic was the already difficult personal relationship, and disputes, between Luigi Casale and René Leprestre. Relations soured dramatically, as evidenced by the correspondence between Luigi and Maria. Luigi had come to despise Leprestre as a person. Leprestre flaunted his new trophy wife Winifred wherever they went, much to the annoyance and embarrassment of Luigi Casale, and even worse had left his two daughters in America without proper supervision. Casale was appalled and nauseated. The "showgirl" was a cause of gossip, and lurid tales. According to Maria, Leprestre was continuously claiming for himself considerable amounts, on account of "various expenses and costs incurred"—not always clearly defined, but no doubt in support of an extravagant lifestyle—causing, in turn, a state of persistent quarreling and fighting between Leprestre and Casale.

Luigi Casale, increasingly disgusted with his partner, advised Maria that Leprestre was a "phony," "insincere," and a "betrayer." Their correspondence reveals as much about Luigi Casale's moral values, who in his background and his faith was committed to his family and the ways of Catholicism, as it does about Leprestre's apparent clear lack of such values. The presence and behavior of Winifred, especially on the trip to Japan, only aggravated the situation. Leprestre had by the fall of 1923 become the antithesis of Casale. This was especially the case after Casale returned from Japan and was annoyed to discover that in his absence Leprestre had taken certain actions—details of which were not placed on record—but must have been so serious that they went against the interests of Ammonia Casale. Tensions between the two men frayed what had been a vital alliance. Their quite different backgrounds had come to the fore. The reservoir of trust was exhausted. The need for fresh faces and investors on the board of Ammonia Casale seemed nothing less than obvious. Among them was a group of bankers and Hippolyte Bouchayer, whose social disposition was more to Casale's liking, as was Hippolyte's important direct contact with Pechiney.

5.2.1 Raising Capital

In order to raise much needed capital a number of options for sale of an interest in Ammonia Casale were investigated. Thus early in 1924, the Belgian Solvay Process Company, an important consumer of ammonia, came close to acquiring a substantial stake in Ammonia Casale, to the extent that it signed an agreement stipulating a penalty of 100,000 Swiss francs if its offer was withdrawn, which is what happened, and Solvay settled the penalty. (In December 1923, as already noted, Belgian Solvay had acquired a stake in Société Ammonia, the French consortium that held licenses for use of the Casale process. This is perhaps related to the fact that the Solvay plan to invest directly in Ammonia Casale was scuttled.)

By the summer, however, Luigi Casale had managed to encourage several bankers and others in Europe to make substantial investments in Ammonia Casale. Most important was the long-established Basel-based bank Carl Gutzwiller through the senior partner Ernst Gutzwiller (who led the bank from 1927, when the name changed to E. Gutzwiller & Cie). As a result, at a special meeting of the shareholders held on 11 August 1924, Ernst Gutzwiller was appointed managing director and head of the board of directors of Ammonia Casale, replacing Leprestre. Luigi Casale remained the technical director, a post that he had held since the founding of the company in 1921.

The paid-up capital was increased 90-fold to nine million Swiss francs, in shares each of 500 francs, through the involvement of the Gutzwillers and other bankers. Three others who also joined the board of directors in August 1924 were: Pierre (Pietro) Lagier, of Vaud, Switzerland, who Casale had met when in the US; Alexander (Alessandro) Bungener, of Myes, Switzerland; and Odon de Lubersac, of Paris. Leprestre's shareholding, and influence, in Ammonia Casale had become considerably diluted [1] (Table 5.1).

Table 5.1 Ammonia Casale, Board of Directors, their occupations and places of residence, from 11 August 1924[a]

Name	Occupation	Place of Residence
Lieutenant-Colonel Giuseppe Albisetti	Representative	Massagno (Lugano)
Alessandro (Alexandre) Bungener	Banker	Myes (Vaud)
Dr Luigi Casale	Industrialist	Rome
Ernesto (Ernst) Gutzwiller	Banker	Basel
Pierre (Pietro) Lagier	Banker	Peroy (Vaud)
René Leprestre	Industrialist	American citizen, resident in New York
Odone (Odon) de Lubersac	Banker	French citizen, resident in Paris
Dr Lodovico-Federico Meyer	Lawyer/advocate	Lucerne

[a] From reference [1]

The corporate changes may not have satisfied Leprestre, but they certainly served to loosen his connection with Luigi Casale and Ammonia Casale. Luigi Casale was now free to plan for the future.

5.3 Società Italiana Ricerche Industriali (SIRI)

In 1925, the Nera Montoro site of SIAS started production of ammonia. Ammonia Casale's Terni workshop continued with pilot-plant and semi-industrial studies, and in collaboration with SIAS provided a comprehensive service in the design of manufacturing equipment. But this was insufficient for future needs. A framework for change, focusing on research and development in order to remain innovative, was much needed. Luigi Casale had planned this right from the start. On 8 July 1925, Casale, as principal investor, on behalf of Ammonia Casale, jointly with the Industrial Consortium of Velino and the Terni Company cofounded the Società Italiana Ricerche Industriali (SIRI) [2, 3]. The Terni workshop, under SIRI, was outfitted in order to greatly expand the research needs of Ammonia Casale. SIRI was Italy's first modern private research and development organization devoted to chemistry, and was certainly superior in its broad range of activities to the important research laboratory of the Societá Bomprini Parodi-Delfino, established around 1920 at Segni, near Rome. Casale subscribed to 5800 of the 10,000 shares, each of 500 lira, which constituted the share capital. SIRI, managed by Luigi Calissanno and Guido Santagostino, served as the research and development division of Ammonia Casale, charged with the design of equipment and installations, as well as catalyst improvements.

Luigi Casale continued to encourage licensees to use Italian-made machinery. Ammonia Casale procured hardware, including converters, compressors, and other machinery from Italian companies, notably Pignone for compressors, and Armstrong-Whitworth for converters. Casale's firm had turned itself into a consulting and engineering contractor, providing design and startup services for its clients. Also in 1925, Casale set up an International Consultative Committee to serve the needs of Ammonia Casale.

Luigi Casale was greatly assisted by his wife, Maria, who had become an expert in the design of catalysts. The opening of SIRI necessitated additional technical experts. Miolati was a consultant and a number of his students undertook research in the new SIRI workshop. Among others taken on to participate in technical studies was the electrical engineer and later famous writer and poet Carlo Emilio Gadda. Another literary figure associated with Ammonia Casale, and Gadda, was Franco Rampa Rossi (see later).

5.4 Global Expansion

By late 1924, Ammonia Casale was a highly respected firm with growing global connections. Bank loans for capital investment, international investors, licensees from within Europe, the success in Japan, in addition to a startup in the US, had enabled it to devote attention to potential new markets. Ammonia Casale was also some way ahead of the competition. The directors decided that the company's future growth lay in consolidation and reinforcing its presence globally, though not in the way that Leprestre had planned. This in part led to a further round of restructuring at the board level.

At an extraordinary meeting held on 18 December 1924, just five months after the major increase in the capital of Ammonia Casale, the shareholders agreed to a further increase in the paid-up capital, then of 9,000,000 Swiss francs, to 10,500,000 francs through the issue of 2700 new shares of 500 francs each and 3000 new shares of 50 francs each. The purpose of the increase in share capital was to acquire through an equity issuance the rights to the Casale ammonia patents according to the concession earlier granted to Pechiney, as listed in an inventory dated 16 December 1924, at a value of 1,500,000 francs. In exchange, Pechiney received, respectively, 2700 shares of 500 francs and 3000 shares of 50 francs, without any additional compensatory payment [4]. This no doubt followed the great success in France, especially the recent contract with the French government for providing Casale ammonia units to the Toulouse factory.

On the same occasion membership of the board of directors was increased to seventeen, including, and arising directly from the transaction with Pechiney, three of its executives, Hippolyte Bouchayer, Emile Boyoud, and Jacques Level. Bouchayer was appointed president of Ammonia Casale. There were also two members that represented UK interests, Sir Halford John Mackinder, and Andrew Henry Beatty, cofounder of the engineering firm Balfour Beatty & Co., Ltd (Table 5.2). Promotion of the Casale process in London by the Irish-born civil engineer John F. Crowley generated the great interest in the UK. Mackinder, an accomplished geographer, academic and former Member of Parliament, was appointed vice president. The availability of synthetic ammonia for ensuring national security in the British Empire and nations that Mackinder strongly favored—Poland in particular—by a process that rivaled the German BASF process, explains his strong attraction to Ammonia Casale.

Balfour Beatty was involved in electric tramway and power distribution systems, in addition to civil engineering projects. In 1922, the directors founded the Power Securities Corporation, which acquired the whole of the issued share capital of Balfour Beatty. Power Securities projects in hand included hydroelectric power schemes in the Scottish Highlands. In June 1924, at the First World Power Conference held at Wembley, just north of London, Crowley together with Eric M. Bergstrom of Balfour Beatty discussed installation of Casale synthetic ammonia factories in the Highlands, taking advantage of abundant hydroelectric power. In addition to low-cost electricity for power and electrolytic production of hydrogen, they drew attention to the several advantages of the Casale process, including the

Table 5.2 Ammonia Casale, new members appointed to the Board of Directors, their occupations and places of residence, from 18 December 1924[a]

Name	Occupation	Place of Residence
Sir Halford John Mackinder	British politician	west London
Andrew Henry Beatty	Co-founder of Balfour Beatty	southwest London
Paul Christ	Banker	Basel
Baron d'Ornellas, Charles Evariste	Civil engineer	Portuguese citizen, resident in Paris
Felix-Sigismond-George Marcel	Engineer	Paudex, near Lutry (Vaud)
Hippolyte Bouchayer	Engineer, executive at Pechiney, and associated with Compagnie de produits chimique Roche la Molière, and Tréfileries et Laminoirs du Havre	Grenoble (Isère)
Emile Boyoud	Executive at Pechiney	Brioude (Haute Loire)
Jacques Level	Executive at Pechiney	Paris
Hugues-Jean-Guillaume Jéquier	Director of Crédit Commercial de France	Neuchâtel
John Noseda	Lawyer	Vacallo, Switzerland

Notes Odon de Lubersac ceased to be a member of the board in December 1924, probably as a result of ill health. He died in June 1928. Hippolyte Bouchayer was associated with Ammonia Casale until his death in 1957, as the then president; he was succeeded by Ernst Gutzwiller. Albisetti remained on the board until 1975

[a] From reference [4]

quick startup time, when compared with the Haber–Bosch process [5]. At around that time, Crowley became involved in the planning for the large Shannon (Ardnacrusha) hydroelectric power scheme in County Clare, in the new Irish Free State (established in December 1922).

Crowley went to considerable trouble, and expense, in his promotion of Ammonia Casale in the UK. On 30 April 1925, he arranged a dinner to honor Luigi Casale and the board of directors of Ammonia Casale at London's Royal Automobile Club's Pall Mall headquarters (Figs. 5.1 and 5.2). Representing Ammonia Casale was its president, Hippolyte Bouchayer, and, apart from Luigi Casale, nine other directors. The more than fifty participants included James McNeill, High Commissioner to London of the Irish Free State, and various British military men, industrialists, and scientists, notably chemists Sir Arthur Cope, Professor J. R. Partington and Sir Robert Robinson, as well as Sir Charles Parsons (inventor of the eponymous turbine), consulting electrical engineer William Henry Patchell, president of the Institution of Mechanical Engineers, and engineering consultant Henry Selby Hele-Shaw. Among the political figures and civic leaders present were the Conservative MP George Balfour, cofounder with Andrew Beatty of Balfour Beatty, Labour MP Sir Henry Slesser, Liberal politician Frederick Ceaser

Linfield, and the Irish Nationalist and journalist Thomas Power O'Connor. The small but influential Irish contingent was present as a result of Crowley's contacts, probably because of his involvement in the Shannon scheme. Attesting to the importance of the event, reporters from the *Journal of the Society of Chemical Industry* and other leading technical journals were also in attendance. In proposing a toast to Luigi Casale, Crowley drew attention to the "strong British financial participation" in Ammonia Casale. George Clayton of United Alkali observed that the Casale process was easily adaptable to small-scale operation, which made it strategically advantageous, because several compact plants could be distributed throughout the country.

Mackinder was unable to attend the dinner. In August 1925, he met in Rome with Luigi Casale and Hippolyte Bouchayer to discuss progress in promoting Casale's process and the recent reorganization of the board of Ammonia Casale. In October, Bouchayer reported to Luigi Casale on Mackinder's involvement in an Anglo-Polish Chemical Syndicate, and the promise of financial assistance from the UK Government for installation of a factory based on the Casale process in Poland. Bouchayer also revealed that a meeting had taken place between Luigi Casale and the top management at BASF, which by this time was concerned about the ambitions of Ammonia Casale [6]. No doubt BASF was alarmed at the rapid spread of the Casale process. By this time BASF was loosening its policy regarding licensing, though only if there was much to be gained, such as, in exchange, access to a process that it needed, or a controlling interest in a new plant based on BASF technology. However, the price was way beyond what its potential suitors were prepared to pay.

At a board meeting held in Paris on 7 November 1925, the reorganization of Ammonia Casale on the lines proposed by Mackinder was discussed, along with other pressing issues. Based on his extensive experience in British government and other committees, Mackinder had suggested the setting up of subcommittees within Ammonia Casale. One was a small technical committee; another was for overseeing activities. A few days later, an agreement was drawn up with a "British Syndicate" involving Power Securities (Balfour Beatty) and United Alkali for promotion of the Casale ammonia process throughout the British Empire [7]. No doubt Mackinder had played a prominent role in bringing about this arrangement. Also, it was decided to transfer all administrative duties, certain of which had been carried out in Paris, to the Rome office.

One proposal at the 7 November board meeting was that in future the royalties were to be paid always in their gold equivalent, to protect against depreciation of weak currencies [8]. This latter practice must have been adopted, because, as retired engineer Ettore Comandini remembered, "Back then Casale enjoyed royalties and in gold!" Moreover, royalties for each ton produced were then standard in licensing arrangements, as was the case in the contracts between Fritz Haber and BASF. (Today this is no longer acceptable. Engineering contractors receive a one-time fee.)

In 1925, Casale-based ammonia facilities worldwide, operational, under construction, or at the planning stage, with converters of rated capacities of up to

Fig. 5.1 Royal Automobile Club, 30 April 1925. Dinner given for Dr Casale and the Members of the Conseil d'Administration of the Ammonia Casale Company. (Casale SA. With permission)

Fig. 5.2 Enlargement of a section of Fig. 5.1, showing, from left to right, Sir Charles A. Parsons (seated), Luigi Casale, John F. Crowley (both standing), Hippolyte Bouchayer, James McNeill, Sir Henry Slesser, and Emil Boyoud. In the foreground is the British chemist and historian of chemistry J. R. Partington. (Casale SA. With permission)

16-mtpd, were claimed to have a total rated capacity of 169,200-mtpy, though actual production was just 30,200 metric tons. In Italy, Montecatini in 1925 placed a break on Casale output. This was through an agreement between the Terni company and Montecatini which provided for a maximum production of 3500 metric tons of ammonia per year for Terni (including SIRI) and 17,500 for Montecatini. From 1927 Terni was able to increase production up to 7000 metric tons per year, while there was no restriction on Montecatini's production of synthetic ammonia.

In October 1925, the Soviet Union would take the first steps in the inauguration of its synthetic ammonia industry when it inked a licensing agreement with a new subsidiary of Ammonia Casale.

5.5 The Russian Connection: Casale Ammonia Società Anonima Italiana (CASAI)

In the mid-1920s, the Casale ammonia process appeared to offer the most viable means for the widespread growth, beyond Germany, of an industry that would ensure national security, through nitrogen products, in agriculture and munitions. Industrial leaders and groups of entrepreneurs drew up plans for forming consortia to promote the Casale process in various geographic regions, including throughout North America and parts of Western Europe, as well as in Eastern Europe and the British Empire. The success in Japan had attracted much attention, notably in the Union of Soviet Socialist Republics (USSR), which had been formed in 1922. Three years later, the USSR was preparing to set on a course of industrialization, based on imported technologies. The leading chemist engaged in the acquisition of ammonia and nitrogen technologies was Vladimir N. Ipatieff, the Russian expert in high-pressure chemistry. In 1925, Ipatieff visited German, French, Italian, and Swedish facilities engaged in the research and development of nitrogen processes. Ipatieff found that the most readily available, and well-proven technology for synthetic ammonia was that of Ammonia Casale. However, the adoption of nitrogen fixation processes had to confront a few hurdles in their acquisition by the communist country. Thus Ipatieff had to convince Leon Trotsky, head of the Soviet electrotechnical board and the scientific-industrial board of industry, as well as chairman of the chief concessions committee, that large-scale removal of nitrogen from the atmosphere would not pose a threat to its composition. Rather than providing concessions to western corporations, as had often been the case prior to 1925, and with which Trotsky was dissatisfied, he preferred the signing of contracts for know-how, licensing, and the supply and installation of machinery (Fig. 5.3).

In the field of synthetic ammonia, the Casale process was the first to be taken up on the lines favored by Trotsky. This required a novel mechanism for undertaking trade with the communist nation, a separate company, Casale Ammonia Società Anonima Italiana (CASAI), established in Terni following a meeting of the directors of Ammonia Casale on 21 September 1925. Hippolyte Bouchayer was appointed president. Among the board of directors of CASAI was Arturo Miolati. The parties to the contract were CASAI and a Russian trust, a horizontally integrated chemical firm, the Severny Chimicheskyi Trust URSS (Severnii Khimicheskii Trest; Northern Chemical Trust), based in Moscow. On the Russian side was Mr Paul Tschekin, the vice president and signatory of the trust. The signatory for CASAI was Luigi Casale, the managing director. According to the contract, CASAI agreed to sell licenses for "several ammonia plants" to be set up "in the northern regions of Russia." The Soviets paid a deposit of US$ 150,000, following the signing of the contract on 13 October 1925. A Russian

Fig. 5.3 Assembling compressors at the Rastiapino (Dzerdzinsk) Casale factory, around 1926. (Casale SA. With permission)

engineer was dispatched to Terni. During 1926, Casale ammonia converters and related equipment arrived in Rastiapino—located in the Central District, 400 km east of Moscow—accompanied by Italian technical staff. (Rastiapino from 1929 was known as Dzerdzinsk.)[1] The ammonia factory, part of the Chernorechenskii (Black River) industrial complex, was the first synthetic ammonia facility in the Soviet Union. It began production of ammonia in 1928. Hydrogen was obtained from water gas by the process designed by Julius Pintsch of Berlin. In 1930, during the First Five Year Plan (1928–1932), additional Casale units were installed, raising the rated capacity from 15-mtpd to 30-mtpd. By then, the American firm Chemico, and Montecatini had each signed contracts for supply of ammonia units to the USSR, and Casale Ammonia's role had declined. In Russia, the great interest in synthetic ammonia led to the creation of a dedicated high-pressure research institute.

In 1940, following the outbreak of World War II, Miolati replaced the Frenchman Bouchayer as president of CASAI. In 1942, when Italy and the USSR were on opposing sides in World War II, it was decided to wind up CASAI, in accord with the terms of the original contract [9].

[1] Dzerdzinsk was near Nizhny Novgorod (Nijni, Nizhiny), birthplace of Maxim Gorky and called Gorky between 1932 and 1990.

5.6 Death of Casale

In late 1926, Luigi Casale's wartime laboratory accident would now take a great personal toll. After a debilitating illness that stretched on for five months, his brilliant but short career in high-pressure industrial chemistry, spanning less than a decade, was brought to an end by his untimely death. On 18 February 1927, Casale died in Vigevano, Lombardy, some thirty kilometers from his birthplace at Langosco.

This was a tremendous blow. In the months before his death, Luigi Casale, who had almost single-handedly revolutionized the international nitrogen industry during the first half of the 1920s, was planning further expansion at Ammonia Casale, in both technology and strategy. One week before his death, at a meeting of the International Consultative Committee, which took place at his bed-side, Luigi Casale, as sharp-minded as ever, continued to radiate optimism by announcing the imminent availability of new, enlarged converters of 30-mtpd capacity. Casale advised his audience that he was ready to produce even larger units [10].

But for Luigi Casale it was too late.

What Casale achieved in his short life was truly remarkable. The impact of the death of this outstanding inventor and entrepreneur was such that it was deemed necessary to make an official announcement in the Italian parliament, duly noted in the minutes of the next meeting of shareholders of Ammonia Casale, held in June. A decade after Luigi Casale's death, Maria Sacchi-Casale was bestowed with the title Papal Countess by Pope Pius XI—with the right of transmission to her (male) descendants—in recognition of Luigi Casale's contribution to society. The title was also confirmed by Italy's King Vittorio Emanuel III in 1938.

5.7 After Luigi Casale

In 1927, total annual Casale rated capacity, including new plants then under construction or on order, was expected to achieve 350,000 metric tons. While this was lower than for ammonia produced by IG Farben (founded in December 1925 as the result of a merger of BASF and other leading German chemical firms), they were greater than corresponding figures for all other competing processes.

The minutes of the June 1927 meeting of Ammonia Casale's shareholders recorded that the first 20-mtpd ammonia units were placed on stream in 1926. The converters—8.5-m in length (with lids, 10 m) and weighing 40 tons—were installed at Nobeoka, in Japan. Also in 1926, several new units, with a combined capacity of 250-mtpd were commissioned; eight new ammonia units had been licensed with a total rated capacity of 180-mtpd, and four plants were under construction with a total capacity of 260-mtpd. The Casale process even posed a minor threat to IG Farben when, in 1927, three Casale units each of rated capacity 20-mtpd were commissioned by a consortium of German coal mining firms that in 1928 established Ruhrchemie AG. In 1929, Ruhrchemie commenced manufacture of ammonia at Oberhausen-Holten. In 1930, Ruhrchemie ordered a fourth unit

(Fig. 5.4). After 1927, 30-mtpd converters were available. Smaller units, including of 2- and 4-mtpd capacity, were also produced. Casale converter output invariably exceeded guaranteed figures. The first Lonza converters at Viège averaged, per unit, 4-mtpd, exceeding Ammonia Casale's guaranteed 3.5-mtpd. Subsequently Lonza received converters of 7.5-mtpd capacity.

Ammonia Casale acquired Luigi Casale's majority stake in SIRI, amounting to 60% of the paid-up capital, corresponding to 6000 shares each of 500 lira [11]. Luigi Calissano took over as manager of Ammonia Casale. Maria Sacchi-Casale was appointed to the board, and became increasingly active in the affairs of the company (which continued up until her death on 29 March 1950 at her home in Rapello). This included research activities at SIRI, where she was already an important contributor in the area of catalysts. Around 1925, high-pressure synthesis of methanol and urea had been investigated by Luigi Casale. In 1927, a methanol pilot plant at Terni produced 2-mtpd, and by the end of 1928 methanol was produced in 5-mtpd units at Nera Montoro, based on Casale's process. In 1931, a 10-mtpd methanol unit for Nera Montero was under construction at Terni [12].

Fig. 5.4 Compressors for mixed gases, Ruhrchemie AG, Oberhausen-Holten, Germany, 1930s. (Casale SA. With permission)

By 1929, there were one hundred and twenty people on the payroll of SIRI, which was directed by the engineer Renato Fabi. In the same year, the first 30-mtpd ammonia converters entered service. During 1930, the first 50-mtpd Casale ammonia converter was under construction. SIRI was undertaking research into various processes, including for generating hydrogen. Worldwide, there were twenty-eight facilities in operation using Ammonia Casale technology [13]. This included the explosives manufacturer Canadian Industries Limited (an ICI affiliate) which started production of ammonia with a 7.5-mtpd Casale converter at Sandwich, Ontario, in June 1930. A Casale converter of 2-mtpd rated capacity was installed at Celulosa Argentinas SA, Juan Ortiz, Argentina, in 1931 (Fig. 5.5). The source of hydrogen was probably the company's chloro-alkali electrolytic cells. By 1938, thirty factories in thirteen countries had installed Casale synthetic ammonia technology (Fig. 5.6). Converters of 75-metric ton daily capacity were available (Fig. 5.7).

Calissano, however, confronted a number of challenges. They included from the mid-1920s the sale of licenses for synthetic ammonia by Air Liquide for the Claude process, after it had overcome problems with high-pressure converters, Montecatini for the Fauser process, Friedrich Uhde for the Uhde/Mont Cenis process, and by the late 1920s the Chemico Nitrogen Engineering Corporation (NEC) process. Moreover new refrigeration technologies enabled efficient production of liquid ammonia from the lower pressure processes. At the end of the decade there was the looming global problem of over-production of nitrogen products, particularly because this impacted the main exporters, IG Farben and ICI (as well as Chilean nitrate interests). Notwithstanding John F. Crowley's efforts—as a member of the Ammonia Casale International Consultative Committee, and in connection with the "British Syndicate"—Casale ammonia units did not appear in the British Empire, nor in the UK. At the close of 1926, United Alkali and Brunner, Mond had been merged into the new Imperial Chemical Industries (ICI). ICI had its own synthetic ammonia process up and running at Billingham, which no doubt put an end to interest in the introduction of the Casale process in the UK. As a result, in January 1928, MacKinder and Beatty ceased to be board members of Ammonia Casale, as did Leprestre, Félix Marcel, Paul Christ, and Pierre Lagier [14]. Over the following years, the declining economic climate, in addition to competition between technologies, and collusion between IG Farben and ICI over control of markets, would place licensing fees under pressure [15]. The reasons for the departure of MacKinder and Beatty included the fact that by 1928, ICI had tied up the British Empire business with exports of Billingham-made ammonium sulfate fertilizer. Ammonia Casale failed in competitive bids to gain business in the Netherlands and China, while Kuhlmann in France showed a preference for the Nitrogen Engineering Corporation (NEC) process. As for Leprestre, his departure was hastened by failure during 1927 of the American Casale enterprise (Chap. 8), and, especially, differences with Maria.

At this point, Leprestre, whose backing of Luigi Casale had been so critical in the early years, disappears from the history of Ammonia Casale. What happened to him next is mainly unknown. In a 1930 census, he described himself as a civil

Fig. 5.5 Casale converter of 2-mtpd rated capacity, Celulosa Argentina SA, Juan Ortiz, Argentina. The factory, located near Rosario, opened in 1931. (Casale SA. With permission)

CAPACITÉ DES USINES À AMMONIAQUE SYNTHÉTIQUE PROCÉDÉ CASALE

EN FONCTIONNEMENT ET EN CONSTRUCTION

DÉCEMBRE 1938

PAYS	LICENCIÉ	LOCALITÉ	Tonnes NH_3 anhydre par jour
ALLEMAGNE	Ruhrchemie A.G.	OBERHAUSEN	240
ARGENTINE	Celulosa Argentina S.A. . .	JUAN ORTIZ	2
BELGIQUE	Union Chimique Belge S.A. .	OSTENDE	80
"	Société Carbochimique S.A.	TERTRE	120
"	S. A. Fabrication Engrais Azotés "S.A.F.E.A." . . .	LA LOUVIERE	60
"	Etablissements Kuhlmann . .	SELZAETE	24
CANADA	Canadian Industries Ltd. .	SANDWICH	8
ESPAGNE	Energia e Industrias Aragonesas	SABIÑANIGO	15
ETATS-UNIS	E.I. Du Pont de Nemours & Co.	BELLE	106
FRANCE	Cie. Alais, Froges & Camargue	ST. AUBAN	12
"	Société des Mines de Dourges	HENIN-LIETARD	15
"	Société Industrielle & Financière de Lens	DOUVRIN	50
"	Cie. des Mines de Vicoigne, Noeux & Drocourt	HENIN-LIETARD	50
"	Cie. des Produits Chimiques Anzin Kuhlmann	ANZIN	32
"	Société Mosellane Industrielle & Financière . .	CARLING	60
"	Sté. des Phosphates Tunisiens	SOULOM	48
"	Gouvernement Français . . .	TOULOUSE	180
"	Compagnie des Produits Chimiques Marles Kuhlmann . .	CHOCQUES	24
FRANCE	Société des Produits Azotés	LANNEMEZAN	10
"	Cie. de Roche-La-Molière. .	FIRMINY	24
"	Compagnie de Béthune. . . .	BULLY-LES-MINES	60
ITALIE	Società Italiana Ricerche Industriali "S.I.R.I." .	TERNI	10
"	"Terni" Società per l'Industria e l'Elettricità . .	NERA MONTORO	130
"	"Vetrocoke" S.A.	VENISE	100
JAPON	Nippon Chisso Hiryo Kabushiki Kaisha.	NOBEOKA	62
"	Nippon Chisso Hiryo Kabushiki Kaisha	MINAMATA	100
"	Nippon Chisso Hiryo Kabushiki Kaisha	KOÑAN	480
RUSSIE	Gouvernement U.R.S.S. . . .	DZERJINSK	64
SUISSE	Lonza Usines Electriques et Chimiques	VIEGE	23
YOUGOSLAVIE	Gouvernement du Royaume de Yougoslavie	—	11
		Total Tonnes	2200

Fig. 5.6 Locations and rated annual capacities of factories that worked the Casale process (or were under construction) in December 1938. (Casale SA. With permission))

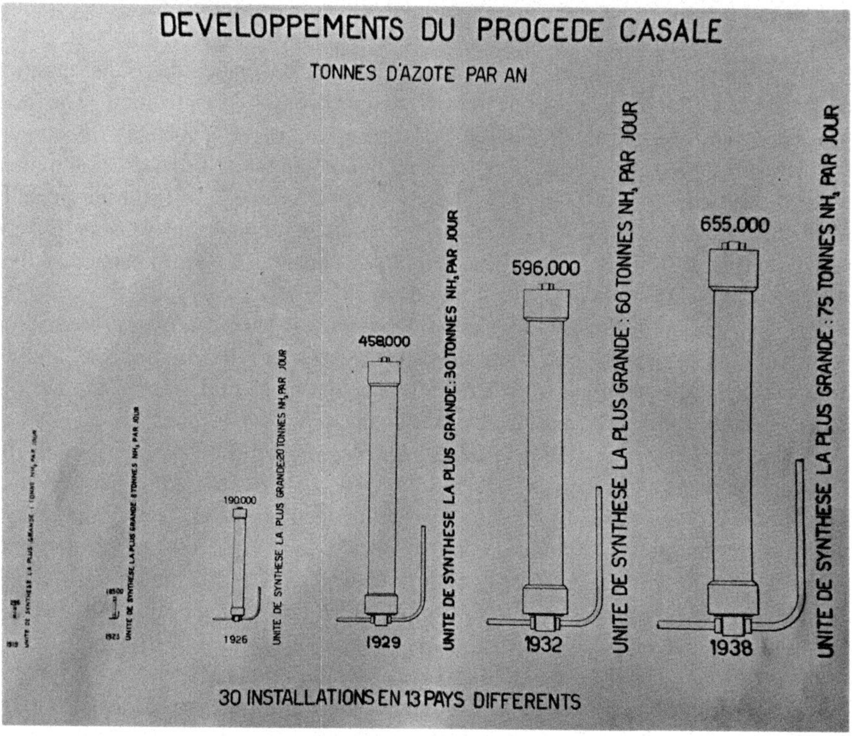

Fig. 5.7 Global rated capacities of Casale ammonia converters, in metric tons, until 1938. By that time, Casale converters of 75-metric ton daily capacity were available. The one-ton per day capacity shown for 1919, as intended at Ruminaca, was not realized. (Casale SA. With permission)

engineer. Around that time, he separated from his "showgirl" wife, and embarked on some family social climbing, at least for his daughter Paule and himself. In 1932, Paule married Henry Watson Cornell Jr. (1899–1957), a great grandson of Ezra Cornell the founder of Cornell University. Later in the 1930s, Leprestre married for a third time. His wife was Henry Cornell's mother, Margaret Feek Cornell (née Bouck) (1892–1975), a granddaughter of former New York Governor William C. Bouck.[2] Genevieve Leprestre, who had taken a distinct dislike to Leprestre's second wife, Winifred, left for Hawaii and became a nurse at the Molokai Leper Colony under Father Damian [16]. A single patent document reveals that by the late 1930s, René Leprestre was involved in research into catalytic hydrogenation of hydrocarbons for use as high-octane fuels. His converter design, adapted to this special purpose, bore a strong resemblance to Luigi Casale's ammonia apparatus.

[2] Margaret (Marjorie) Leprestre was a graduate of Cornell University (1914), she spent much of her career with the American Red Cross, working on disaster relief, and from 1927 in education. She was in France, with the Red Cross in 1918.

The patent claimed operation at up to around 900 atm, not far off Casale's pressure [17].

In summary, during 1924–1926, Luigi Casale had overseen the transformation of Ammonia Casale into a well-funded, research-based organization. This was the time when a small number of synthetic ammonia processes began to compete against one another. The threat of overproduction lay in the future. Larger and more efficient versions of the Casale converter were introduced. Investors, especially from the UK, were attracted by plans for opening up new markets in Europe and Asia. These relationships did not continue, but the reorganization that they stimulated ensured survival in the difficult times ahead.

In 1932, just after the peak of the world recession, global nitrogen production was represented by: synthetic ammonia, 70.4%; town gas and coking oven ammonium sulfate, 10.8%; Chilean nitrate 10.5%; calcium cyanamide, 7.6%; and electric arc and nitrate (potassium nitrate) from Bengal, India, 0.7%. The two factories of IG Farben, the leader in synthetic ammonia, represented 33% of global capacity, followed by users of the Casale process, at 17% [18]. In 1935, after the worst of the economic crisis was over, the total world nitrogen production capacity (mainly as synthetic ammonia), including Chilean nitrate, was 2,991,000 metric tons (as nitrogen). The total world consumption amounted to 2,400,000 metric tons, which included 217,000 metric tons of Chilean nitrate. Egypt was the main importer of ammonium sulfate (462,348 metric tons), followed by the US (452,619 metric tons), and Spain (236,521 metric tons) [19]. Global production in the mid-1930s was reduced, partly as a result of the international nitrogen cartel, the Compagnie Internationale de l'Industrie de l'Azote, established in 1930 mainly under the auspices of IG Farben and ICI, the two main producers of synthetic ammonia [20].

5.8 Ammonia Casale as an Incubator of Culture. The Engineers Gadda and Rossi

Here we get a hint at a broader context of the Ammonia Casale story, in which people were knitted with the company in recognized cultural developments. The emphasis is on the 1920s and 1930s in Italy, a decade of great political divisions and upheavals.

The Italian author Carlo Emilio Gadda (1893–1973) studied engineering at the Milan Polytechnic (Politecnico di Milano). When World War I broke out, he volunteered to join the Alpini Corps, and was appointed a lieutenant. He was taken prisoner during the Battle of Caporetto in October 1917. In the early 1920s, Gadda made his name as a journalist who wrote essays on technical subjects for a popular audience. He joined Ammonia Casale in 1925 and worked on the ammonia process in the new Terni research facility. After Luigi Casale's death, in 1927, Gadda worked under the supervision of Luigi Calissano, the then manager of Ammonia Casale. In January 1931, Gadda resigned from Ammonia Casale in order to devote himself to writing. Initially, the going was rough, and he even asked Calissano if it would be possible to rejoin the firm—a request that was met with a polite but firm refusal. This did not stop Gadda from penning impassioned pieces on the significance of the technology of nitrogen fixation, through the high-pressure synthetic ammonia and the electrothermal calcium cyanamide processes (cyanamide was an important commodity in Italy). He stressed the superiority of these processes, especially as carried out in Italy, where they contributed towards economic autarky, when compared with the Norwegian Birkeland-Eyde electric arc process, with its total reliance on inexpensive hydroelectric power, and the former importation into Europe, especially to the UK, of nitrogen-containing mineral products (nitrates) from South America. In time, Gadda would become one of Italy's literary greats. His unique poetic approach to nitrogen chemistry could also be discerned in his later writings. However, that was not before he was taken on again by Ammonia Casale as an external consultant during 1937–1940 [21, 22].

Francesco (Franco) Rampa Rossi was born in Rapallo (near Genova), and studied chemistry at the University of Turin, where he graduated in 1923. In his younger years, he was a communist and an adherent of the avant-garde Futurist ("Futurismo") movement, which in the early 1920s received recognition from

Fig. 5.8 Franco Rampa Rossi (at left), the Impressionist painter Fedele Azari, and the Futurist artist Fortunato Depero, probably March 1922. (Courtesy of the Museum of Modern and Contemporary Art of Trento and Rovereto)

Fig. 5.9 Rossi (at right), Azari, and Depero, on the same occasion, in the cockpit after distributing leaflets by air in order to advertise the "Esposizione Futurista Internazionale," held from 27 March to 27 April 1922 at the Winter Club della Galleria Subalpina, in Turin. Azari and Depero were proponents of "Aeropittura," or "flying painting." (Anonym, Fortunato Depero, Fedele Azari and Franco Rampa Rossi, vintage silver print, post card, 8.7 × 13.9 cm, courtesy PRISKA PASQUER, Cologne)

the fascists, and on which the communists were making claims. Rossi is credited with having composed several poems and organizing the important "Esposizione Futurista Internazionale," held in the spring of 1922 at the "Winter Club" Galleria Subalpina in Turin (Figs. 5.8 and 5.9). This is considered the most important contact between communists and Futurists. From Rossi's extensive correspondence with Gadda, it seems that for a time he gave up his youthful leftist ideals and became a fascist. In 1924, Rossi joined the Rome office of Ammonia Casale. He was described as a polyglot, particularly well-versed in Russian. He translated works by Ivan Turgenev (*The Watch*), Vadeem Andreevich Sofonov, and Clara Milic. Rossi befriended Gadda, with whom he shared a passion for literature. His fluency in several foreign languages was no doubt put to good use in Ammonia Casale's commercial department, especially after a contract for know-how and supply of machinery for manufacture of synthetic ammonia was signed with the Soviet Union in October 1925. He served as a member of the editorial committee of *Materie Priome per L'italia e L'Impero* ("Raw Materials for Italy and its Empire"), and contributed articles to several journals. He also indulged in writing recipes for the 1932 book "Futurist Cooking" under the pen name Sirocofran. Through Rossi, Gaddi ghost-wrote, for the industrialist Pietro Ferrario, *La nuova fabbrica italiana del magnesio*. After 1945, Rossi penned occasional pieces for *L'Unità*, the official newspaper of the Italian Communist Party, on technical subjects, including electricity, and penicillin and antibiotics [23–25].

References and Notes

1. Anon (1924) Schweitzer Handelsamblatt/Foglio Ufficiale Svizzero di Commercio, no. 207, 4 Sept pp 1473–1474. The paid-up capital, until 11 August 1924 amounting to 100,000 francs, split into 200 registered shares each of 500 francs nominal value, was increased to 9,000,000 francs, through the issuance of an additional 17,800 ordinary (common) shares of 500 francs each, fully paid up. All new shares were bearer shares. The stated objectives of the company were to purchase, exploit, and defend existing and future Casale patents for the production of synthetic ammonia and the hydrogen-nitrogen mixture, according to the Casale process.
2. Agreement between Ammonia Casale and Luigi Casale, 7 November 1925. According to the preamble to this agreement, the contract between Casale and Leprestre dated 2 August 1919, as ratified by the Conseil d'Administration on 5 June 1922, included the establishment of a laboratory where Casale would undertake research into problems of industrial chemistry. Casale would receive 20% of the benefits arising from any future inventions patented and commercialized. Research and development work, that had not been possible at the existing facilities, would be undertaken by SIRI, under Casale's leadership. Casale's share of the SIRI paid-up capital amounted to 3,000,000 lira, out of 5,000,000 lira, while Terni contributed the balance of 2,000,000 lira in the form of a fully equipped workshop and the ammonia machinery formerly owned by SIAS. Casale's initial outlay towards the paid-up capital was 900,000 lira in cash, with Ammonia Casale responsible for making up the balance of 2,100,000 lira. In addition, Casale opened a bank account for SIRI with a deposit of 2,000,000 lira as working capital, treated as a loan to Ammonia Casale. Ammonia Casale S.A. Massagne-Lugano. Conseil d'Administration, Réunion du 7 Novembre 1925, pp 11–13. (Casale SA) The founding individuals in SIRI, in addition to Luigi Casale, were the engineer Arturo Bocciardo, managing director of the Terni company (Terni Company for Industry and Electricity) and its affiliate Terni-Vickers, and Carburo, and president of the Consortium of Velino; and Commendatore Vincenzo Nava, managing director of SIAS. SIAS, with the approval of the Terni company, conferred on SIRI the property of the former Ferriera di v. Campofregoso with all the tools and machinery. The registered office of SIRI was in Terni. Its duration was until 1960, later extended to 2060.
3. Stentella D (1995) Dalla Ferriera Pontifica alla Zecca di Terni all SIRI. Terni
4. Anon (1925) Schweitzer Handelsamblatt/Foglio Ufficiale Svizzero di Commercio, no. 76, p 551
5. Crowley JF, Bergstrom EM (1924) The development of national water power resources. In: The transactions of the First World Power Conference, London, June 30th to July 12th, 1924, vol. I. Power resources of the world available and unutilized. Percy Lund Humphries & Co., Ltd, London, pp 349–368, on 359–363
6. Hippolyte Bouchayer to Luigi Casale, 6 October 1925. (Collection of Francesco Casale)
7. Luigi Casale, and another (a legal representative), to Balfour Beatty, London, 12 November 1925: "Further to our discussions a few days ago in Paris, with various representatives of the British group of shareholders of Ammonia Casale, we are pleased to advise that, in our capacity as attorney for Ammonia Casale, we have signed on May 1st in London the agreement with Power Securities concerning the foundation of the British Syndicate which will be in charge of developing the Casale plants throughout the British Empire. It is our understanding that Power Securities was not acting for itself but only on behalf of the Syndicate currently being incorporated, of which the various members will be at the same level as Power Securities." (Casale SA)
8. Ammonia Casale S.A. Massagne-Lugano. Conseil d'Administration, Réunion du 7 Nov 1925, pp 13–14, 15. (Casale SA)
9. CASAI (Casale Ammonia Società Anonima Italiana) file; Ammonia Casale S.A. Massagne-Lugano. Conseil d'Administration, Réunion du 7 Novembre 1925, pp 13–14. (Casale SA) In 1943, the registered office of CASAI was Via Campofregoso 72, in Terni. Surviving correspondence suggests that previously it had been at no. 42, and before that at no. 36A

10. Crowley JF (1927) Dr Luigi Casale. J Soc Chem Ind 46(11 Mar):220–221
11. Minutes of the meeting of shareholders, Ammonia Casale, 2 June 1927. (Casale SA)
12. Minutes of the meeting of shareholders, Ammonia Casale, 24 June 1931. (Casale SA)
13. Minutes of the meeting of shareholders, Ammonia Casale, 12 June 1930. (Casale SA)
14. Anon (1928) Schweitzer Handelsamblatt/Foglio Ufficiale Svizzero di Commercio, no. 39, p 307
15. Kwan MB (2017) Patriots' game: Yongli Chemical Industries, 1917–1953. Brill, Leiden, pp 85–105
16. Elizabeth Burrows to author, emails December 2021
17. Leprestre R (1938) US patent, 2319508, filed May 1938, granted 18 May 1943; assigned after his death to his executors, Margaret C. Leprestre, H. Douglas Madden, Ralph C. Tobin, and Joseph Dannenberger
18. Toniolo C (1934) Le industria dell'azoto sintetico nella crisi mondiale. Associazone Italiana di Chimica, Rome, pp 164, 166
19. Anon (1936) International trade in certain raw materials and foodstuffs. League of Nations, Geneva, 1936-II, A 26
20. Reader WJ (1975) Imperial Chemical Industries: a history, vol. II. The first quarter-century 1926–1952. Oxford University Press, London
21. Falkoff RR (2015) Carlo Emilio Gadda's professional alchemy, or how to make bread from air and literature from science. Roman Stud 33(3–4):232–244. https://doi.org/10.1080/02639904.2015.1124219
22. Italia P (2013) Gadda e Roma prima del acciaccio. Con un articolo 'disperso.' I quaderni dell'ingenere: Testi e studi gaddiani no 4, new series, pp 234–270
23. Salaris C (1996) Dizionario del Futurismo: Dizionario del futurismo: Idee, provocazioni e parole d'ordine di una grande avanguardia. Biblioteca tascabile/Editori riuniti
24. Ialongo E (2015) Filippo Tommaso Marinetti: the artist and his politics. Farleigh Dickinson University Press, Madison
25. Marinetti FT, Fillia [Luigi Colombo] (1932) La cucina futurista. Sonzogno, Milan

Casale's Process Described

6

Abstract

The technical novelty of Luigi Casale's synthetic ammonia process, which operated at 850 atmospheres pressure, arose from several significant innovations, most especially the design of the converter (reaction tube), the ability to control overheating within the catalyst bed, and a device called an ejector which improved the recirculation of unreacted gases. As with the Haber–Bosch process, the Casale process was based on a closed loop, or train. The converter was made of ordinary steel, which posed problems of failure in the other early ammonia processes. Success was made possible by the design of the internals: an arrangement of concentric tubes which ensured that the severe reaction conditions, brought on by pressure, heat, and a catalyst, did not place undue stress on the shell of the converter. The ammonia was a pure, anhydrous liquid, a feature made possible by the extremely high operating pressure. The ejector, introduced in the mid-1920s, reduced the number of moving parts.

6.1 Introduction

In September 1922, the London-based engineer John A. Harker made his way by boat and train to Terni. He had come to visit Luigi Casale's workshop, to see for himself the newly installed apparatus for producing synthetic ammonia. Its centerpiece was a novel chemical reactor, a converter (synthesis tube), that at a pressure of 850 atmospheres brought about the formation of ammonia from nitrogen and hydrogen. The converter was claimed to be capable of producing two metric tons of ammonia every day.

The converter, a steel tube resembling the barrel of a naval gun, was made of ordinary steel, which was known to respond poorly to reactive hydrogen under even far milder conditions. This usage of ordinary steel, such that decarburization did not take place, was made possible by the design of the internals: concentric tubes which ensured that the severe reaction conditions, brought on by pressure,

A. S. Travis, *The Global Synthetic Ammonia Industry*, Perspectives on the History of Chemistry, https://doi.org/10.1007/978-3-031-87978-4_6

heat, and a catalyst, did not place undue stress on the shell of the converter.[1] Large bolts held the lid of the converter tightly closed. The design was elegantly simple and remarkably compact. The mixture of gases—hydrogen produced from electrolyzers and nitrogen produced from the burning of oxygen in the air with hydrogen—entered together at the bottom of the converter through a narrow pipe. They were pre-heated via internal heat exchange. The electrical wiring for the internal heater was inserted at the top. Harker's curiosity was particularly piqued by important advantages of the new process over the Haber–Bosch process. The ammonia, leaving by another pipe at the base of the converter, was a pure, anhydrous liquid, a feature made possible by the extremely high operating pressure. After observing the apparatus at work, Harker could barely contain his enthusiasm. And he was taken aback by the confidence of his hosts in the safety of the process. This response must have been not untypical of those who saw Casale's novel ammonia process at work.

This chapter discusses more technical aspects of Luigi Casale's synthetic ammonia process, and its development during the 1920s (Fig. 6.1). Specifically, it will cover the converter, the control of overheating within the catalyst bed, and a device called an ejector which improved the recirculation of unreacted gases (and also enabled a reduction in the pressure).

6.2 The Casale Converter

Luigi Casale's converter was, with modifications, the standard model for over four decades. The flow of compressed mixed gases was axial. By flushing the inner wall of the outer tube, or shell, of the converter with cold synthesis gas, it was maintained at a temperature which greatly reduced the stress on the tube. This was the feature which permitted the use of ordinary steel rather than special alloys. As described in patents "the reacting gases take a tortuous path before entering the innermost tube" [1]. This innermost tube, containing the catalyst bed, surrounded the electrical heating element of the converter. Heat from the catalyst bed warmed up the incoming gas mixture.

The "tortuous path" began with entry of the compressed nitrogen–hydrogen gas-feed at the bottom of the converter. From there, it flowed up in the annular space between the inner wall of the converter shell and a corrugated partition. At the top, the gas reversed its direction and flowed down the next annular space, where it came into contact with the outer wall of the inner catalyst tube. Here it absorbed heat by heat exchange from the exothermic reaction. At the bottom it again changed direction, this time entering into the catalyst tube where, within the catalyst bed, the ammonia was formed (Fig. 6.2). In the words of John Crowley,

[1] Decarburization of steel in the shell of an ammonia converter results from the reaction between the carbon in the steel and hydrogen under the conditions of a high pressure. As a result of the loss of carbon the strength and hardness of the steel decreases.

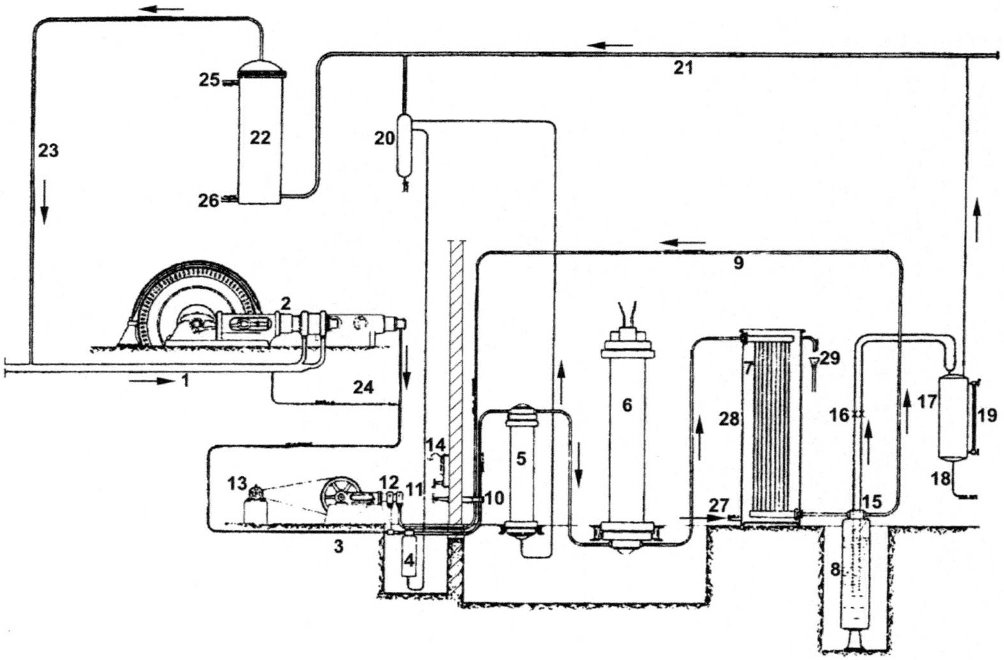

Fig. 6.1 Scheme for the original Casale synthetic ammonia process. From an original Ammonia Casale diagram. (Casale SA. With permission) 1. Low-pressure nitrogen-hydrogen make-up gas suction piping; 2. Six-stage make-up gas compressor (final delivery pressure 750-850 atm); 3. Circulation pump; 4. Oil scrubber for removal of compressor lubricating oil; 5. Gas purifier (removal of final traces of lubricating oil); 6. Ammonia converter; 7. Ammonia condenser tube bundles; 8. High-pressure ammonia separator (catchpot tank); 9. Recirculation gas piping; 10. Circulation pump bypass; 11. Suction valve; 12. Compression valve; 13. Circulation pump motor; 14. Control panel; 15. Liquid ammonia (suction) pipes; 16. Control valves, also acting as pressure reducers. (The internal pressure of the collection tank is more than sufficient to expel the liquid by means of the suction pipes); 17. Low-pressure ammonia separator (catchpot tank, or flash tank), at 15 to 20 atm; 18. Liquid ammonia discharge pipe; 19. Level indicator; 20. Compressor oil separator; 21. Flash gas return pipe; 22. Water scrubber (to recover the ammonia vapors present in the flash gas); 23. Gas recirculation return pipe; 24. Gas recirculation control valve; 25. Water pipe; 26. Aqua-ammonia discharge pipe; 27. Cooling water inlet; 28. Ammonia condenser (external vessel); 29. Cooling water discharge

"The synthesis tube [converter] is constructed so as to ensure efficient heat interchange, and the pressure container [outer tube] itself never attains a temperature higher than 200 °C. This low temperature enables special steels to be dispensed with in the design of the catalyst bomb" [2].

As in the Haber–Bosch process, a gas recirculation pump, for return of the unreacted gas mixture to the converter, was incorporated into the early synthesis loop.

The important advantage of the use of very high pressures, as Casale had calculated, was the increased yield, 15–18%, compared with around 5% for the Haber–Bosch process. The yield, in fact, was greater than expected by theory. This meant that less gas was pumped and circulated to produce the same volume of ammonia as in the Haber–Bosch process. In the synthetic ammonia processes at the lower pressure range, at 200–350 atm, the magnetite catalyst had to be specially

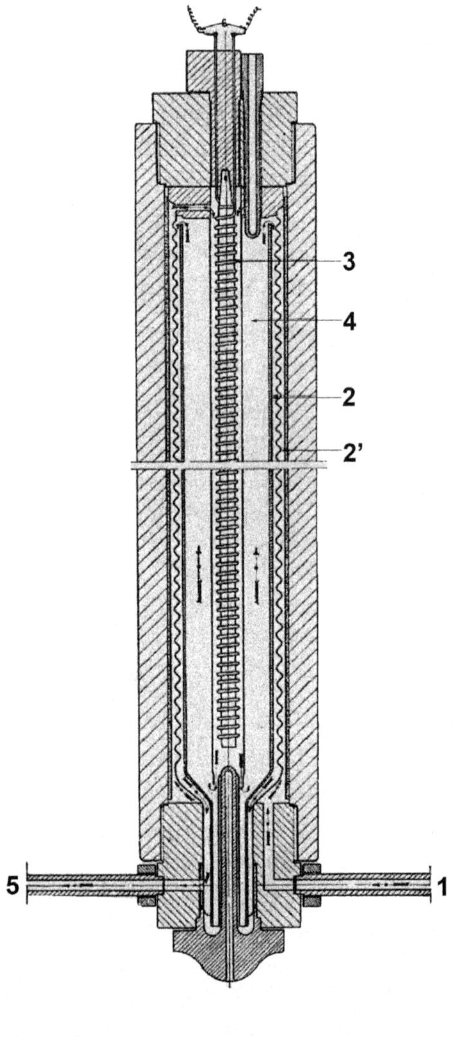

Fig. 6.2 Casale ammonia converter, 1920s (Casale SA. With permission). 1. Gas (nitrogen–hydrogen) inlet; 2. 2,2', concentric annular spaces, divided by a corrugated partition; 3. Electrical heating element; 4. Catalyst bed; 5. Ammonia and unreacted nitrogen–hydrogen outlet. Figure redrawn from Miolati A (2009 [1927]) Synthetic ammonia and the Casale process. Amplified edition of a lecture delivered the 27th February 1927 at the Institute of Chemistry of the Polytechnic School of Prague. Lugano, Ammonia Casale SA, p. 24

formulated with the addition of promoters, or activators, which were oxides of certain metals, to ensure high activity. At the very high pressures, a less active catalyst, though still incorporating promoters, was used, as was the case in the Casale process, which required around one-seventh the amount of catalyst when compared with the Haber–Bosch process. At Casale's working pressure of 800–850 atm, it was possible to separate out anhydrous liquid ammonia from the gas mixture by simply cooling with water. Thus, because of the high working pressure, there was no need for a refrigeration compressor for separating out liquid ammonia; the condenser operated at the water-cooling temperature. This was a major advantage over the Haber–Bosch process, for which expensive refrigeration equipment was necessary in order to produce anhydrous, liquid ammonia from the gaseous product mixture. Moreover, direct production of liquid ammonia under Casale's conditions involved more compact units than were required in the Haber–Bosch process. John Francis Crowley, a partner of John Harker—both were the UK representatives of Ammonia Casale—believed that there were strategic advantages in the use of small units. As Crowley put it, "The development of processes such as that of Casale … permit of the manufacture of small units, which because of their simplicity and ease of operation, can be run with sufficient economy to compete with the large units hitherto essential" [2, p. xxii]. Another important feature was the quick startup time, when compared with the Haber–Bosch process.

In 1922, a converter of one metric ton per day (mtpd) rated capacity was installed at Terni. It was followed in September by a converter of 2-mtpd rated capacity, the one examined by Harker. In July 1923, a converter of 3-mtpd capacity was installed at Terni. By mid-1923, converters of 5-mtpd capacity were available, and converters of 7.5-mtpd (sometimes referred to as 8-tpd) capacity were under construction, for delivery to Japan [3].

At pressures of 800–850 atmospheres, the life of the catalyst was short. However, Casale's iron catalyst, said by Arturo Miolati to be prepared from scrap iron turnings, mixed with promoters, was inexpensive. The need for more frequent replacement than with the catalyst material used in the Haber–Bosch process, which was far more costly, did not impact on the economics of the process. Casale's catalyst also suffered less from carbon monoxide poisoning. Until the widespread introduction of coking oven hydrogen, from the mid-1920s, hydroelectricity was employed for the electrolytic production of hydrogen, as well as for the supply of power, in small and medium-size ammonia facilities. By 1927, closing of the lids of the larger converter tubes was through a screw fitting, rather than with bolts, an innovation that made the apparatus safer, and enabled easier and more rapid assembling and dismounting, particularly when spent catalyst was removed and replaced by a new charge.

6.2.1 Overheating in the Catalyst Bed

Once the ammonia synthesis apparatus was in operation, the reaction could be continued in the absence of internal heating. The temperature was monitored and

Fig. 6.3 Head of Casale ammonia converter, showing electrical heater and thermocouple leads (Casale SA. With permission))

adjusted as required, using remote instruments. However, unless carefully controlled, the enormous heat generated in the catalyst bed, the area of exposure of mixed gases to the catalyst, would cause the apparatus to explode. In order to prevent this, the reaction was slowed down by allowing some product ammonia to be retained in the recirculated gas mixture. Regulation of the process could also be achieved by varying the temperature, through adjusting the electrical heating element, or varying the mass and speed of gases circulating in the synthesis tube (Fig. 6.3).

6.3 The Ejector

As in all other early synthetic ammonia processes, gas recirculation with fresh make-up gas within the synthesis cycle, or loop, was brought about by means of piston-driven, or reciprocating, pumps. This, however, required lubrication of the moving parts, and necessitated stuffing boxes and devices for separating out lubricating oil in order to prevent its entry into the converter, where, as a result of charring, it would poison the catalyst. These various drawbacks were overcome by Luigi Casale with the introduction of an ejector into the circuit (Figs. 6.4 and 6.5). This piece of equipment, with no working parts, was invented in 1923, was in use by 1924, and replaced the circulation pump in the Casale process (Figs. 6.6 and 6.7).

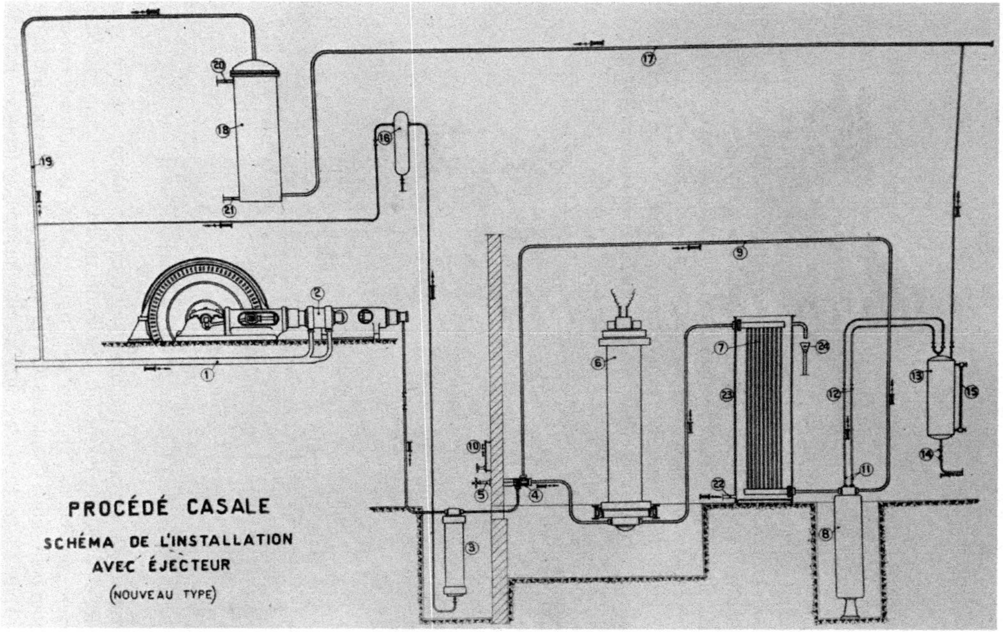

Fig. 6.4 Scheme of Casale ammonia process, with ejector for recirculation of unreacted gas with a new charge of nitrogen-hydrogen gas-feed (make-up gas). From an original Ammonia Casale diagram. (Casale SA. With permission)1. Low-pressure nitrogen-hydrogen make-up gas suction piping; 2. Six-stage make-up gas compressor (final delivery pressure 750-850 atm); 3. Compressor oil removal; 4. Ejector; 5. Ejector needle control wheel; 6. Ammonia converter; 7. Ammonia condenser tube bundle; 8. High-pressure ammonia separator (catchpot); 9. Recirculation gas piping; 10. Control board; 11. High-pressure ammonia discharge piping; 12. Let-down valves; 13. Low-pressure ammonia separator (catchpot tank, or flash tank); 14. Liquid ammonia discharge pipe; 15. Level indicator; 16. Compressor oil separator; 17. Flash gas return pipe; 18. Water scrubber (to recover the ammonia vapors present in the flash gas); 19. Scrubbed flash gas to compressor suction; 20. Scrubbing water inlet nozzle; 21. Aqua-ammonia discharge nozzle; 22. Cooling water inlet; 23. Ammonia condenser (external vessel); 24. Cooling water discharge

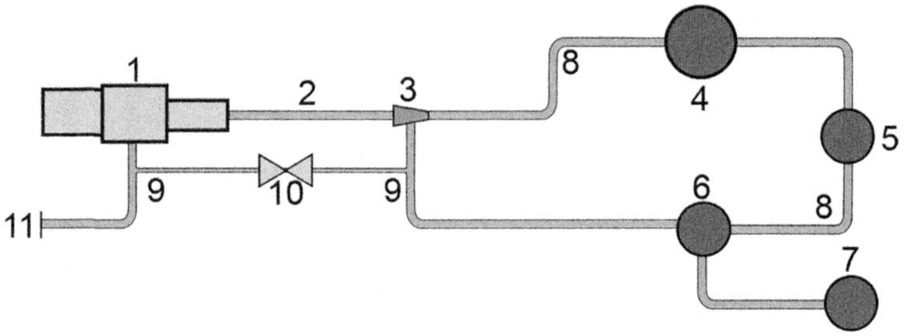

Fig. 6.5 Scheme for operation of Casale's ammonia synthesis process with ejector (Casale SA. With permission). 1. Make-up gas compressor; 2. Make-up gas to the loop; 3. Ejector; 4. Ammonia synthesis converter; 5. Ammonia condenser; 6. High-pressure ammonia separator (catchpot); 7. Low-pressure ammonia separator (catchpot); 8. Recirculating gas piping; 9. Bypass; 10. Flashing (control) valve; 11. Pipe leading gas mixture to compressor (fresh make-up gas suction). *Note*: The bypass (9) is important during the startup, when the gas mixture must circulate through the circuit but the reaction has not started yet. Because there is no withdrawal from the synthesis loop (in the form of ammonia produced), all circulating gas must be kicked back to the compressor and, on account of the lower suction pressure compared to that in the loop, there must be a flashing valve in between (valve 10). During this stage, in order to avoid continuous accumulation of gas in the loop (and consequent overpressure), there is no admission of fresh gas to the compressor

Fig. 6.6 Casale ejector. Luca Bianchi explains: "The larger pipe carries circulating gas in the synthesis loop, while the smaller pipe, the one on the left, carries the motive, or make up gas. Note the peculiar very high-pressure connections, typical of the original Casale design, made of thick square-forged pieces with four bolts. Luigi Casale devised this kind of connection because in the 1920s there were no standard pipe flanges (nowadays they are all cylindrical), or fittings in general, for such very high pressures." Each ejector replaced two reciprocating circulation pumps. Reciprocating equipment was cumbersome, costly, and required frequent maintenance. The entry of any leaking lubricating oil into the gas poisoned the catalyst, and meant shutdown of the synthesis loop. This injector was installed in a synthesis loop at Société Carbochimique, Tertre, Belgium, where ammonia production began in 1932 (Casale SA. With permission)

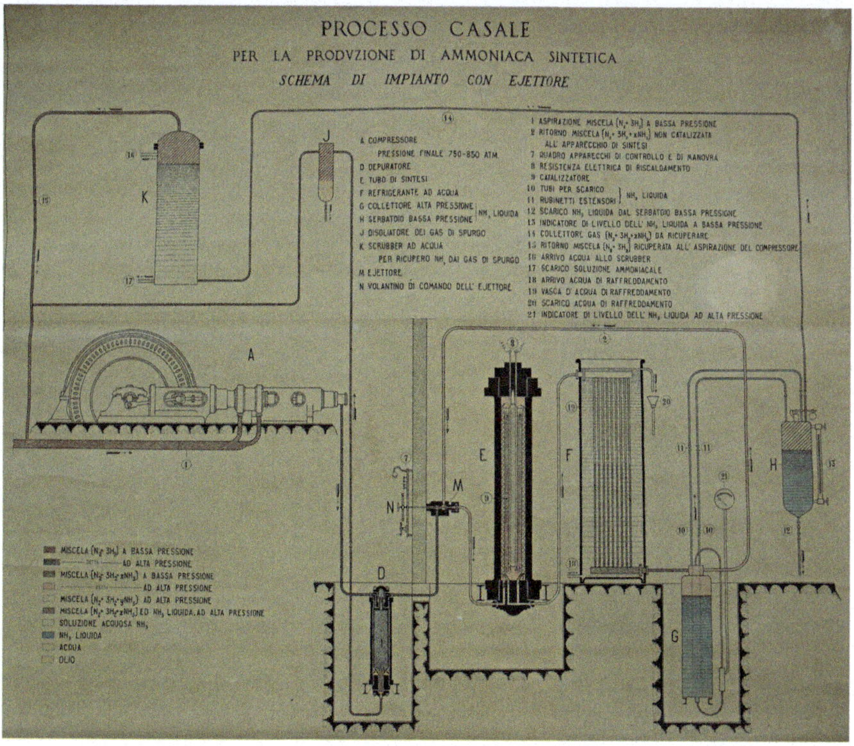

Fig. 6.7 Diagram for early Casale ammonia process with ejector. (Casale SA. With permission)

By means of the ejector, a mixture of unreacted gas with a fresh charge of gas was pulled through the loop and into the converter [4]. The ejector replaced not only the circulation pumps but also the oil separators and scrubbers. It had distinct advantages over pumps then in use which were also difficult to maintain as well as costly. It also enabled the conversion to be undertaken at a lower pressure. In addition, because the synthesis cycle was now completely static, there was a further reduction in the space taken up by the apparatus. The ejector was a standard feature of Casale ammonia units until the 1960s. Control panels for the ammonia process were provided by the Swiss firm Brown Boveri & Cie (Fig. 6.8). The foregoing is a brief outline of developments in Casale's ammonia process until the time of his death in early 1927. Little is known of the design of his experimental converters prior to 1920. But we can assume that, like Carl Bosch, whose early experiments were conducted with tubes of 30-mm-wall thickness, Casale started out with pouring his iron catalyst into a thick-walled tube. This gives credence to the story that at Terni he used the barrel of a redundant naval cannon.

Fig. 6.8 Console-type panel, manufactured by Brown-Boveri, fitted with process-control instruments of both electrical and pneumatic types. Variables such as pressure, temperature, and flow rate are monitored with pressure gauges, thermometers and flow meters, respectively. The diagram above the panel shows the flow scheme of the Casale ammonia synthesis. This arrangement required minimum operating personnel. Compressor controls were located in the compressor rooms (Casale SA. With permission)

References

1. Casale L (1923) Canadian patent, CA 230616, filed 1 May. See also Casale and Leprestre UK patents, 176144, 2 Dec 1920, and 193789, 24 Aug 1922 (addition to patent 185179 of 24 May 1921); and US patent 1408937, 7 Mar 1922 (application 7 Oct 1920)
2. Crowley JF (1926) In: Waeser B (ed) The atmospheric nitrogen industry: with special consideration of the production of ammonia and nitric acid (transl. Fyleman E). P Blakiston's Son & Co., Philadelphia, vol I, p xvii
3. Anon (1925) Dinner to Dr Luigi Casale. J Soc Chem Ind 44(8 May):477–478
4. Casale L (1923) UK patent 221956, 20 Sept

Pure Gases, Compressors, Catalysts, and Special Steels

7

Abstract

The Haber–Bosch process of BASF when introduced in 1913 was achieved by a team of chemists and engineers headed by Carl Bosch. Less than a decade later, Luigi Casale almost single-handedly introduced a slew of new features in order to design a more effective variant for synthesizing ammonia from its elements. However, the type of challenges faced by both Carl Bosch and Luigi Casale were not unalike. They included: pure and cheap to produce nitrogen and hydrogen, the availability of compressors, the formulation of catalysts, and materials of construction. Economic sources of hydrogen, the main input, were determined by the cost of electricity, in the case of electrolysis, or the availability of fossil fuels. The larger factories relied on water gas, made from coal. By the early 1930s, coking oven gas processes were also important, particularly in Europe. Investigations into the use of natural gas led to the early development of catalytic steam methane reforming.

7.1 Pure Gases

"When the management at the time entrusted me with the task of expanding the conversion of this high pressure synthesis to an industrial scale with a view to its economic application, it was obvious that there were three main problems which had necessarily to be settled before the construction of a plant could be undertaken. These were: supply of raw materials, i.e. of the gases hydrogen and nitrogen, at a lower price than hitherto possible, the manufacture of effective and stable catalysts, and lastly the construction of the apparatus. These problems were dealt with simultaneously." From Carl Bosch, Nobel Lecture, 1932 [1].

The Haber–Bosch synthetic ammonia process of BASF was, among industrial and physical chemists, as well as engineers, recognized to be a formidable and truly groundbreaking development when it was started up at Oppau in 1913. This

A. S. Travis, *The Global Synthetic Ammonia Industry*, Perspectives on the History of Chemistry, https://doi.org/10.1007/978-3-031-87978-4_7

was achieved by a team of chemists and engineers headed by Carl Bosch, who, for his triumph, shared the 1931 Nobel Prize in Chemistry. A decade earlier, in 1921, Luigi Casale had almost single-handedly introduced a slew of new features in order to design a more effective variant for synthesizing ammonia from its elements. The type of challenges faced by both Carl Bosch and Luigi Casale were not unalike. Here we briefly place greater emphasis on some of these challenges: pure and cheap to produce pure nitrogen and hydrogen, the availability of suitable compressors, the formulation of catalysts, and the materials of construction of ammonia converters.

7.2 Nitrogen

For the early nitrogen fixation processes, nitrogen could be obtained from the atmosphere by three processes that had been developed originally for isolation of oxygen. First was the Brins process, in which barium oxide was heated in air, and that emerged in the 1880s. By 1910, it was modified for the isolation of nitrogen at American Cyanamid's calcium cyanamide factory at Niagara Falls. By this time, the Linde and Claude (Société de l'Air Liquide) processes had been adapted to the separation of nitrogen from the air. Whereas the availability of oxygen involved distribution networks to remote users, the nitrogen for the fixation processes was produced and used at the site of manufacture. The Linde and Claude processes for nitrogen were used at electric arc and cyanamide factories. However they were costly to install and operate. A less expensive source, for the large-scale manufacture of synthetic ammonia, was producer gas, as used at Oppau. Early on, Linde liquefaction for nitrogen was used to adjust the composition of the nitrogen–hydrogen gas mixture. There was also Luigi Casale's burning of the oxygen from air process, in which hydrogen reacted with oxygen in a special furnace to form steam, leaving almost pure nitrogen. Casale developed such a process around the time that he was working on his ammonia synthesis. During the burning, the oxygen of the air could be enriched with oxygen obtained from electrolyzers [2].

7.3 Hydrogen

Pure hydrogen has always been the most costly input in synthetic ammonia plants. This was emphasized by both Carl Bosch, of BASF, and Charles O. Brown, of the Atmospheric Nitrogen Corporation, a part of Allied Chemical-Solvay. Bosch in his 1932 Nobel Prize lecture, drew attention to the fact that "…the procurement of hydrogen is the largest item in the prime cost, at least nowadays since because of the efficiency of the high pressure synthesis the conversion of the ready-made gas mixture into ammonia is only a minor cost factor" [1]. Brown, in a similar vein, and at around the same time, observed: "[T]he differences [among the synthetic ammonia processes] are, in general, those required because of the different sources for obtaining hydrogen… The development and purification of the hydrogen is still the outstanding feature of cost and extensiveness of plant equipment. The synthesis equipment is obviously a smaller and less expensive part, this being one of the advantages of the use of high pressure synthesis" [3].

Prior to the development of the synthetic ammonia industry, the reactive, lighter than air gas was required for airships, and from the early 1900s in hydrogenation of fats and for manufacture of anhydrous hydrogen chloride. However, the scale of production was not great. That changed with the introduction of synthetic ammonia.

7.3.1 Electrolysis

The early laboratory and pilot-plant research into synthetic ammonia processes drew on electrolytic hydrogen generated in cells (electrolyzers). This was developed for large-scale use by Luigi Casale, Montecatini, and others (Fig. 7.1). The controlled combustion of the hydrogen with air gave the mixture of nitrogen and hydrogen required for the Casale ammonia synthesis.

Production of pure hydrogen on a large scale by electrolysis suited Japan and Italy, and regions such as Scandinavia, where cheap hydroelectric power was available.

Fig. 7.1 Electrolyzers for production of hydrogen, for the Casale process, at Sabiñánigo, Spain. (Casale SA. With permission)

Occasionally, hydrogen was produced by fermentation, as was the case in the late 1920s at Commercial Solvents, of Terre Haute, Indiana, which adapted its ammonia units to the synthesis of methanol.

7.3.2 Water Gas

Outside of regions where cheap hydropower was available, coal-based processes were important. This followed work at BASF on water gas for production of hydrogen and was associated with the larger ammonia facilities. The water gas process involved the action of steam on coke or lignite, to give hydrogen and carbon monoxide, known as synthesis gas (syn gas), followed by the water–gas shift reaction which converts carbon monoxide, a catalyst poison, into carbon dioxide, and gave more hydrogen (Chap. 2) [4]. Manufacture of water gas by coal-based processes also included from low-volatile coal, using oxygen by the Fritz Winkler (BASF) or Lurgi processes; and the reaction of coking oven gas with steam, with or without a catalyst.

In the US, catalyst and other studies related to the production of hydrogen from water gas were conducted in the 1920s at the Fixed Nitrogen Research Laboratory. Apart from at BASF the water gas process was used at Toulouse, the first Soviet ammonia factory, in the US (by both Allied Chemical-Solvay and Du Pont), and in the UK at ICI.

An estimate of hydrogen production for synthetic ammonia during 1926–1927, indicated that 89% of hydrogen then used in the manufacture of ammonia came from water gas (in part reflecting the great output of German synthetic ammonia), 3% from coking oven gas, 6.4% from electrolysis, and 1.6% from other processes [5]. In 1932, 57.5% was from water gas, 21% from coking oven gas, 15.8% from electrolysis, and 9.7% from natural gas [6].

The water gas process required very complex, costly plant, which is why it was used in the larger facilities (Fig. 7.2). BASF at Merseburg (Leuna) drew on locally available brown coal, or lignite, which also provided all the electricity for the factory: "[E]nergy used daily at the factory is generated out of 12,000 t of brown coal….The electric current required at the factory is generated at a power plant of 85,000 KW" [7]. In 1926, Merseburg started up Winkler's continuous fluidized-bed generator, in which a six-foot layer of finely divided lignite (generally the waste material) was heated and kept in motion by a blast of air. The heating was stopped and steam was introduced to give the water gas.

The German engineering firm Bamag, which worked closely with BASF, developed an apparatus for producing hydrogen using iron and a high temperature to decompose steam. A similar iron-steam reaction was studied by Pechiney for its use with Casale synthetic ammonia units.

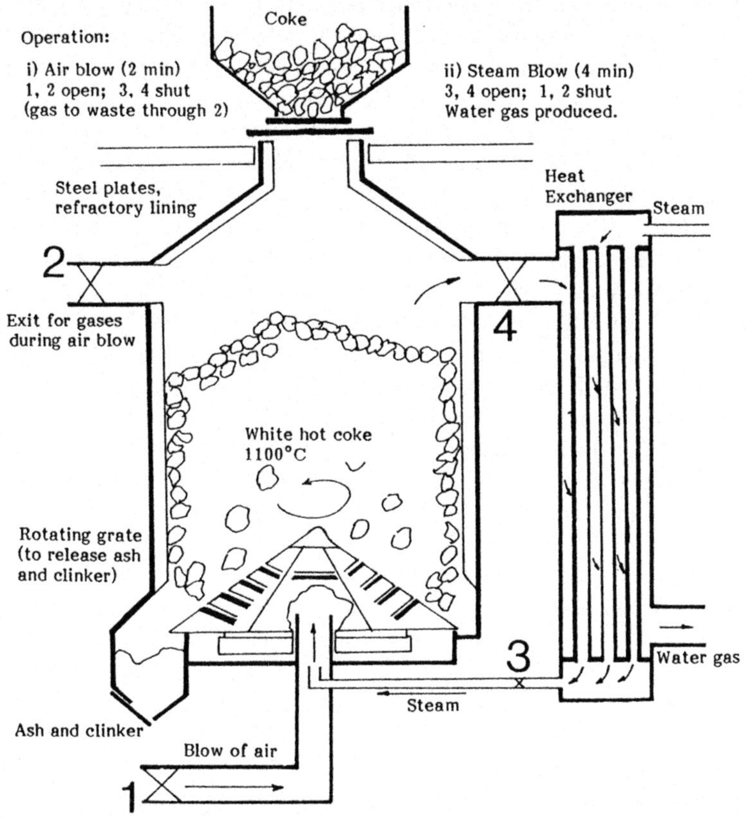

Fig. 7.2 Manufacture of water gas (carbon monoxide and hydrogen), from coke, steam, and air by the Bamag (Berlin-Anhaltische Maschinenbau AG) process, as employed at BASF

7.3.3 Hydrogen from Coking Oven Gas: Casale in Belgium

During the mid-1920s, coking oven gas became a source of low cost hydrogen following research into gas liquefaction at the Claude (Air Liquide) and Linde concerns. Purification included removal of: hydrogen sulfide over iron oxide; carbon dioxide by scrubbing with water and alkali under pressure; and carbon monoxide and nitrogen by liquefaction. The constitution of coking oven gas varied according to the source, but in general it contained around 50% hydrogen, in addition to methane, carbon monoxide, and nitrogen. The gas mixture could be converted to hydrogen and carbon monoxide by reaction with steam in the presence of a catalyst. It was this rich source of hydrogen that drove expansion in synthetic ammonia from the late 1920s until the 1940s. Coking oven gas for hydrogen was well-suited to locations where coal was in abundant supply. This enabled Japanese corporations engaged in coal mining to enter the ammonia industry by producing hydrogen that was less costly than from electrolytic processes. They become strong rivals of Noguchi's enterprise that continued to rely on electrolytic hydrogen.

Luigi Casale's synthetic ammonia process was integrated into one of the first manufacturing units that demonstrated the viability of coking oven hydrogen. This took place in Belgium, at a coking oven plant located at Zanvoorde, in the city of Ostend, which supplied the coastal resort with town gas. During 1923–1924, Casale ammonia units were installed at Zanvoorde by SA des Fours à Coke Semet-Solvay & Piette, a part of the Solvay group. It was the first facility to combine ammonia synthesis at an ultra-high pressure with the Linde system of coking oven gas fractionation for hydrogen, and the Semet-Solvay & Piette gas treatment process [5]. Linde units produced both nitrogen from the atmosphere and hydrogen from coking oven gas. The Semet-Solvay & Piette coking-oven gas treatment process as applied to bulk production of pure hydrogen was based on an experimental setup built in collaboration with Linde at Oberhausen in the Ruhr Valley, and designed by Jegor Israel Bronn, an engineer at Rombacher Hüttenweke AG. Following by-product recovery, the gas produced at Ostend was extracted by means of exhausters, and subjected to washes with ammonia, water, sulfuric acid, and caustic soda.

The associated coal carbonization plants were arranged for both generation of power and utilization of gaseous by-products. In 1928, the year in which Fours à Coke Semet-Solvay & Piette was merged into Union Chimique Belge, the rated annual capacity of the Zanvoorde factory was 10,000 tons. Production of pure hydrogen for the Casale ammonia process was described by F. A. F. Pallemaerts, research manager, division of coking ovens and ammonia, at Union Chimique Belge, before the second International Conference on Bituminous Coal held at Pittsburgh in November 1928. The Linde system of gas liquefaction for hydrogen, based on the Joule–Thomson cooling effect, was chosen in preference to the Claude system, in which an expansion engine liquefied gases. The reason probably had to do with Claude's own involvement in high-pressure synthesis of ammonia. In the Casale-based process, carbon monoxide in the hydrogen was reduced to methane, which could be removed from the circuit.

The original experimental ammonia facility at Ostend consisted of three Casale ammonia units, each one with a converter of 8-tpd capacity, and a Linde gas liquefier. One unit was kept in reserve. For the first few weeks of operation production was low and irregular due to impurities in the coking oven gas. Following careful purification of the hydrogen, the output was 50% above that for which the plant was designed. By around 1928, four more Casale ammonia units had been installed, bringing daily production to 40–55 tons. Most carbon monoxide was removed by liquefaction. Final traces of carbon monoxide were removed when liquid nitrogen, obtained by Linde's process, was added at the nitrogen wash stage in the purification of hydrogen. Some nitrogen was already present in the coking oven gas. Additional nitrogen was added to bring about a mixture of hydrogen and nitrogen in the proportions required for the ammonia synthesis. According to Pallemaerts, the ammonia was produced at a very low cost, and in 1928, there were already "about a dozen plants" working the Linde-Casale processes, with a total daily capacity of one thousand tons of ammonia [8]. However, the overall capacity was restricted because of the limited output of existing coke-oven plants. A considerable expansion in the production of hydrogen was anticipated from construction of central carbonization plants in the southern Belgian coalfield (Borinage).

By 1928, the circulation pumps installed in the ammonia plant at Ostend were undergoing replacement by Casale ejectors. This enabled further savings in energy. The six operating compressors each handled 1200 m^3 of the nitrogen–hydrogen mixture per hour. Despite the high working pressure of 750–800 atmospheres, it was found that Casale energy requirements for compression of the gas mixture were not much greater than those required at 250–300 atmospheres (1.2 kWh per kg of ammonia, when compared with 1.05 kWh, respectively). Since the Ostend coking oven plant also supplied gas to the town—a major seaside resort, with fluctuating demand—water gas plant was also installed, though the cost of hydrogen was twice that of coking oven gas. According to Pallemaerts: "Thanks to the cheapness of the hydrogen [recovered from the coking-oven gas] and also to the general economy of the Casale synthesis process, ammonia can be produced at the cost of about 9 American cents per kilogram of nitrogen in the form of ammonium sulfate (including the cost of the sulfuric acid and total operation costs, amortization, etc.)" [8, on p. 22]. During 1931–1932, Ammonia Casale supplied the Ostend facility with a newly developed catalytic purification system for hydrogen. This enabled a significant increase in the production of ammonia.

The move to coking oven gas as a source of hydrogen for ammonia synthesis was rapid. In 1928, it was reported that Linde had provided 7 units with rated capacities of 1600 m^3 of coking oven gas per hour, and 19 units of 5000 m^3 per hour, all for ammonia synthesis. Hydrogen was purified using processes developed by both Linde and Claude. This enabled coking oven processes to be taken up on a large scale from the late 1920s, with enormous implications for the global spread of the synthetic ammonia industry. It encouraged the setting up of ammonia factories in all countries with extensive reserves of coal, and coke producing plants, especially France, Belgium, Holland, and the Soviet Union, but not in the US. Pechiney's promotion of the Casale process focused on the use of coking oven gas.

Claude, whose research into producing hydrogen from coking oven gas took place at Béthune, in the Pas-de-Calais department, participated in formation of Société Belge de l'Azote in 1923. From around 1924, Claude's coking oven process was taken up in Spain by Sociedad Ibérica del Nitrógeno, at Flix, Catalonia, and, by 1930, in Asturias, at a site close to the large coal mining and engineering firm Sociedad Metalúrgica Duro Felguera.

In order to overcome the disadvantages of the batch operation of coking-based plants, continuous gasification processes using oxygen or oxygen-enriched air and steam in the same reactor, similar to Winkler's process, were developed. This was important because both coal and hydrocarbons could be used as feedstocks. The liquid nitrogen by-product was used as the washing agent in the final purification step of the synthesis gas, during which the correct amount of nitrogen for the ammonia synthesis was allowed to evaporate into the gas.

In 1934 coking ovens were responsible for 25% of the hydrogen used in the synthesis of ammonia. By then, Ammonia Casale had received orders for five more of its hydrogen purification units. In Europe, coking ovens were the main source of hydrogen until after World War II.

7.3.4 Synthesis Gas and Hydrogen. Steam Reforming

The most notable change in ammonia-related technology was in the production of synthesis gas from hydrocarbons such as methane. The process had been developed during the 1930s, and took off from around 1940.

Though methane is a component in the residual gas from the coking oven process, converting it to hydrogen (and carbon oxides) was originally considered too cumbersome and costly. The first industrial investigation of the reaction was undertaken by Alwin Mittsach at BASF in 1912, using a method described a decade earlier by Paul Sabatier (1854–1941) and Jean-Baptiste Senderens (1856–1937). Mittasch prepared the methane from the reaction between carbon monoxide and hydrogen. Synthesis gas reactions are particularly associated with the work of Franz Fischer and Hans Tropsch who in the mid-1920s catalytically produced hydrocarbons from the gas mixture. Their gas source was coal or coke, and the method was placed on an industrial footing in Germany in 1936 [9, 10]. Around 1950, continuous coal gasification methods, based on the work of the US Bureau of Mines and Standard Oil-Pittsburgh Consolidation, were considered viable for production of synthesis gas suited to production of both methanol and ammonia. There was also an interest in underground gasification, as carried out in the Soviet Union [11].

The ammonia synthesis, however, and especially at the present time, draws on hydrogen from hydrocarbons, in particular the methane in natural gas. In the 1920s, there was already a growing interest in using natural gas and other hydrocarbons as a source of hydrogen. Pallemaerts in his 1928 lecture referred to a recent development, steam reforming of methane, the "methane-steam" reaction that gave a mixture of hydrogen and carbon monoxide (synthesis gas). This is steam reforming.

It was this novel way of producing hydrogen that provided the first opportunities for petroleum companies to enter into the manufacture of synthetic ammonia. However, the first use of natural gas in the synthesis of ammonia was based not on steam reforming but on cracking. In the usual cracking processes larger hydrocarbon molecules are broken down into light hydrocarbons, as suited to production of fuels. Cracking was originally carried out by heating petroleum oil at a temperature above that at which it normally distils. This is thermal cracking. The American firm Universal Oil Products was a pioneer, during 1919–1920, in thermal cracking, based on the improved Dubbs process, invented by Jesse Dubbs, and perfected by his son Carbon Petroleum Dubbs. Cracking of coking oven gas was carried out in Germany to give a hydrogen-rich gas suitable for enriching water gas.

Thermal cracking of natural gas followed the acquisition in April 1928 by Royal Dutch-Shell's subsidiary Bataafsche Petroleum Maatschappij of world rights, apart from Germany, to the Mont Cenis process for producing hydrogen from natural or cracked gases. Another father and son partnership, chemical engineer Daan Pijzel (1877–1972), at Royal-Dutch Shell, and Dan Pijzel, produced hydrogen by thermal cracking of natural gas. In 1931, Royal Dutch-Shell's Shell Point facility (Shell Chemical), at Pittsburg, California, began producing ammonia using hydrogen from their cracking process. Linde technology was used for gas and air separation. In 1937, Vladimir Shukov's thermal cracking process for producing hydrogen from natural gas was used at a new alkali and nitrogen fertilizer plant near Baku in the Soviet Union.

Reforming of gaseous hydrocarbons for production of hydrogen, as distinct from cracking, and quite different from catalytic reforming used to produce high-octane gasoline, was taken up by Standard Oil Development Company, following exchange of information with BASF. In 1930 it was adopted at the Bayway refinery, New Jersey, for oil hydrogenation. At this time, both natural gas and petroleum gases were being investigated as sources of hydrogen by methods that borrowed from petroleum technologies. The main emphasis was on hydrogen for BASF's Bergius process that involved coal hydrogenation in the manufacture of synthetic fuels. A steam reforming process, similar to the process mentioned by Pallemaerts, was developed in the US in the mid-1930s to produce synthesis gas from light hydrocarbons, instead of from coal.

Interest in novel methods of producing hydrogen for hydrogenation of coal and oil for use in the synthetic fuel processes in the US declined in the 1930s, following the discovery of large deposits of gasoline. However, steam reforming for manufacture of hydrogen, from off gases of coal hydrogenation, was installed at ICI's Billingham ammonia factory in 1936. What was special about ICIs work in relation to synthetic ammonia was that the reforming was carried out in two steps, the second of which introduced nitrogen from the air in the amount required for the ammonia synthesis. This was the first stage in the development of modern processes for producing hydrogen and synthesis gas.

The reforming process consists of heating the gas to between 700–1100 °C in the presence of steam and a nickel catalyst. Hydrogen and carbon monoxide are formed.

$$CH_4 + H_2O \rightleftharpoons CO + 3H_2 \quad \Delta H + 206\,\text{kJ/mol}$$

For ammonia, the mixture is then subjected to a further treatment, in the presence of air, which introduces the nitrogen. The ICI process for reforming of natural gas was first taken up during World War II in North America and Australia (Chap. 9).

7.4 Compressors

Among the most important items of machinery investigated by Carl Bosch and colleagues while developing the synthetic ammonia process were compressors. Those available around 1910 were of the reciprocating type, used to provide compressed air for mine locomotives and Linde air liquefaction plant. The problem was that these machines were of limited output, often stopped, and their stuffing boxes could not prevent some leakage. In the ammonia synthesis, loss of gas had to be eliminated, because of its cost and the possibility of escaping hydrogen causing explosions. It was also essential to prevent lubricating oils from entering the compressor and contaminating the gas mixture.

Uninterrupted operation on a much larger scale than hitherto required at very high pressures was essential if the ammonia process was to prove economical. The outcome was the design of a new compressing system capable of continuously providing the mixed nitrogen–hydrogen gases, fed from a gasometer, to the converter. This was achieved stepwise, with three separate reciprocating compressors: first a low-pressure compressor, to 10 atm; then an intermediate two-stage compressor, to 50 atm; and finally a high-pressure compressor to 250 atm. Though very efficient, the limited capacity of centrifugal compressors meant that batteries of compressors were needed to drive the individual ammonia loops in larger facilities. At Merseburg (Leuna), forty-five gas engines, of 3000–4000 horsepower, were required to drive the compressors used in the synthesis of ammonia [7].

In 1922, Ammonia Casale had access to new a compressor, the first of its kind, and a major innovation in its own right. For Casale, the firm of Fonderia del Pignone, which had manufactured compressors since around 1900, produced a high-pressure machine capable of bringing the pressure stepwise to 1000 atm. For the Casale process, six stages of compression raised the pressure, to 3, 9, 27, 90, 270, and finally 850 atmospheres (Fig. 7.3). Later, compressors were manufactured by other engineering firms, including the French SACM (Société Alsacienne de Constructions Mécaniques), for Société Houillère de Sarre & Moselle, at Carling.

Compressor improvements emerged with the need for larger equipment required in handling and processing gases for the ammonia synthesis [12].

7.5 Catalysts

In November 1909, Alwin Mittasch and co-workers at BASF succeeded to formulate an iron catalyst containing promoters, or activators, that was well-suited to the Haber–Bosch process. They continued to undertake screening experiments until around 1920. Bosch and Mittasch, sometimes with co-workers, filed several patents for catalysts from 1910. In all, there were 20,000 trials, but there were no major improvements to the original catalyst, based on magnetite promoted with small amounts of unreducible oxides, typically of aluminum, potassium, and chromium. Research into catalysts was very much by trial and error, based on well-reasoned hunches. British, American, and French investigators had the benefit of samples of

Fig. 7.3 Compressors for mixed gases for the Casale process at the ammonia factory of the Société Carbochimique in Tertre (Belgium) (Casale SA. With permission)

the valuable BASF catalysts, for both ammonia and nitric acid processes, which had been taken on their inspections at Oppau—probably surreptitiously, or aided by bribery—and of recipes, obtained by industrial espionage.

Luigi Casale's iron catalyst did not need to be as active as those used under Haber–Bosch conditions. Manufacture of the catalyst involved burning steel turnings in oxygen in the presence of the metal oxides that served as promoters. According to one Casale patent: "Iron or iron oxide catalysts are prepared from ordinary iron or such alloys as molybdenum steel by placing the metal in the form of turnings in a heated magnesium crucible and then suddenly bringing it into violent ebullition by a jet of oxygen under very high pressure." On cooling, the solid mass was broken up and crushed. "This process can be applied to the regeneration of catalysts, the efficiency of which has become impaired by poisons or otherwise" [13]. The method was subsequently modified "by adding and dissolving in the mass, during the oxidation process, over 10% of an oxide or mixture of oxides, not reducible by hydrogen" [14]. Another Casale patent described heating of metal oxides or iron, or metals belonging to the "iron group," with compounds "possessing a high heat of combination with oxygen, such as carbon, magnesium, aluminium, and calcium" [15]. Magnetite and iron served as the basis of all synthetic ammonia catalysts until almost the end of the twentieth century. In the 1920s, the life of a catalyst was measured in months.

7.6 Steels for Converters

The design of Luigi Casale's converter, or pressure tube, was such that ordinary steel could be used in its construction. Cooling of the inner wall of the shell by the incoming gas mixture was an important design feature. The converter was sometimes described as being not unlike a big naval gun, which was probably the original inspiration. The 2-mtpd converter at Terni installed in September 1922 was after one year in operation capable of withstanding a pressure of 1500 atm on testing. According to Miolati, chemical and metallographic analysis "of the inner wall of a catalyst tube, carried out exactly in the hottest point and after one year of operation, showed that the metal's physical and chemical characteristics were unchanged" [16]. After 1922, it was transferred to the factory of Pechiney at Saint-Auban, in the Alpes-Maritimes department, southeastern France. In 1927, it was still capable of producing two metric tons of ammonia every day. Success in its use, probably as a semi-commercial plant, no doubt contributed towards Pechiney's licensing and other arrangements with Ammonia Casale.

Developments elsewhere focused on novel steels. The steel industry undertook extensive metallurgical studies and developed a range of novel alloys well-suited to high-pressure processes. The problems encountered with the Claude converters, which until the mid-1920s often exploded, were overcome with the use of nickel–chromium-iron, with some tungsten, alloys (BTG steels) for the converter body and certain other parts, as introduced by Vickers in England. However, the ammonia industry was not without its risks (Fig. 7.4). Austenitic stainless steels, alloys containing chromium, nickel, and molybdenum, overcame the technical limitations of ordinary steels used in chemical processing equipment. General Chemical in the US from around 1916 had demonstrated the advantages of chrome-vanadium steels over a range of other steels. In the 1920s, research at the Midvale Steel Works, in Philadelphia, and elsewhere, included into the increase of creep strength under conditions of high pressures and high temperatures, notably by addition of 0.35% of molybdenum to ordinary steel. The availability of large forgings enabled improvements in the production of synthetic ammonia and in reactions involving hydrogen to be carried out safely at high pressures. In 1941, as an alternative to forged steel converters, IG Farben used the 1938 method of its engineer Julius Schierenbeck, based on strip-wound coiling (filament winding), for manufacture of large converters. Such synthesis tubes, of 1200-mm diameter and 18 ft in length, produced 250 tons, or more, of ammonia every day [17–19].

Fig. 7.4 On 7 November 1930 this Casale converter at the factory of Société de Produits Chimiques-Anzin-Kuhlman, at Anzin (France) exploded. The cause of the failure was probably decarburization of the steel of the outer shell. As Luca Bianchi comments: "It's a reminder that making ammonia in those pioneering times was sometimes a risky business!" Early Claude converters were especially notorious for developing fractures, at least until a suitable steel for their construction was invented in the UK during the mid-1920s. (Casale SA. With permission)

References

1. Carl Bosch—Nobel Lecture. NobelPrize.org. Nobel Media AB 2020. https://www.nobelprize.org/prizes/chemistry/1931/bosch/lecture/
2. Casale L (1924) UK patent 233040, 31 Jan
3. Brown CO (1933) High pressure synthesis: basis of new chemical engineering industries. In: Kirkpatrick SD (ed) Twenty-five years of chemical engineering progress. Silver anniversary volume. American Institute of Chemical Engineers/D. van Nostrand Company, Inc., New York, pp 152–168, on 167
4. Evans RM, Newton WL (1926) Hydrogen from water gas: catalysts for its production. Ind Eng Chem 18(May):513–517
5. Tongue H (1934) The design and construction of high pressure chemical plant. Chapman & Hall, London, p 109
6. Toniolo C (1934) Le industria dell'azoto sintetico nella crisi mondiale. Associazone Italiana di Chimica, Rome, p 165
7. "Visit to the Ammoniakwerk Merseburg G.m.b.H. in connection with the World Power Conference at Berlin. Some technical data given by Direktor Dr Schneider, when addressing the visitors. Leunawerk, June 25th 1930." Papers of Herbert Alfred Humphrey, College Archives and Corporate Records Unit, Imperial College London, item M7
8. Pallemaerts FAF (1929) Synthetic ammonia plant at Ostend. Ind Eng Chem 21(January):22–29
9. Stranges AN (2024) Petroleum from coal: a century of synthesis. Leiden, Brill
10. Weil BH, Lane JC (1948) Synthetic petroleum from the synthine process. Remsen Press Division, Chemical Publishing Co., New York

11. Martin PT (1950) Synthesis gas from coal. Chemical Industries, 66(3):365–372
12. Anon (1929). Recent designs in compressing plant. High pressures in the manufacture of synthetic ammonia. J Soc Chem Ind 48(14 June):591–598
13. Casale L (1922) UK patent 197199, 16 May
14. Casale L (1923) UK patent 218237, 11 Dec
15. Casale L(1923) UK patent 227491, 20 Sept
16. Miolati A (2009) Synthetic ammonia and the Casale process. Amplified edition of a lecture delivered the 27th February 1927 at the Institute of Chemistry of the Polytechnic School of Prague. Lugano, Ammonia Casale SA, p 25. Translation by Impallomeni G, of Miolati A (1927) L'ammoniaca sintetica ed il processo Casale. Ammonia Casale SA/L'Universale, Tipografia Poliglotta, Rome
17. Witsch W (1968) Large pressure vessel manufacture. Chemical and Process Engineering, pp 63–66, on 65
18. Timm K (1960) The synthesis of ammonia. The contribution to the development of modern chemical technology and its economic importance for the nitrogen supply of the world. Chem Ind, no. 11, Mar, pp 274–281
19. Eggers R (ed) (2012) Industrial high pressure applications: processes, equipment and safety. Wiley-VCH, Weinheim

American Synthetic Ammonia, 1920–1940

8

Abstract

Luigi Casale's synthetic ammonia process played a significant role in the development of the US synthetic ammonia industry. The manner in which this came about initially involved Ammonia Casale's creation of a company for the purpose of manufacturing ammonia on American soil. However, the company, Niagara Ammonia, founded in 1923, faced trading difficulties. Among the several reasons was the fact that US nitrogen demand was met through the availability of ammonium sulfate; calcium cyanamide; and imported Chilean nitrate. The Casale venture not only faced marketing problems but also a threat to Luigi Casale's lead high-pressure patent in the US. The subsequent acquisition of Casale technology by Du Pont enabled the corporation to overcome major problems with its Claude synthetic ammonia process. This was an important contribution to the growth of the US synthetic ammonia industry.

8.1 Introduction

Luigi Casale's ammonia process played a significant role in the development of the synthetic ammonia industry in the US. The manner in which this came about differed considerably from other countries. It initially involved Ammonia Casale's creation of a company in the US for the purpose of manufacturing ammonia on American soil. However, the company, Niagara Ammonia, founded in New York City in 1923, faced trading difficulties. Among the several reasons was the fact that the US synthetic ammonia industry developed far slower than in Europe and

Part of this chapter is reused from the author's article, First steps: synthetic ammonia in the United States, in *Substantia* 5(2) (September 2021):55–77. Open Access under a CC BY 4.0 license, https://creativecommons.org/licenses/by/4.0/.

A. S. Travis, *The Global Synthetic Ammonia Industry*, Perspectives on the History of Chemistry, https://doi.org/10.1007/978-3-031-87978-4_8

Japan. Moreover, there were also technical problems, and a threat to Luigi Casale's lead high-pressure patent in the US. To understand the mix of forces at play, we need to consider the US nitrogen industry in the early 1920s.

The most important feature of the industry was the near total dependence on supplies from three well-entrenched and competitive industries whose products enabled the US to become practically self-sufficient in nitrogen compounds for use as fertilizers. These products were ammonium sulfate, produced from the modern by-product coking ovens that were rapidly replacing the beehive ovens from which valuable by products were lost; calcium cyanamide, as manufactured by the American Cyanamid Company at Niagara Falls; and imported Chilean nitrate, widely used as fertilizer and for producing nitric acid. In 1922, Chilean nitrate represented 39% of nitrogen consumed in the US. And well into the 1920s, there was the possibility that an additional source of nitrogen might come from the government's Muscle Shoals calcium cyanamide factory in Alabama that had been constructed during the recent war and mothballed following the end of hostilities.

Nitric acid is relevant here. It was an important product, used extensively in the manufacture of synthetic dyestuffs and explosives. Just prior to World War I it had been made in the US by catalytic oxidation of ammonia derived from calcium cyanamide and coking ovens, by, respectively American Cyanamid, at Warners, New Jersey, and the Solvay Process Company, at Syracuse, New York State. However, the earlier method of acid making starting with Chilean nitrate was generally favored for both economic and technical reasons. This is why during 1924–1925 the import of nitrate into the US grew considerably, including through the involvement of the increasingly diversified Du Pont corporation [1]. A few years earlier, in 1920, Du Pont agreed to supply Mississippi farmers, who were facing a severe shortage of fertilizer, with 50,000 tons of Chilean nitrate. By that time, the US was emerging as a major exporter of ammonium sulfate. In general, then, America's nitrogen requirements were satisfied without the need for a synthetic ammonia industry.

In many ways this explains why it was not until 1928 that there emerged two major US producers of synthetic ammonia for fertilizer, Allied Chemical & Dye, and Du Pont. The Allied Chemical process was similar in operating conditions to the Haber–Bosch process. The Du Pont process was at first based on the Claude process but was beset with technical problems. It was the acquisition in 1927 of Niagara Ammonia and rights to the Casale process that enabled Du Pont to achieve success in large-scale, high-pressure industrial chemistry. This was almost a decade after Germany's much vaunted synthetic ammonia industry had stimulated interest in both US government and industrial laboratories.

8.2 The Fixed Nitrogen Research Laboratory and Its 3-Ton Units

Following American failure to produce synthetic ammonia for munitions production during World War I, research into high-pressure processes and general nitrogen fixation was taken up at the state-sponsored Fixed Nitrogen Research Laboratory (FNRL) in Washington DC, opened in 1919 under the aegis of the U.S. Army's Nitrate Division. In the same year, American military officers visited the BASF ammonia facility at Oppau, then in the French zone of occupation. Invariably unwelcome, they did manage to make useful observations, and procured samples of catalysts, which ended up at the FNRL [2]. Within a few years its scientists resolved most of the chemical problems related to the ammonia synthesis. Thus, Alfred T. Larson developed a catalyst similar to that used in the Haber–Bosch process, based on analysis of samples obtained from Oppau. This contributed to the success of the FNRL in producing synthetic ammonia, at around 200 atm pressure, drawing on pure hydrogen obtained by electrolysis of water. The rated capacity of the FNRL unit was just 3-tons per day (US tpd). It was adopted at a handful of factories. As elsewhere, the most expensive input was hydrogen, in this case mainly the by-product gas from electrolytic processes. In 1921 the Nitrate Division was suspended, and administration was transferred to the Department of Agriculture. In 1926, the FNRL was absorbed into the laboratories of the Bureau of Soils.

8.3 The Atmospheric Nitrogen Process of the General Chemical Company

In the private sector, the introduction of catalytic high-pressure synthetic ammonia processes followed the attempt by the General Chemical Company to master a Haber–Bosch type process just prior to the outbreak of war in Europe. At General Chemical, Frederick W. de Jahn and colleagues developed a process similar to that of BASF, and discovered an iron catalyst incorporating sodamide. The company put together a small pilot plant during 1915–1916. Subsequent government-supported wartime experiments and trials were not successful. However, by 1919 these studies apparently enabled General Chemical chemists to overcome several technical problems, including a suitable steel for the converter shell, and to devise an improved catalyst. No doubt the visits to Oppau and the assistance of the FNRL played a role in these developments. Towards the end of 1919, General Chemical and Solvay Process combined their interests in synthetic ammonia and oxidation of ammonia to afford nitric acid, respectively, through the Atmospheric Nitrogen Company. Further development of the ammonia process benefited greatly from the free exchange of technical information between Solvay Process and the UK firm Brunner, Mond & Company, in accord with arrangements dictated by the Belgian licensor of the Solvay ammonia-soda process, Solvay et

Cie. This included the results of analyses of catalysts acquired by Brunner, Mond investigators during their visits to Oppau in 1919 [3].

Late in 1920, General Chemical, Solvay Process, and five other firms, including the National Aniline & Chemical Company (a merger of four firms dating from 1917), were absorbed into a new conglomerate, Allied Chemical & Dye Corporation. In August 1921, Atmospheric Nitrogen (then a part of Allied Chemical), launched the General Chemical synthetic ammonia process at Syracuse. The engineer in charge of the ammonia program was Walter H. Knishern. The initial rated output was 9-tpd. The source of hydrogen was water gas, obtained by the reaction between steam and red hot coke in Semet-Solvay coking ovens. Nitrogen was obtained by air liquefaction [4]. The liquefied ammonia was used in refrigeration.

Published figures of the rated annual output of Syracuse ammonia suggest growth from 3000 tons in 1921 to 7750 tons in the mid-1920s [5, 6]. Further expansion followed the decision of the head of Allied Chemical, Orlando Weber, to invest heavily in synthetic ammonia and inorganic products at the expense of dyes and other organic chemicals [7]. The ammonia converter was improved, with the synthesis pressure increased from the original 100 to 350 atmospheres. The catalyst was cooled by heat exchange using an arrangement of tubes within the converter. By 1925, Allied had started work on a large synthetic ammonia facility, using the modified General Chemical process, at the new Hopewell, Virginia, factory of Solvay Process which opened in late 1928. Water gas made from coke provided the hydrogen. The first product made at Hopewell was anhydrous ammonia, soon followed by the fertilizers ammonium sulfate and sodium nitrate. By the end of the decade, economies of scale enabled synthetic ammonia to compete in nitrogen products for the agricultural fertilizer market with Chilean nitrate, coking oven and gas works by-products, and calcium cyanamide.

8.4 Ammonia Casale in America

After the filing of certain key patents on his inventions, Luigi Casale, encouraged by Leprestre, set about establishing an American branch of Ammonia Casale that would not only engage in licensing arrangements, but would also undertake the manufacture of ammonia in the US using converters, compressors and other machinery imported from Italy.[1] As we have seen, in late July 1923, Leprestre, then in New York City, established two American entities: The Ammonia Corporation and Chemical Investment Corporation. They were associated with,

[1] Luigi Casale US patent 1,348,428 of 12 July 1921 (burning of air; process for generating from air and hydrogen, either nitrogen and hydrogen mixtures or nitrogen; filed 7 July 1920); and 1,408,987 of 7 March 1922 (catalytic apparatus for the synthesis of ammonia; filed 7 October 1920). See also patents 1,447,123 (self-making of ammonia), 27 February 1923, and 1,478,549 (catalytic apparatus for the synthesis of ammonia; filed 31 March 1921) and 1,478,550, both of 25 December 1923. In 1924, Casale received a US patent for preparing a catalyst, and on 28 July 1925 patent 1,547,362, for electrolysis of water (filed 19 July 1922).

Table 8.1 Ammonia Casale SA (Casale Ammonia Company) in the United States

Holding corporation	Affiliate	City/state of foundation	Date	Fate
The Ammonia Corporation		New York City	1923	To Du Pont 1927. Bankrupt 1928
	Niagara Ammonia Company, Inc.	New York City	1923	To Du Pont 1927. Bankrupt 1928
Chemical Investment Corporation		New York City	1923	To Du Pont 1927. Bankrupt 1928
	Hydro-Electric Chemical Company	Maine	July 1923	To Du Pont 1927

respectively, the Niagara Ammonia Company, Inc., and the Hydro-Electric Chemical Company (Table 8.1). Share certificates reveal cross holdings, such as a holding of Hydro-Electric, in the form of preferred capital stock, in The Ammonia Corporation. The Ammonia Corporation retained all rights to the Casale patents in the US [8]. Around four-fifths of both preferred and common stock in Niagara Ammonia were held by The Ammonia Corporation, while around one-fifth of the stock was held by Electric Bond and Share, in the name of nominee Frank L. Smiley, a dealer in utilities stocks. The Ammonia Corporation assigned patent rights and provided funding to Niagara Ammonia; they were to share any licensing compensation. Hydro-Electric was also founded with the close involvement of Electric Bond and Share. A director, and the holder of a single share, was René Leprestre.

8.4.1 The Business Plan

The business plan was that The Ammonia Corporation would benefit from the growing demand for nitrogen products, including in the long term the fertilizer market, ensuring, it was anticipated, vast royalties following the signing of licensing agreements. This differed from arrangements elsewhere whereby licenses for designated countries or geographical regions were granted to manufacturers, without, it appears, direct investments from the Swiss Ammonia Casale firm. Electric Bond and Share and Ammonia Casale embarked on an ambitious venture aimed at dominating certain novel and emerging sectors of the US chemical industry, based on electrochemicals and high-pressure technologies.

8.4.2 Colonel Frederick Pope

The president of both The Ammonia Corporation and Niagara Ammonia was the Harvard educated chemical engineering consultant and entrepreneur Colonel Frederick Pope (1877–1961) [9]. In 1919, he was one of the technical experts on the Allied side who visited German chemical factories, in his case twice, to follow up developments in organic chemical and dye manufacture and the production of war gases. Pope's brief published account does not indicate the sites that he visited, though they probably included the synthetic ammonia facility of BASF at Oppau [10]. His Chemical Warfare Service colleague Theodore Sill visited Oppau, which he described as a wonder of modern technology. What Pope saw or heard about Oppau must have aroused in him an interest in the potential for high-pressure industrial chemistry [11].

8.5 Niagara Falls

The factory of the Niagara Ammonia Company was located at Niagara Falls, the hub of the US electrochemicals industry. There, firms such as Hooker Electrochemical manufactured chlorine, alkali and chlorinated organic compounds, and also made available by-product hydrogen. In November 1923, a contract with the Electrolabs Corporation for purchase of a hundred and sixteen of its 5000 amp cells (electrolyzers), including their installation, was approved, as was a contract with the General Electric Company for "motor generators, exciters, switchboards, etc., being the complete apparatus to furnish direct current at the proper voltage to the cells, for the sum of $45,308:00." Also, it was reported that the Casale ammonia plant in Japan was so successful that an order had been placed for two additional converters, each with a rated daily capacity of seven and a half tons of ammonia. The source of nitrogen for the ammonia process at Niagara Falls was based on Casale's invention, namely, the removal of atmospheric oxygen from air by its reaction with hydrogen, the burning of air.

In December, Frederick Pope stated that progress in construction work at the Niagara Falls site was satisfactory. The first of three consignments of Italian-made machinery had arrived. Director and investor A. E. Bonn, of A.E Bonn & Company, Inc., advised that there might be some delay in receiving a second seaborne shipment of machinery from Italy, and that there "is a substantial amount of machinery on the third boat" [12]. Each shipment included Casale converters and Pignone high-pressure compressors and circulation pumps.

It was during May 1924 that Niagara Ammonia first received by-product hydrogen from Hooker [13]. Though Niagara Ammonia intended to generate its own supply of hydrogen from the in-house electrolyzers, the availability of hydrogen from Hooker was a useful reserve supply.

On 30 July, Pope reported that this was the first occasion when he could report actual production, "the plant having operated off and on during the month of July, making about 72,000 pounds of ammonia, which at 15¢ a pound, amounts to

$10,800." The monthly bill for hydrogen was approximately $5000, payroll was about $4000, and cost of supplies $600. Pope believed that "the plant profit [per month], excluding overhead and sales expenses, could be about $1000" [14].

However, Niagara Ammonia was forced to delay the sale of its anhydrous ammonia until quality issues had been dealt with. Pope drew attention to the fact that for use as refrigerant the anhydrous ammonia was unacceptable to an important potential customer, Armour & Company, "as it contained more than the allowable percentage of water and foreign gases." This difficulty could be overcome by the installation of a distilling apparatus, costing between $10,000 and $15,000, a sum that was approved. Funds were also approved for completion and "tuning up" of the first ammonia unit. Pope observed that Henry J. Pierce, as a member of the Niagara Ammonia board, "has been of great service to the company while abroad, in urging Dr Casale to come to the United States the middle of August, and in following the delivery of the Italian machinery for the third unit of our plant" [14].

On 22 August 1924, Luigi Casale and two of his engineers, Renato Fabi and Iacopo Volpi, departed from Naples, bound for New York in order to deal with business interests, and especially certain technical matters, including the quality of anhydrous ammonia. They arrived at Ellis Island, on 1 September. According to Casale's professor, Arturo Miolati, Casale also attended an event with Fritz Haber in New York, which must have been around the time of the Franklin Institute centenary celebrations, held in Philadelphia during 17–19 September, and in which Haber participated [15].

8.6 Changes and Challenges at Niagara Ammonia

8.6.1 The Departure of Frederick Pope

On 14 November 1924, Colonel Frederick Pope tendered his resignation as president of Niagara Ammonia, and apparently of involvement in other Casale interests, and was replaced by Henry Pierce. Pope perhaps foresaw difficulties, including the great expense of generating hydrogen from the electrolyzers at the Niagara Falls operation [16].[2] That may have suited Europe and Japan, where strong political and strategic commitments and local and state support were significant factors in the adoption of hydroelectric power. In the US, however, electricity, though widely available, was expensive. The actual output of synthetic ammonia at Niagara Ammonia never exceeded 8-tpd, which was considerably less than the rated capacity of the three converters. In August 1925, according to data supplied by

[2] Pope later sold (what probably included his own) holdings in The Ammonia Corporation. Note of sale of 350 preferred (at $100) and 652 2/3 common (at $25) stock from Pope & Company, 60 Broadway, New York, to Tucker, Anthony & Co., 120 Broadway, New York, 22 September 1925. Pope appears to have been involved in various enterprises as a consultant. During the war he became a partner in the consulting engineers Moses, Pope & Trainer, Inc., of New York.

ammonia manufacturers, Niagara Ammonia's rated capacity was 13-tpd, to be increased to 18-tpd by the end of year. Allied Chemical's daily rated capacity at Syracuse was then 30 tons.

Lured away by new challenges, Pope's experience at Niagara Falls, and of the new synthetic ammonia technology of Luigi Casale, was put to good use. He most likely became involved in the Nitrogen Engineering Corporation (NEC), founded in the mid-1920s, which offered an ammonia converter based on a process that was far better adapted to large-scale production than the so-called "American" process of the FNRL. The co-inventors of the NEC process, Louis C. Jones and Major Charles O. Brown, had previously been employed by Allied Chemical. Unlike the Allied Chemical ammonia process, the NEC process was available for direct licensing, which is where Pope's input was no doubt invaluable. In the US, at least, there was far more interest in synthetic ammonia in the late 1920s than had been the case earlier in the decade when Niagara Ammonia was established. Pope also became associated with the Chemical Construction Corporation (Chemico), an engineering firm founded by Peter S. Gilchrist, Ingenuin Hechenbleikner, Thomas C. Oliver, and A. Mangum Webb, in December 1914. Chemico, based in Charlotte, North Carolina, held patents on novel catalytic processes, including the Selden vanadium catalyst for sulfuric acid, and was working on oxidation of ammonia to give nitric acid. In 1929–1930, NEC and Chemico were taken over by the American Cyanamid Company.

American Cyanamid's Chemical Construction Corporation (still referred to as Chemico), offered design and construction services for high-pressure plant and other equipment. Around 1930, Pope was involved in the introduction of the NEC process in the Soviet Union, perhaps after convincing the Russians that it was better suited to their conditions than the Casale process (which had been introduced in Russia in 1928). In June 1932, at a time when the Soviet Union was anxious to gain diplomatic recognition from the US, the Russians managed to gain his assistance [17]. However, Pope's relationship with the Russians was not smooth and he ceased dealing with them in 1934 [18]. In 1931, Pope was in China, promoting, on behalf of Chemico, the NEC process [19]. Later, he was associated with attempts to introduce Chemico technologies in Mexico, Egypt, and elsewhere.

8.6.2 The Hydro-Electric Chemical Company and the Liljenroth Process

In 1924, an interest in production of mixed nitrogen-phosphorus fertilizers, and at the same time a new source of hydrogen, led Hydro-Electric to back the electrothermal phosphoric acid-hydrogen method of Frans Georg Liljenroth (1881–1960), of Stockholm, Sweden. Phosphorus, produced by reduction of phosphate rock in an electrically heated furnace, was reacted with steam to afford phosphoric acid and hydrogen. However, Liljenroth's process was not taken up in practice.

Hydro-Electric, like Niagara Ammonia, also promoted the Casale ammonia process. One item in its correspondence file is of interest for revealing the royalty rates, and also the difficulty in finding licensees. During 1925, the royalty on ammonia from the Casale process was fixed at 5 dollars per ton. On 9 December 1925, the board of Hydro-Electric discussed a reduced royalty, to be "granted by this Company at any time within two years from and after the date of this meeting, the royalty rate to be fixed ... at not less than 0.4¢ per pound of NH_3." Board member Charles J. Hardy voted against this proposal because he considered that it was not good for business to reduce the royalty rate from 5 dollars per ton until the latter "had been refused by some of the prospective licensees" [20].

The discussion on royalty rates had been stimulated by the abrupt 50% fall in the price of anhydrous ammonia late in 1925. The price of aqueous (aqua) ammonia also fell, though less sharply. This did, however, encourage widespread oxidation of ammonia into nitric acid from 1926, in particular using processes developed by Du Pont and Chemico [21].

The fall in price hardly helped Niagara Ammonia and other companies that relied on the sale of synthetic ammonia for the industrial refrigeration market. They had a hard time matching the price of anhydrous ammonia from the other sources [22]. By 1926, the ammonia supply industry in general was suffering from severe competition and overproduction. The Niagara Ammonia Company, operating at less than half capacity, confronted an uncertain future, and not just as a result of an increasingly difficult trading environment.

By March, Niagara Ammonia was facing heavy losses. For various reasons, the daily output target could not be met. Though there were plans to enter into the nitrogen fertilizer business at the Niagara Falls site, the facility was too small to compete with the major producers of nitrogen products for agriculture. As with the operators of FNRL units, the scale of working with electrolytic hydrogen in the US was both limiting and expensive.

8.7 Du Pont Ammonia

Du Pont took up an interest in synthetic ammonia early in 1923, following somewhat exaggerated claims concerning the technological capability of the Claude process, as made by a representative of Air Liquide to Fin Sparre, director of Du Pont's Development Department, and head of much of the corporation's diversification program. Sparre had previously been against embarking on manufacture of synthetic ammonia, though he expressed a strong interest in oxidation of ammonia to nitric acid. What impressed him now was the high yield achieved at pressures far greater than used in the Atmospheric Nitrogen/General Chemical process of Allied Chemical, possibly sufficient to make the latter obsolete, and the potential of catalyzed high-pressure technology [23]. In the summer of 1924, Du Pont acquired exclusive US rights to the Claude process from Air Liquide. Claude's first ammonia patent in the US had been filed in March 1920 [24].

On 21 May 1925, jointly with Air Liquide, Du Pont incorporated in West Virginia the firm Lazote, Inc. Air Liquide provided 25% of the capital. Construction began on an ammonia factory at Belle, East Kanawha County, West Virginia. Soon after, Du Pont announced that Lazote had on 30 June 1925 received letters patent 1,544,373, assigned to Georges Claude, "covering broadly a method of conducting the synthetic operation wherein the pressure-sustaining wall is cooled by the incoming gases which are preheated by heat exchange with outgoing gases from the zone of reaction." Tellingly, "Lazote, Inc., is prepared to enforce its rights under this patent and prevent infringement thereof" [25].[3]

A catalyst testing laboratory, under the supervision of chemical director Roger Williams, and modeled after that of the FNRL, was installed at Du Pont's Wilmington Experimental Station. Alfred Larson, previously at the FNRL, conducted further investigations at Wilmington. The Belle facility, where hydrogen was produced from water gas, and nitrogen from producer gas (blow-run gas, or blow gas), both starting with coke, opened on 1 April 1926. Five 20-tpd Claude ammonia converters were installed at Belle. Du Pont announced that it was represented by its distributor, the National Ammonia Company [26].

The Claude process did not come up to expectations [27]. There were problems with the very high-pressure, or hyper-, compressors, often called "hypers," water gas and gas purification equipment, liquefaction plant, and especially the converters [23]. The rated capacity of the converters was difficult to achieve. Aware of the difficulties, and dangers, including explosions and casualties, Du Pont began to take an interest in the Casale process, no doubt in part because, having invested in hypercompressors, it wished to continue working with pressures way beyond those used in other synthetic ammonia processes (all of which Du Pont intended to outperform). Apart from these considerations, Casale's process was still the only independent, well-tried process suited to large-scale production available for licensing in the US in 1926. While access to Casale's process could have been obtained through a licensing arrangement with The Ammonia Corporation, Du Pont chose another way, a direct challenge to the validity of the Casale patents.

This involved a lawsuit in which it was argued that the Casale patents infringed certain Claude patents. In 1926, Lazote initiated proceedings for infringement against The Ammonia Corporation and Niagara Ammonia, in Buffalo, New York State. Because the claims in the US for both processes were construed as broad, the wording of certain of the Casale patents were open to attack. This included the high pressure, approaching that of the Claude process. Du Pont, through Lazote, pushed for an injunction against Niagara Ammonia's operations, in addition to claiming damages [28]. The outcome went in favor of Du Pont. This and the technical difficulties that Niagara Ammonia was facing, brought about the downfall of existing financial interests in the Casale process in the US and opened the way for Du Pont's acquisition of Casale technology in 1927. A press release from Du Pont's

[3] During 1926, Claude's US ammonia patents were assigned to Lazote.

Publicity Bureau revealed only the acquisition of patent rights from The Ammonia Corporation [29, 30]. However, the arrangements brought about Du Pont's total control of Niagara Ammonia, The Ammonia Corporation, and Hydro-Electric.

Du Pont's Jasper E. Crane, a member of the executive committee responsible for international affairs (and its vice president from 1929), was appointed chairman of the board of directors of Lazote. On 26 January 1928, at a meeting held in the Du Pont Building, Wilmington, the voluntary petition of bankruptcy of the Niagara Ammonia Company was sanctioned [31, 32].Two weeks later, The Ammonia Corporation, unable to pay its debts (as a result of the failure of Niagara Ammonia) was also declared bankrupt [33]. The *New York Times* reported that the liabilities of Niagara Ammonia were $947,748, and assets $107,899. The principal creditor was The Ammonia Corporation, with liabilities of $839,249, and assets of $498,467. The corporation was indebted to Lazote for the sum of $821,714. The principal listed assets of The Ammonia Corporation were rights to "foreign" patents, namely those relating to the Casale synthetic ammonia process, valued at $350,000 (this sum probably included the Phosphorus-Hydrogen Company's patents). These valuable patents were now in the hands of Lazote. Plant, stock and fixtures of Niagara Ammonia were valued at $97,579, which in the newspaper report were assigned to The Ammonia Corporation, as creditor [34].

Lazote, now the owner in the US of the Casale ammonia process and the useful Italian-made high-pressure converters and machinery of Niagara Ammonia, shipped the Casale converters and machinery from the defunct Niagara Falls factory to Belle. There, the Claude and Casale processes were each confined to a separate production area. Du Pont developed its "much modified Claude-Casale system" by applying design features from the Casale converters to its Claude converters [35]. The Du Pont process, it was noted, included "decidedly new and advantageous features and represents a radical departure from the original Claude process" [36, 37]. Jasper Crane of Du Pont late wrote that the "best features" of the French and Italian processes "were adapted to American conditions" [38]. The success with ammonia encouraged Roger Williams to develop a high-pressure methanol process, using the same converters as employed in the ammonia synthesis.

8.7.1 Expansion at Belle

Over the winter of 1927–1928, Du Pont authorized three-fold expansion in output of synthetic ammonia at Belle, based on the introduction of the Casale process (Fig. 8.1). In 1929, following the increased reliance on Casale technology, Du Pont bought out Air Liquide's holdings in Lazote. The problems with the Claude

Fig. 8.1 Du Pont's synthetic ammonia factory, Belle, East Kanawha County, West Virginia, 1930s. (Casale SA. With permission)

process brought to an end the contract with Air Liquide.[4] Du Pont's ammonia interests became the Du Pont Ammonia Corporation. While Du Pont may have forced an end to the Casale enterprises in the US, it did not shy away from purchasing converters from Ammonia Casale during 1929–1930. In fact it was so heavily reliant on Casale technology that in June 1930, just three months after one of Ammonia Casale's new 30-tpd converters was installed at Belle, and found to be successful in operation, four more were ordered [39]. This brought the facility to a rated daily capacity of around 225 tons, with 100-tpd produced by the Casale process, and 125-tpd by the modified Claude process [40].

Expansion at Belle during the 1930s brought the annual rated capacity of synthetic ammonia to around 146,000 tons in 1940. Single pass conversion to around 20% of ammonia through a single converter was claimed (also for the Nitrogen Engineering Corporation process). Ammonia was converted into ammonium sulfate fertilizer, nitric acid, and other industrial products. Contrary to the early expectations at Du Pont, its very high-pressure process did not make Allied Chemical's and similar processes carried out below 300 atm obsolete. By the mid-1930s, Allied was responsible for around 60%, and Du Pont for around 40%, of synthetic

[4] Independently, Air Liquide later modified its Claude process, with, in the case of the Mississippi Chemical Corporation of Yazoo City, Missouri, two converters in series, and recirculation of unreacted gases.

Table 8.2 Rated annual synthetic ammonia capacities, United States, for 1940

Company	Year manufacture began	Process	Source of hydrogen	Capacity, tons
Allied chemical	1928	General Chemical/ Allied/Solvay	Water gas	250,000
Du Pont	1926/1928	Claude-Casale	Water gas	146,000
Shell	1931	Mont Cenis	Natural gas	27,000
Midland	1930	American	By-product	9,000
Mathieson	1925	NEC	By-product	5,800
Pennsylvania Salt	1932	Mont Cenis	By-product	4,900
Du Pont R&H Division, from 1932	1926	American	By-product	4,500
Great Western (Dow)	1927	NEC	By-product	1,100
Hercules	1940	Claude	Natural gas	10,900

Notes NEC = Nitrogen Engineering Corporation. American = based on the Fixed Nitrogen Research Laboratory method. By-product = from electrolytic processes. Source: Cope WC (1949) Ammonia. Part I. US production facilities. Chemical Industries 64(6) (June):920–925

ammonia produced in the US. Notwithstanding the entry of other firms into the manufacture of ammonia, they maintained a clear lead until the end of the decade (Table 8.2). As a powerful duopoly, they collaborated in control of the fertilizer market, partly through Allied's Barrett Company (Division from 1940), as distributor of ammonium sulfate (Fig. 8.2). One important outcome of the availability of inexpensive synthetic ammonia was that during the 1930s it almost completely replaced Chilean nitrate as the source of the world supply of nitric acid.

Though synthetic ammonia was certainly low in the list of Du Pont's original innovations, and at first hardly met the criteria for investment, it had a tremendous impact on the long-term standing of a technology giant. High-pressure chemistry was the key to one of the most profound transformations at Du Pont. It required a new understanding of methods employed in chemical manufacture and processing, and ensured that the corporation remained relevant in commercialization of new discoveries for decades to come. Du Pont's Ammonia Department, drawing on Casale inspired technology, helped propel the corporation to first place as the leading innovative chemical firm in the US, enabling a proliferation of new of products, such as nylon and other polymeric materials.

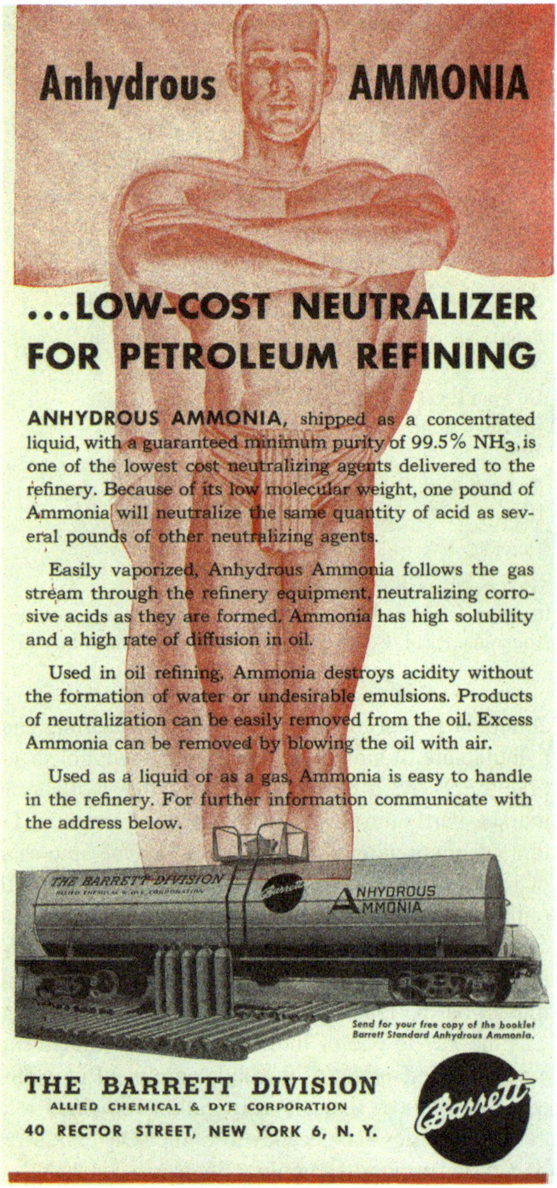

Fig. 8.2 Barrett advertisement for anhydrous ammonia, 1946. Until around 1940, Barrett distributed the nitrogen fertilizer products of Allied Chemical and Du Pont. (Petroleum Refiner 25(4):264) (Barrett Industries. A Colas Company. With permission)

References

1. Brunet P (1925) Fixation de l'azote aus Etats-Unis et les usines de Muscle Shoals. Revue d'Electrochimie et d'Electrometallurgie 19(129)(April):2–54

2. Clarke MJ (1976). The federal government and the fixed nitrogen industry 1915–1926. PhD dissertation, Oregon State University
3. "No. 77, Nitrogen prod. Synth amm. Haber proc. catalysts. Analysis of samples received from Germany between February and August 1919. H. E. Jones, Brunner, Mond & Co., Northwich, February-August 1919"; and "No. 78, Nitrogen prod. Synth amm. Haber proc. catalysts and catalysers. Examination of Oppau catalysts from vanadium. H. E. Jones, Brunner, Mond & Co., Northwich, May 1920." Solvay Process Company archives, microfilm reel 2, the Sidney M. Edelstein Library, National Library of Israel, Jerusalem
4. Anon (1921) American Chemical Society favors selective embargo. Oil, Paint and Drug Reporter 12 Sept, p 23
5. Haynes W (1945) American chemical industry: a history, vol. II. The World War I period: 1912–1922. D Van Nostrand, New York, p 122
6. Ernst FA (1928) Fixation of atmospheric nitrogen. Chapman & Hall Ltd, London, pp 131–132
7. Steen K (2014) The American synthetic organic chemicals industry: war and politics, 1910–1930. The University of North Carolina Press, Chapell Hill, p 254
8. Records of E. I. Du Pont de Nemours & Co. Absorbed companies. Hagley Museum and Library, Wilmington, Delaware, DE 19807, Accession 500, Series II, part 1
9. Anon (1961) Frederick Pope, engineer dead. Set up synthetic ammonia plants here and abroad. New York Times, 13 Aug, p 88
10. Pope F (1919) Condition of chemical plants in Germany. Ind Eng Chem 11(June):512
11. Sill TW (1919) The dyestuff plants and their war activities. Ind Eng Chem 11(June):509–512
12. Regular monthly meeting of the directors, Niagara Ammonia Co., 21 December 1923. Records of E. I. Du Pont de Nemours & Co. Absorbed companies. Hagley Museum and Library, Wilmington, Delaware, DE 19807, Accession 500, Series II, part 1. Niagara Ammonia Co., box 1366
13. Haynes W (1949) American chemical industry: a history. vol VI. The chemical companies. D Van Nostrand, New York, p 394
14. Minutes of meeting of board of directors, Niagara Ammonia Co., 30 July 1924. Records of E. I. Du Pont de Nemours & Co. Absorbed companies. Hagley Museum and Library, Wilmington, Delaware, DE 19807, Accession 500, Series II, part 1. Niagara Ammonia Co., box 1366
15. Miolati A (2009) Synthetic ammonia and the Casale process. Amplified edition of a lecture delivered the 27th February 1927 at the Institute of Chemistry of the Polytechnic School of Prague. Lugano, Ammonia Casale SA, p. 34. A translation by G. Impallomeni of Miolati A (1927) L'ammoniaca sintetica ed il processo Casale. Ammonia Casale SA/L'Universale, Tipografia Poliglotta, Rome, p 34
16. Minutes of meeting of board of directors, Niagara Ammonia Co., 14 November 1924. Records of E. I. Du Pont de Nemours & Co. Absorbed companies. Hagley Museum and Library, Wilmington, Delaware, DE 19807, Accession 500, Series II, part 1. Niagara Ammonia Co., box 1366
17. Pope F (1932) Envoy needed in Russia. New York Herald, 14 Nov
18. Sutton AC (1971) Western technology and Soviet economic development, 1930–1945. Hoover Institution, Stanford, pp 99–100
19. Kwan MB (2017) Patriots' game: Yongli Chemical Industries, 1917–1953. Brill, Leiden, pp 83–105
20. Minutes of meeting of the board of directors, Hydro-Electric Company, 9 December 1925. Records of E. I. Du Pont de Nemours & Co. Absorbed companies. Hagley Museum and Library, Wilmington, Delaware, DE 19807, Accession 500, Series II, part 1. Niagara Ammonia Co., Hydro-Electric Company, box 847
21. Fairlie AM (1936) Sulfuric acid manufacture. Reinhold Publishing Corporation, New York, pp 150–151
22. Anon (1926) Why synthetic ammonia prices are lower. Chemicals 26(25)(20 Dec):12–13
23. Hounshell DA, Smith JK Jr (1988) Science and corporate strategy. Du Pont R and D, 1902–1980. Cambridge University Press, Cambridge, pp 183–189
24. Georges Claude, US patent no. 1332460 of 2 Mar 1920

25. Anon (1925) Claude patent for a method for direct synthesis of ammonia. Ind Eng Chem 17(Sept):985
26. Anon (1927) Glancing back over 1926. The Du Pont Magazine 21(Mar):1
27. Varnes SK (2001) Pioneering in high pressure. In: Denham CJ (ed) Sentimental history: the Du Pont Belle Works. A 75 year history. Charleston, Du Pont, p 14
28. Haynes W (1948) American chemical industry: a history, vol IV. The merger era. D Van Nostrand, New York, p 88
29. Anon (1927) Du Pont obtains American rights on Casale process. Chem Metallur Eng 34(6):388
30. Anon (1927) Du Pont gets the U.S. rights to Casale patents. Oil, Paint, Drug Reporter, 6 June, p 21
31. Du Pont, Wilmington, letter regarding meeting to be held on 26 January 1928 to consider voluntary petition of bankruptcy of Niagara Ammonia Company, Inc., 19 January 1928. Records of E. I. Du Pont de Nemours & Co. Absorbed companies. Hagley Museum and Library, Wilmington, Delaware, DE 19807, Accession 500, Series II, part 1. Niagara Ammonia Co., box 1366
32. Board of directors, meeting no. 41 (special meeting), Ammonia Corporation, 26 January 1928. Records of E. I. Du Pont de Nemours & Co. Absorbed companies. Hagley Museum and Library, Wilmington, Delaware, DE 19807, Accession 500, Series II, part 1. Ammonia Corporation, box 67
33. Board of directors, meeting no. 42 (special meeting), Ammonia Corporation, 10 March 1928. Records of E. I. Du Pont de Nemours & Co. Absorbed companies. Hagley Museum and Library, Wilmington, Delaware, DE 19807, Accession 500, Series II, part 1. Ammonia Corporation, box 67
34. Anon (1928) Niagara Ammonia bankrupt. Ammonia Corporation creditor, also is insolvent. New York Times, 16 Feb, p 37
35. Shreve RN (1945) The chemical process industries. McGraw-Hill Book Company, New York, pp 402–404
36. Anon (1930) Pressure-synthesis operations of the Du Pont Ammonia Corporation. Ind Eng Chem 22(May):433–437, on 433.
37. Crane JE (1930) Development of the synthetic ammonia industry in the United States. Ind Eng Chem 22(July):795–799
38. Tongue H (1934) The design and construction of high pressure chemical plant. Chapman & Hall, London, pp 377–378
39. Annual report to shareholders, Ammonia Casale, 12 June 1930 (Casale SA)
40. Stewart C (2001) A glance at 25 years. In: Denham CJ (ed) Sentimental history: the Du Pont Belle Works. A 75 year history. Du Pont, Charleston, pp 16–18 (from an account written in May 1950)

The United States Emerges as the Global Leader

9

Abstract

Government-sponsored expansion of the synthetic ammonia industry during World War II enabled the US to achieve world leadership in ammonia production after 1945. An important wartime innovation was the reliance at certain of the new factories on natural gas, rather than coal, as feedstock for hydrogen. The outcome was a nationwide dispersal of the ammonia industry. Sale of government factories brought new entrants to the industry. The needs of the Korean War gave a further boost to expansion and production. By the mid-1950s, there were around 50 synthetic ammonia factories in the US. Significantly, by 1960, through its American agent, the Casale ammonia process was adopted at several sites. Another European process introduced into the US was the redesigned Claude process which was taken up by a major farmers' cooperative. Manufacture of methanol from synthesis gas and based on ammonia technology achieved considerable importance. The growing use of natural gas as a feedstock encouraged petrochemical firms to undertake the manufacture of ammonia.

9.1 The Government Ammonia Plants

The threat of war in the late 1930s transformed the nitrogen situation in national markets from one of surpluses to one of potential shortages. In the case of synthetic ammonia in the US, Du Pont, Allied-Solvay, Chemico, and the engineering firm M.W. Kellogg contributed to American preparedness for war after the government, through its military authorities, authorized construction of nitrogen factories for munitions production. In 1940, funding became available for the construction of the first ordnance synthetic ammonia factories. With the entry of the US into World War II on 7 December 1941, this program was considerably enhanced. By 1945, the government had funded ten factories, with a combined installed annual capacity

A. S. Travis, *The Global Synthetic Ammonia Industry*, Perspectives on the History of Chemistry, https://doi.org/10.1007/978-3-031-87978-4_9

of around 800,000 tons of ammonia (Table 9.1). They contributed towards the wartime production of around 2.25 million tons of TNT, alone, by early 1945.

The first government-owned contractor-operated ammonia factory to be completed, in 1941, was built by Du Pont under the supervision of the Quartermaster Corps. This was the Morgantown Ordnance Works, in West Virginia, and based on coal as feedstock. The rated capacity was 180,000 tons per year (tpy). The factory used the modified Claude (Claude-Casale) technology.

Most of the wartime ordnance works were equipped with one, two or three trains, with converters of 150-tons per day (tpd) rated capacity [1]. Part of the program involved the erection of new manufacturing plants away from the

Table 9.1 Synthetic ammonia processes and locations of US government 1940s wartime ordnance works[a]

Process	Location	Notes
Du Pont (Claude-Casale)	*Morgantown*, West Virginia	Ordnance Department of the Army. Built by Du Pont, 1941
Allied-Solvay (Atmospheric Nitrogen Corp.)	*Buckeye*, South Point, Ohio	Ordnance Department of the Army
Allied-Solvay (Atmospheric Nitrogen Corp.)	*Ohio River*, Henderson, Kentucky	Ordnance Department of the Army. Operated by Solvay Process Division, September 1942-August 1945
Chemico	*Ozark*, El Dorado, Arkansas	Ordnance Department of the Army. Operated from May 1943
Chemico	*Cactus*, Etter, Texas	Ordnance Department of the Army. Plant completed 1945
Chemico	*Jayhawk*, Pittsburg, Kansas	Ordnance Department of the Army. Commenced production 1943, operated by K. A. Spencer
Kellogg	*Dixie*, Sterlington, Louisiana	Ordnance Department of the Army. Operated from April 1943
Kellogg	*Mathieson*, Lake Charles, Louisiana	Sponsored by the War Production Board. Built by Defence Plant Corporation. Operated 1944-September 1945
Claude (L'Air Liquide)	*Missouri*, Louisiana	Ordnance Department of the Army. Built and operated by Hercules Powder Co., 1941–1942. Operated from November 1942.
TVA	Muscle Shoals, Alabama	Tennessee Valley Authority. Operated from 1942

[a] From reference [1]

Pennsylvania-Kentucky-West Virginia coal regions. Four plants were based on coal: Buckeye, South Point Ohio; Muscle Shoals, Alabama; Ohio River, Henderson, Kentucky; and Morgantown, West Virginia [2]. Significant to the future history of ammonia production was the reliance of the six other plants in the war program on natural gas (methane) as feedstock, drawing in five cases on the Imperial Chemical Industries (ICI) two-stage steam reforming technology, and in one case on the single-stage Hercules Powder Shapleigh reforming process. During 1940–1941, ICI provided the engineering contractors M.W. Kellogg and Chemico with data on its steam reforming process. ICI engineers assisted Chemico to install steam reforming units at the Ozark, Jayhawk, and Cactus ordnance works [3]. Subsequently, Kellogg used the ICI steam reforming process at the Dixie and Mathieson ordnance works. The Dixie works was authorized in August 1941, for a single train unit with a capacity of 150-tpd, subsequently increased, with three additional units, to 600-tpd. However, only one unit was completed and began production in April 1943.

The Ozark Ordnance Works, later owned and operated by the Lion Oil Company (Lion Chemical Corporation) at El Dorado, Arkansas, "led the Ordnance plants and perhaps the industry throughout the world in low production cost for the manufacture of ammonia. Furthermore, this plant was largest producer of ammonia [from natural gas] among the various Ordnance facilities." The Ozark plant was authorized by the War Department on 10 October 1941, and began operating on 13 May 1943 (Fig. 9.1). Ozark's rated annual production capacity in 1944 was 155,000 tons of ammonia. "It was unique in the fact that the design of the plant included not only the use of natural gas as a source of hydrogen but also as the prime source of fuel … for steam production, as fuel for gas engine driven refrigeration and process compressors, as fuel for gas engines in electric power generation, and as fuel for driving process water pumps" [4].

Hercules Powder, at Hercules, California, had in December 1938 selected the Claude process for manufacture of anhydrous ammonia from natural gas. Hercules Powder subsequently developed the catalyst for James H. Shapleigh's reforming process, which, at Hercules, was the first steam reforming process used in an ammonia facility. Production of ammonia began in March 1940, at 40-tpd, later increased to 90-tpd, with two trains. During 1941–1942, Hercules, which received 40 million dollars in government funding, installed five 30-tpd Claude ammonia units at the Missouri, Louisiana, ordnance works. Again, the feedstock was natural gas [5].

Together, the six ordnance works that relied on natural gas represented the first large-scale use of steam reforming of natural gas as feedstock for synthetic ammonia production.

In 1942–1943, the Tennessee Valley Authority (TVA), a federal agency created by Congress in May 1933 as part of Franklin D. Roosevelt's New Deal, promoted an ammonia converter working at 250–350 atm, originally with a rated capacity of 140-tpd. It was designed for the Office of Chemical Engineering of the TVA by Charles Brown, of the Nitrogen Engineering Corporation, and a member of the Chemical Referee Board of the War Production Board [6]. The TVA design was

FIGURE 1
Ammonia Plant—Lion Oil Company, Chemical Division, El Dorado, Arkansas

AMMONIA
Its Production
From Natural Gas

GUY S. MITCHELL, Chief Chemist
Chemical Division—Lion Oil Company

Fig. 9.1 The Ozark Ordnance Works at Eldorado, Arkansas, where ammonia was made by the Chemico (Nitrogen Engineering Corporation, NEC) process. After the war the facility was acquired by the Lion Oil Company (From Mitchell GS (1946) Ammonia. Its production from natural gas. Petroleum Refiner 25(June):245–258/97–110) (Hydrocarbon Processing. With permission)

used at the Muscle Shoals site of the former World War I calcium cyanamide facility, and was made freely available. In addition, Muscle Shoals supplied ammonium nitrate for agricultural use, in accord with the War Production Board's directives. However, the farmers had to be convinced of the suitability of ammonium nitrate because it had a reputation for caking. Use of Chilean nitrate and ammonium sulfate from coke ovens remained significant. The War Food Administration favored a high level of imports of Chilean nitrate, a product with which farmers were long familiar.

The wartime program in many ways compensated for the slow uptake of high-pressure synthetic ammonia processes in the US prior to the 1930s. As a result, by 1945, not only was the US almost self-sufficient in its needs for ammonia, it was also the leading global ammonia producer. The industry had a strong showing in the Gulf Coast region.

American engineering contractors were also active in Canada, where, in order to mainly support the UK's wartime demand for explosives, government-backed expansion in ammonia took place, especially in the western provinces. In 1941, Kellogg assembled its first steam methane reforming unit near Calgary, Alberta, for Alberta Nitrogen Co., using locally available natural gas to produce hydrogen by the ICI process. The converter was based on Kellogg's first design for manufacture of synthetic ammonia. It operated at a pressure of 300 atm (as was also the case at Dixie and Mathieson in the US). At Trail, British Columbia, the Consolidated Mining and Smelting Company of Canada, Ltd, using Casale technology, had a rated capacity of 38,000-tpy around 1940. Engineers from ICI assisted in increasing production at Trail [7]. Ammonium nitrate was produced at both Alberta Nitrogen and Consolidated Mining; the nitrate made at Trail was exported to the US for agricultural use. However, the Canadian chemical industry remained concentrated in the east. Thus, Chemico's Nitrogen Engineering Corporation converter was used at Welland, Ontario, which relied on coal as feedstock.

The US wartime production of synthetic ammonia was an integral part of a program for production of munitions carried out on an unprecedented scale [8]. Among the novel products was RDX explosive, made by reacting ammonia with formaldehyde and nitrating the product.

Ammonia production, as in other sectors, relied on technological collaboration between the UK and US, most notably, in the case of ammonia, through transfer of ICI's technology for producing hydrogen from natural gas. This would have far-reaching consequences for the post-war growth of the US synthetic ammonia industry, including through the involvement of the petroleum and petrochemicals firms (Sect. 9.2.2).

9.2 Further Expansion in the United States

By 1950, five of the ten US government's ordnance works had been sold, or leased out and then sold, by the War Assets Administration, in some cases to the corporations that had constructed and operated them on behalf of the government, in other cases to new players in the synthetic ammonia industry. This at least encouraged greater competition among manufacturers (Table 9.2).

New ammonia plants were built by Dow (using its by-product electrolytic hydrogen) and Du Pont. Dow, at Freeport, Texas, produced refrigeration (99.95%) and technical grade (99.5%) ammonia. The intended markets for anhydrous ammonia were the west coast and the Delta region (Arkansas, Louisiana, and Mississippi). Mathieson Chemical Corporation, at Lake Charles, Louisiana, close to the northern part of the Gulf of Mexico, and Spencer Chemical expanded their new and existing facilities. Commercial Solvents purchased the Dixie works for $5,850,000 in September 1946. The single ammonia unit was retained. The rest of the site, which was incomplete, was used for production of synthetic methanol [9].

Buckeye, at South Point, Ohio, acquired in 1946 by Allied Chemical-Solvay, was assigned to its Nitrogen Division for production of ammonia, urea, and methanol. For this division the engineering contractor Catalytic Construction Company of Philadelphia erected an ammonia-urea plant at La Platte, Nebraska. It came on stream in 1954. At that time, the newer ammonia units, with few equipment changes, could be used for both ammonia and methanol synthesis. In 1952, alone, 65% of methanol produced in the US was from the high-pressure process starting with natural gas. Typical conversion pressures for methanol were 250 to 300 atmospheres, similar to the pressures used in ammonia units. Other sources

Table 9.2 Post-1945 disposal of US government wartime ordnance works, by sale, and leasing or contractual arrangements[a]

Location	Source of hydrogen	Notes
Morgantown, West Virginia	Coke (water gas)	Producer for the U.S. Army fertilizer program. Later leased to and then sold to Heyden Chemical Co.
Buckeye, South Point, Ohio	Coke (water gas)	Sold to Solvay Process Division, Allied Chemical & Dye Corporation/Allied Chemical-Solvay, Nitrogen Division, June 1946
Ohio River, Henderson, Kentucky	Coke (water gas)	Wartime operator Allied Chemical-Solvay, and then producer for the U.S. Army fertilizer program. Spencer Chemical Co., the only bidder, acquired Ohio River Ordnance Works, and the lease of nitric acid facilities at Indiana Arsenal, Charleston, Indiana, for $3 m; a further $3 m was invested in the two plants
Ozark, El Dorado, Arkansas	Natural gas	Leased to Lion Chemical Corporation/Lion Oil Company, May 1946, and sold to Lion in March 1948. Capacity expanded to 580-tpd, followed with a plant at Luling, near New Orleans, Louisiana, of 350-tpd capacity
Cactus, Etter, near Amarillo, Texas	Natural gas	Producer for the U.S. Army fertilizer program. Leased to Phillips Chemical, a subsidiary of Phillips Petroleum Co., for 15 years, in 1948
Jayhawk, Pittsburg, Kansas	Coke (water gas)	Wartime lessee K. A. Spencer; sold to Spencer Chemical Co., May 1948. Produced ammonia and from 1950 methanol
Dixie, Sterlington, Louisiana	Natural gas	To Commercial Solvents Corporation, September 1946
Mathieson, Lake Charles, Louisiana	Natural gas	To Mathieson Chemical Corporation
Missouri, Missouri, Louisiana	Natural gas	In 1948, two of the five 30-ton units were moved to San Jacinto Ordnance Depot, Houston, Texas, and in 1950 leased to San Jacinto Chemical Corporation; during 1947–1949 the three remaining units were used for coal-to-oil conversion at what became the U.S. Bureau of Mines Experimental Fuels Plant
Muscle Shoals, Alabama	Natural gas	Tennessee Valley Authority. Retained by the government. Fertilizer sold to farmers' cooperatives

[a] From reference [1]

of methanol were, at around 30%, wood alcohol (methanol from distillation of wood), and, at 6%, oxidation of hydrocarbons. Around 75% of the methanol was used as antifreeze or converted to formaldehyde, for use in production of Bakelite and urea and melamine resins. Both Allied Chemical-Solvay at Buckeye, and Lion, at Ozark, expanded their new acquisitions, raising total US annual rated capacity of anhydrous ammonia to 1,500,000 tons.

Spencer Chemical owed its origins to the Kansas coal mine owner Kenneth Aldred Spencer, who in 1941 responded to a War Department request to manufacture ammonium nitrate at Jayhawk. This facility commenced production of ammonia during 1943 at 14,500-tpy. After the war, with J.H. Whitney & Company, Spencer founded Spencer Chemical and leased and then in 1950 purchased Jayhawk and Ohio River. Spencer Chemical converted both plants from coal to natural gas as feedstock [10, 11]. Spencer concentrated on the manufacture of nitrogen fertilizer. Jayhawk and Ohio River (originally 56,000-tpy; increased to 68,000-tpy), together made Spencer the fourth largest US producer of ammonia, after Allied Chemical-Solvay, Du Pont, and Lion [12]. Spencer, by modifying the wartime ammonia installation at Jayhawk, also moved into methanol [13]. In January 1954, Spencer opened an ammonia factory at Vicksburg, Mississippi.

In the 1950s, Du Pont at Belle changed over from water gas to natural gas for production of hydrogen. Armour & Co., which in the 1920s had an interest in the Casale enterprise in the US, produced extremely pure hydrogen from ammonia (the reverse of the ammonia synthesis reaction) for use in hydrogenation processes.

9.2.1 Farmers' Cooperatives

The passing of the 1922 Capper-Volstead Act and the 1926 Cooperative Marketing Act led to the formation of thousands of farm cooperatives in the US. The cooperatives received federal support; among their leading activities were distribution of seeds and fertilizers. From 1945, the TVA took on an important role by supplying nitrogen fertilizers to the cooperatives, several of which had consolidated their operations during the war. However, the supply of synthetic nitrogen fertilizers could not satisfy the tremendous growth in demand. Even Chilean nitrate, which was still popular with cotton farmers in the Delta region, was in short supply. There was an increasing interest among American farmers, through the now large cooperative ventures, in maintaining self-sufficiency in ammonia, and locating new nitrogen and fertilizer production plants close to markets [14].

A cooperative of farmers in Mississippi founded the Mississippi Chemical Corporation, in November 1948, in order to erect at Yazoo City the first farmer-owned synthetic ammonia facility. The feedstock was natural gas. The synthesis process used modified Claude technology, with recycling of unreacted gas, as was then common in the US, and the manufacturing plant was purchased on a "turn-key" basis. For the high-pressure Claude process, worked at 1000 atm, the Yazoo City plant incorporated four Ingersol-Rand HHE-type compressors, of which two 4-stage units were used for primary compression, and two 2-stage units were used as boosters. The Girdler Company (a division of National Cylinder Gas Company) of Louisville provided the ammonia catalyst. This was one of the first plants to employ methanation, again using a Girdler catalyst, rather than a copper liquor scrubbing, in removal of carbon oxides during gas purification (Sect. 10.5.1). Under Claude conditions the catalyst life was short, about three months. The change of catalyst cartridge was timed to coincide with maintenance work at the plant. For the first fiscal year, 1951–1953, Yazoo produced 94,963 tons of anhydrous ammonia. The facility was considered "an excellent example of what is being done to bring industry to the people of a long industrially backward state" [15, 16]. The ammonia at Yazoo was stored in two refrigerated Horton spheres, pressure vessels of 48-foot in diameter, each of 100-tons capacity. Storage of ammonia was an important consideration because farmers in the south required ammonia for no more than three months every year. They were encouraged to install their own ammonia storage facilities. Mississippi farmers, especially in the Delta region, applied ammonia directly to the mineral rich soil, an application introduced by Shell Chemical Corporation in California during the 1930s. As required, ammonia was converted into urea by a process designed by Girdler. Mississippi Chemical later formed a partnership with the major cooperative Farmland Industries, Inc., for the supply of ammonia [17].

9.2.2 Entry of the Petroleum Industry

The petroleum industry, increasingly engaged in production of hydrogen from liquid hydrocarbons and natural gas, also emerged as a major producer of ammonia, rivalling Allied Chemical, Du Pont, Lion, and Spencer. This involved creation of specialty chemical divisions and subsidiaries. Shell Chemical Corporation, one of the four major producers of ammonia in the petroleum industry, manufactured 350-tpd at each of its two west coast plants in California, Ventura, opened in 1953, and the existing prewar plant at Pittsburg. Phillips Chemical Company, which by the mid-1950s was producing 500-tpd of anhydrous ammonia at the former government Cactus plant, supplied about 200 industrial operators, including marketers of petroleum products, with anhydrous ammonia. Phillips converted ammonia into prilled ammonium nitrate fertilizer and nitrogen solutions for direct application to the soil. Other applications of Cactus ammonia included in production of nylon and urea. In late 1953, Phillips opened a second anhydrous ammonia factory with a daily capacity of 450 tons at its Adams Terminal (a site purchased in 1948) on the Houston Ship Canal. The Chemico (Nitrogen Engineering Corporation, NEC) ammonia process, as at Catcus, was used. Natural gas was delivered through a 30-km pipeline from the Phillips Chocolate Bayou gasoline facility. Part of the ammonia was converted into ammonium sulfate [18, 19]. By 1940, liquid anhydrous ammonia had found new uses, including in the petroleum industry for protecting refining equipment from corrosion, and case hardening of steel.

Brea Chemicals, Inc., of Brea, California, was, from its opening in May 1954, another important producer of ammonia. C.F. Braun & Co., was the lead engineering contractor. Annual production eventually reached 150,000 tons of ammonia and 85,000 tons of urea. Around 1960, the Brea facility was sold to Collier Carbon and Chemical Corporation, a subsidiary of Union Oil Company of California (later known as Unocal). Collier was a leading producer on the west coast, and in the 1960s constructed an integrated ammonia-urea unit on the Kenai Peninsula in southcentral Alaska.

Some North American manufacturers did not easily give up on coal, or electrolyzers, as sources of hydrogen. In the mid-1950s, Ketona Chemical Corporation, at Birmingham, Alabama, opened the first ammonia factory in the US designed around coking oven gas as the source of hydrogen. The location was near the Tarrant coke plant of Alabama By-Products, the joint owner, with Hercules Powder Company, of Ketona. Engineering and construction were carried out by The Fluor Corporation, Ltd. American Air Liquide supplied units for recovery of hydrogen and the low temperature air separation of pure nitrogen. The annual capacity was 45,000 tons of anhydrous ammonia. In Canada, Consolidated Mining and Smelting (Cominco, from 1966), at Trail, continued to generate hydrogen by electrolysis.[1] The oxygen from the electrolyzers was used to gasify coke to generate more hydrogen, also for ammonia. However, this facility later relied mainly on natural gas as feedstock. Consolidated Mining acquired the Alberta Nitrogen wartime ammonia plant near Calgary, which was converted to fertilizer production.

9.3 The Korean War

Defense and security issues returned with the Korean War (1950–1953). Military targets included one of the largest synthetic ammonia factories in the world: United Nations bombers severely damaged the Hûngnam (Konan) plant in North Korea that since around 1930 had operated the Casale ammonia process.

In September 1950, the US Congress passed the War Production Act, which gave the government control over mobilization of American industry throughout the conflict. This included a much increased output of ammonia for both munitions and food production. Early in 1951 the Department of Agriculture calculated that farmers would need over two million tons of nitrogen fertilizer by 1955, twice the amount consumed in 1950 (and quite apart from nitrogen required for military purposes). This did much to encourage farmers to increase consumption of fertilizers based on synthetic ammonia. Apart from ammonia, methanol production in the US was increased during 1950–1951: at Du Pont, by adapting ammonia converters; at Dow, Freeport, Texas; and at the Mississippi Chemical Company, Yazoo City. In May 1952, American Cyanamid started work on its Fortier complex, at Avondale, on the Mississippi River, near New Orleans. Its subsidiary Chemico designed and constructed the facility, which included acetylene, ammonia (NEC process), hydrogen cyanide, and acrylonitrile units (Fig. 9.2). Hydrogen for ammonia was derived from off gases available from the acetylene unit. Ammonia was

[1] Consolidated Mining & Smelting produced zinc at Trail using an electrochemical process. From 1943, Trail produced heavy water for the Manhattan Project (P-9 project). In 1993, Cominco Fertilizers merged with Calgary Energy Company to form Agrium.

Fig. 9.2 Chemical Construction Corporation (Chemico) advertisement (1950) (Author's collection. Chemical Industries 67(2)(August):301)

converted into ammonium sulfate fertilizer. This large and diversified petrochemical plant, drawing on natural gas, began full production in 1955. The climatic conditions favored outdoor construction of most units, which was also a feature of new chemical manufacturing facilities [20]. Moreover, journals dealing with the petrochemical industry, such as *Petroleum Refiner*, increasingly featured articles dealing with ammonia and related technologies.

By 1954, as a consequence of the military requirements, around fifty ammonia factories were in operation or under construction in the US, and a further twenty-one were either under construction or proposed (Fig. 9.3). Between 1950 and 1955, the US increased annual production of anhydrous ammonia from 1.56 million tons to 3.25 million tons. Though the US had become the world's leading producer of synthetic ammonia, it was still a major importer of nitrogen fertilizers, in part because they carried no import duties. During 1953–1954, 400,000 tons of nitrogen entered the United States.

The momentum created by the Korean emergency situation continued after the cessation of hostilities. In 1954, in addition to the startups at Spencer's Vicksburg facility and Brea Chemicals, ammonia units went on stream for, among others, National Distillers, and Salt Lake Chemical. Also in 1954, the engineering contractor M.W. Kellogg started work on a 300-tpd ammonia plant for the Standard Oil Company of Ohio, at Lima, Ohio; and Fluor Corporation Ltd was awarded a contract to erect an ammonia plant for Mississippi River Fuel Corporation. As a result, between 1945 and 1955, US ammonia capacity doubled, with the heaviest concentration of the industry now in the southern states [21]. No less significant, the Du Pont-Allied duopoly had been broken [22].

Soon, however, there arose a problem of overproduction, at least as regards fertilizers. The increased production of ammonia during the emergency, and a farm recession during 1955 and early 1956, led to a substantial glut, accompanied by a fall in prices, and the start of a slowdown in the ammonia industry. Plans for new and expanded facilities were shelved. In 1956, Electric Bond and Share Company (EBASCO) acquired Chemico (which in 1969 was sold to the Boise Cascade Corporation, and in 1973 to the Aerojet-General Corporation.)

By 1957, the market situation was improving, with consumption at about 70% of ammonia production capacity. The estimated annual capacity, as nitrogen, of the fifty-two operating ammonia plants in the US was 3.9 million tons.

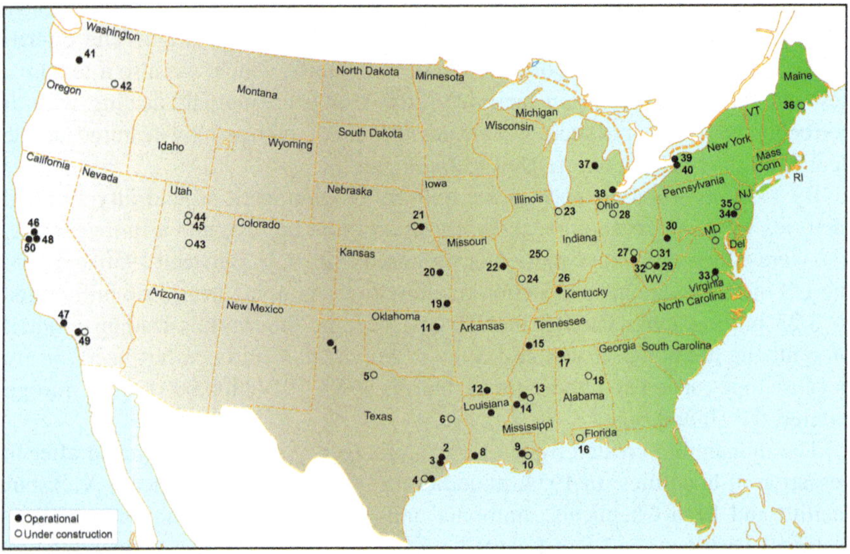

Fig. 9.3 Map of ammonia facilities operating and under construction in the United States, 1955

Key

State	Company	Location	Map code	Process	Source of hydrogen
Southern					
Texas	Phillips Chemical	Etter	1	NEC	NG
	Phillips Chemical	Port Adams	2	NEC	NG
	San Jacinto Chemical	Houston	3	Claude	NG
	Dow Chemical	Freeport	4	ANC	BPC
	Chillicothe Petrochemicals, Inc	Chillicothe	5	Casale	NG
	United Chemical Co.	Timpson	6	?	NG
Louisiana	Commercial Solvents	Sterlington	7	Kellogg	NG
	Olin Mathieson Chemical	Lake Charles	8	Kellogg NEC	(NG) (BPC)
	Lion Oil Company	Luling	9	NEC	NG
	American Cyanamid	Fortier	10	NEC (Kellogg)	NG

(continued)

(continued)

State	Company	Location	Map code	Process	Source of hydrogen
Oklahoma	Deere & Company (Grand River Chemical Division)	Pryor	11	Casale/Texaco	NG
Arkansas	Lion Oil Company	El Dorado	12	NEC	NG
Mississippi	Mississippi Chemical Co.	Yahoo City	13	Claude	NG
	Spencer Chemical Co.	Vicksburg	14	ANC	NG
Tennessee	W.R. Grace & Co.	Memphis	15	Casale/Texaco	NG
Florida	Gulf Chemical Co.	Penascola	16	?	NG
Alabama	TVA	Muscle Schoals	17	TVA	NG
	Ketona Chemical Co.	Ketona	18	?	NG
Mid-Continent					
Kansas	Spencer Chemical Co.	Pittsburg	19	NEC	NG
	Cooperative Farm Chemicals Association	Lawrence	20	Casale/Texaco	NG
Nebraska	Allied Chemical & Dye (Nitrogen Division)	La Platte	21	ANL	
Missouri	Hercules Powder	Louisiana	22	Claude	NG
Indiana	Standard-Sinclair Oil	Hammond	23	?	PRG
Illinois	Mississippi River Fuel	Crystal City	24	Texaco	NG
	National Distillers Products Corporation	Tuscola	25	Kellogg	BPE
Kentucky	Spencer Chemical Co.	W. Henderson	26	ANC	NG
Ohio	Allied Chemical & Dye (Nitrogen Division)	South Point	27	ANC	CWG/NG
	Standard Oil of Ohio	Lima	28	Kellogg	PRG
West Virginia	Du Pont	Belle	29	Claude-Casale (Du Pont)	CWG
	Olin-Matheson	Morgantown	30	Claude-Casale (Du Pont)	CWG
	Columbia-South Chemicals	Natrium	31	NEC	BPC
	Westvaco (Food Machinery and Chemical Corporation)	South Charleston	32	?	BPC
Virginia	Allied Chemical & Dye	Hopewell	33	ANC	CWG/NG

(continued)

(continued)

State	Company	Location	Map code	Process	Source of hydrogen
Pennsylvania	Atlantic Refining	Point Breeze	34	HR	PRG
	Sun Oil Company	Marcus Hook	35	?	PRG
Northeastern					
Maine	Northern Chemical Industries	Sears Port	36	Texaco	FO
Michigan	Midland Chemical Co.	Midland	37	FNRL	BPC
	Pennsylvania Salt Co.	Wyandotte	38	Mt Cenis Casale	BPC
New York	Olin-Mathieson	Niagara Falls	39	NEC	BPC
	Du Pont	Niagara Falls	40	FNRL	BPC
Western					
Washington	Hooker Electro-Chemical	Tacoma	41	NEC	BPC
	Columbia River Chemical	Pasco	42	Claude	FO (NG later)
Utah	Utah Chemical Co.	Mt Pleasant	43	?	NG
	Salt Lake Chemical Co.	Salt Lake City	44	Texaco	NG
	U.S. Steel Corporation	Geneva	45	?	CWG
California	Shell Chemical Co.	Pittsburg	46	Mt Cenis	NG
	Shell Chemical Co.	Ventura	47	Kellogg	NG
	Dow Chemical Co.	Pittsburg	48	NEC	BPC
	Brea Chemicals Inc. (Union Oil Company of California)	Brea	49	MMC	NG
	Hercules Powder	Hercules	50	Claude	PRG

Processes: Claude (Air Liquide); Claude-Casale (Du Pont); ANC, Atmospheric Nitrogen Corporation (Allied Chemical & Dye); FNRL, Fixed Nitrogen Research Laboratory; HR, Hydrocarbon Research; Mt Cenis, Mont Cenis; MMC, modified Mont Cenis; NEC, Nitrogen Engineering Corporation (Chemico); Texaco, partial oxidation process for hydrogen, often used with the Casale synthesis (FW, or Foster Wheeler, process), as developed jointly by the Texaco Development Corporation and Hydrocarbon Research; TVA, Tennessee Valley Authority; ? not known.
Hydrogen from: BPC, by-product from chlorine cells; BPE, by-product from ethylene manufacture; CWG, coke water gas; FO, fuel oil; NG, natural gas; PRG, Petroleum refinery gas; ? not known.
Map redrawn, and table adapted, from reference [21].

9.3.1 Foster Wheeler and Ammonia Casale

Among the US engineering contractors engaged in promoting ammonia production was Foster Wheeler. This corporation was from around 1953 the main agency outside Italy for Ammonia Casale.[2] Foster Wheeler promoted its "FW Sequence" process for synthetic ammonia worked in conjunction with the partial oxidation process of the Texaco Development Corporation (of the Texaco Company) for producing hydrogen from natural gas (methane) (Fig. 9.4; see Chap. 10, Sect. 5.4). The partial oxidation step was followed by a liquid nitrogen wash to remove carbon monoxide, methane and argon, leaving pure hydrogen; then introduction of nitrogen to make up the gas mixture; and, finally, ammonia synthesis by the "simplest, most dependable method," the Casale process [23]. During the 1950s, Foster Wheeler was responsible for eight new ammonia plants that added 20% to the ammonia capacity of the US. Of these eight plants, seven used the Casale process [24].

Many of the older US facilities, some dating from the 1940s wartime expansion, were in the Gulf Coast, near gas wells or were served by natural gas pipelines. Their competitiveness in the 1950s had come under pressure from the general trend for newer plants to be located close to consumers [25]. However, the construction of lengthy ammonia pipelines during the following decade enabled the Gulf producers to offer highly competitive prices. This was important to Central Farmers Fertilizer Company (CF Industries, from 1970), one of three interregional federations of farmers' cooperatives created between 1943 and 1946 as enhanced distribution networks with the encouragement of the TVA. Central Farmers, founded in 1946, held a 25% interest in the Cooperative Farm Chemicals Association, which commenced manufacture of ammonia in 1954 at Lawrence, Kansas. This was one of the first facilities built by Foster Wheeler using Ammonia Casale technology. The ammonia was converted into urea by the second Girdler urea unit to be installed in the US. (Foster Wheeler mainly promoted the Pechiney urea process.)

Manufacture of ammonia by Central Farmers began in 1966, with the installation of novel high-capacity ammonia units (see Chap. 11), at Donaldsonville, Louisiana, on the west side of the Mississippi River, around 80 km north of New Orleans. Soon after its opening, the Central Farmers facility was connected to the Gulf Central Pipeline, which distributed liquid, anhydrous ammonia directly to the Corn Belt. The fertilizer industry in Louisiana also served the needs of the sugar cane growing region, the Sugar Bowl, mainly in the southern part of the state. The first pipeline for delivery of liquid ammonia from Texas to Iowa, owned

[2] Foster Wheeler of New York, with main branches in London, Paris and St Catharines, Ontario, was created in 1927 by the merger of the Power Specialty Company (previously American Water Works Supply Company established by Pell and Ernst Foster in 1884), and Wheeler Condenser & Engineering Company (established by Frederick Merriam Wheeler in 1891). In 2014, Foster Wheeler merged with AMEC plc, and in 2017 Amec Foster Wheeler was acquired by Wood Group plc.

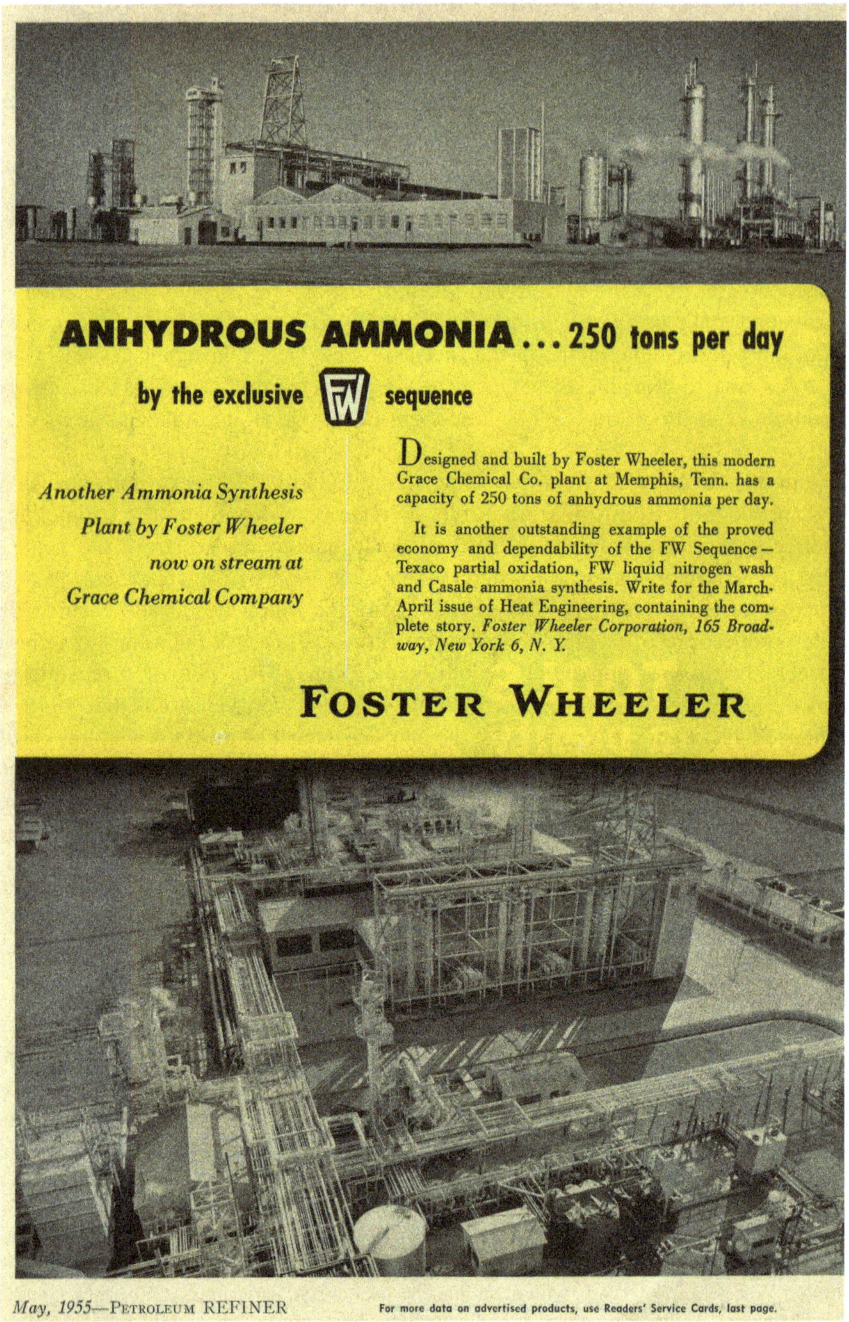

Fig. 9.4 Foster Wheeler's "FW Sequence" for manufacture of anhydrous ammonia, at the rated capacity of 250-tpd, was installed at the Chemical Division of W.R. Grace & Co., Memphis, Tennessee. (Foster Wheeler advertisement (1955) Petroleum Refiner 34:(May)) (Wood Group plc. With permission)

by the Mid-America Pipe Line Co., was laid down around this time; it delivered anhydrous ammonia to distribution terminals in Kansas and Nebraska.

Computer control of ammonia synthesis was used in the US from 1960 at Monsanto's facility, in Luling, Texas, and, around a year later, at Allied Chemical's South Point works, Ohio, at which time it was claimed to be the world's largest ammonia facility, with a rated capacity of 1000-tpd [26]. Significantly, in 1960, Chemico claimed that its 350-tpd ammonia converter installed for U.S. Phosphoric Products, at Tampa, Florida, was "the largest capacity converter in the world" [27].

Finally, in the early 1960s petroleum companies and their subsidiaries joined the trend that had started a decade earlier in which they moved downstream into synthetic ammonia, fertilizers, and other chemicals (Fig. 9.5). From 1960, Sun Oil's Marcus Hook refinery, Pennsylvania, produced 300 tons of ammonia daily as part of a joint project with Olin Mathieson to manufacture urea at North Claymont, Delaware; it was the first major urea plant on the east coast. This move was favored by the then important role of hydrocarbons, mainly natural gas, as feedstocks in the manufacture of hydrogen. In 1964, Gulf Oil Chemicals purchased Spencer Chemicals: "Petroleum concern is sixth in industry to move into mushrooming market," noted the *New York Times* when reporting the intended acquisition in September 1963 [28]. Foster Wheeler, the agent for Ammonia Casale, catered to the growing market and during 1964–1965, could claim in its advertisements that "More ammonia is produced throughout the world using the Casale process than any other" [29, 30]. That would change during the second half of the 1960s with the introduction of high capacity ammonia units (Chap. 11).

Fig. 9.5 For petrochemical manufacturing companies, "3 more ammonia plants by Foster Wheeler." (Foster Wheeler advertisement (1956) Petroleum Refiner 35(August):31) (Wood Group plc. With permission)

References

1. Cope WC (1949) Ammonia. Part I. US production facilities. Chemical Industries 64(6)(June):920–925
2. Thomson HC, Mayo L (1991, 1960) United States Army in World War II. The technical services. The ordnance department: procurement and supply. Center of Military History. United States Army, Washington DC, pp 109–110
3. Allen JB (1965) Ammonia manufacture. Chem Process Eng 46(Sept):473–483
4. Mitchell GS (1946) Ammonia. Its production from natural gas. Pet Refin 25(June):245–258
5. Reed RM, Eriksen A (1948) Hydrogen and synthesis gas production. Pet Process 3(Oct):1011–1016, on 1011
6. Hein LB (1952) Synthesis of ammonia at 350 atmospheres. Chem Eng Prog 48(Aug):412–418
7. Evenden M (2015) Allied power: mobilizing hydro-electricity during Canada's Second World War. University of Toronto Press, Toronto, pp 141–146
8. Kristensen TE (2024) New chemical technology at a time of national emergency: wartime mass production of "Super Explosives" RDX and MDX. Bull Hist Chem 49(2):139–166
9. Kastens ML, Dudley JF, Troeltzsch J (1950) Synthetic methanol production. In: Modern chemical processes: a series of articles describing chemical manufacturing plants. Reinhold, New York, pp 72–82, on 75
10. Martin PT (1950) Synthesis gas from coal. Chemical Industries 66(3)(Mar):365–372
11. Anon (1950) Spencer grows. Chemical Industries 66(5)(May):669
12. Anon (1950) Ammonia from natural gas. Chemical Industries 67(1)(July):4
13. Anon (1950) Methanol muddle. Chemical Industries 66(12)(Dec):873
14. Anon (1990) Cooperative agrichemical and seed operations: farmer cooperatives in the United States. Cooperative information report 1. Section 22 (1990). US Department of Agriculture. Agricultural Cooperative Service, Washington DC
15. Thompson HL, Guillaumeron P, Updegraff NC (1952) Ammonia synthesis at 1000 atmospheres: the present-day Claude process. Chem Eng Prog 48(9)(Sept):468–476
16. Shearon WH, Thompson HL (1952) Ammonia at 1000 atmospheres. Chem Ind Eng 44(Feb):254–264, on 260
17. Prichard JG (1998) Making things grow: the story of Mississippi Chemical Corporation. University Press of Mississippi, Jackson
18. Anon (1954) Starting up a new ammonia plant. Pet Refin 33(Dec):168–170
19. Anon (1957) Phillips gears operations for wide diversification. Pet Refin 37(Sept):194–196, 424–425
20. Strelzoff S (1956) The Fortier plant of American Cyanamid. Pet Refin 35(Mar):167–170
21. Davenport CH (1955) Boom still on for U.S. ammonia. Pet Refin 35(Apr):141–148
22. Martin WH (1959) Public policy and increased competition in the synthetic ammonia industry. Quart J Econ 73(3)(Aug):373–392
23. Foster Wheeler advertisement (1955) Chemical Week 26(Feb):82
24. Foster Wheeler advertisement (1960) Pet Refin 39(Oct):76–77
25. Messing RF, Perkins S (1957) Ammonia: future looks promising. Pet Process 14(Sept 1957):108–109
26. Gutzon J, Killick RD (1962) Computer control of ammonia production. Chem Eng Prog 43(July):324–329
27. Chemico advertisement (1960) Chemical and Process Engineering 41(Nov):530
28. Rutter R (1963) Gulf Oil seeking fertilizer maker; plans acquisition of Spencer Chemicals in a 120-million purchase of shares. 80% of stock sought. Petroleum concern is sixth in industry to move into mushrooming market. Gulf shares up 3/8. Target for rumors. New York Times, 11 Sept, p 57
29. Foster Wheeler advertisement (1964) Hydrocarb Process 43(Aug)
30. Foster Wheeler advertisement (1965) Hydrocarb Process 44(July):94

Methanol and Urea, Food Security, New Feedstocks, Steam Reforming, Fertilizers, Post-war Recovery, and Regional Expansion

10

Abstract

During the 1920s, high-pressure processes were applied to new uses and new needs. Synthetic methanol in particular drew directly on the technology of synthetic ammonia, as did to a large extent synthetic gasoline. These developments were driven by both industrial planning and export programs. However, there were anomalies in the ammonia industry. Thus whereas self-sufficiency in ammonia for nitrogen fertilizers was a political priority in Italy, elsewhere in Europe the leading manufacturers of ammonia from the late 1920s grappled with problems of overproduction, especially for export markets. The situation changed in the 1930s when ammonia for both fertilizers and national security increasingly dominated national planning. By the end of the decade, growth in both Europe and the US was driven by the threat of war. From 1945, growth was driven by food shortages. Around 1950, production of synthetic ammonia was still based on first generation technologies dating from the 1920s. The main process changes took place in production of hydrogen-containing synthesis gas, subsequent to the convergence of industrial hydrogenation and high-pressure chemistry; this encouraged ongoing development of methods for the generation of synthesis gas from hydrocarbons, by steam reforming, of both methane and naphtha, for use in the synthesis of ammonia.

10.1 Introduction

During the 1920s, high-pressure processes were applied to new uses and new needs. Synthetic methanol in particular drew directly on the technology of synthetic ammonia, as did to a large extent synthetic gasoline. These developments drove both self-sufficiency programs and exports. However, there were anomalies in the industry. Thus whereas self-sufficiency in ammonia for nitrogen fertilizers was a political priority in Italy, elsewhere in Europe the leading manufacturers

A. S. Travis, *The Global Synthetic Ammonia Industry*, Perspectives on the History of Chemistry, https://doi.org/10.1007/978-3-031-87978-4_10

of ammonia from the late 1920s grappled with problems of overproduction, especially for export markets. As for the US, it had hardly entered into the manufacture of synthetic ammonia. The situation changed in the 1930s when ammonia for both fertilizers and national security increasingly dominated national planning in most industrialized countries. By the end of the decade growth was driven in both Europe and the US by the threat of war. From 1945, growth was driven by food shortages. Around 1950, production of synthetic ammonia was still based on first generation technologies dating from the 1920s. The synthesis loop, or train, driven by reciprocating compressors, remained fundamentally unchanged. The main process changes took place in production of synthesis gas, subsequent to the convergence of industrial hydrogenation and high-pressure chemistry; this encouraged development of methods for the generation of synthesis gas from hydrocarbons, by steam reforming, of both methane and naphtha, in the synthesis of ammonia.

10.2 Synthetic Methanol and Urea

Developments in high-pressure industrial chemistry based on ammonia led in the 1920s to a process for synthesis of the important solvent methanol, which also found use in the manufacture of organic chemicals. Oxidation of methanol gave formaldehyde, which was required in the manufacture of the phenol–formaldehyde resin Bakelite. During the 1920s and 1930s, formaldehyde was reacted with urea to give other important resins. From the 1930s, methanol was used as a radiator antifreeze. Methanol was the second major application of high-pressure chemistry.

Previously, methanol had been obtained by the destructive distillation of wood. A method for producing synthetic methanol by the reaction between carbon monoxide and hydrogen in the presence of a catalyst had been suggested in 1905 by chemist Paul Sabatier, who earlier had achieved hydrogenation of fats. In 1914, Alwin Mittasch and Christian Schneider at BASF, drawing on the success of synthetic ammonia, attempted to obtain a single product from this reaction, under catalyzed high-pressure conditions, as suggested by studies of the thermodynamics and reaction kinetics.

$$CO + 2H_2 \leftrightharpoons CH_3OH \quad \Delta H = -91\,\mathrm{kJ\,mol^{-1}}$$

However, the BASF workers did not succeed. In 1922, Matthias Pier, also at BASF, took up the research and managed to produce the desired single product in a reaction that took place at 300 atm and at around 700 °C, using a chromium oxide-zinc oxide catalyst. Manufacture at BASF's Oppau site began in 1923. The conditions of the synthetic methanol process were so similar to those for synthetic ammonia that both products could be produced in the very same converter, according to demand, and with little modification. Methanol was made by adjusting the conditions for processing the synthesis gas (carbon monoxide and hydrogen).

Because synthetic methanol became available as the result of developments in high-pressure catalytic chemistry, similar processes, using the same basic chemistry, were worked in the US by Du Pont, at Belle, West Virginia, and Commercial Solvents Corporation, at Peoria, Illinois, from 1927, and in the UK at the ICI Billingham facility, from 1928 to 1929. The first production of methanol at Du Pont drew on a single synthesis gas stream that was passed through both methanol and ammonia converters [1]. For Montecatini, the academic chemist Giulio Natta at Milan (and later Pavia), undertook research from 1927 into the high-pressure synthesis of methanol and higher alcohols and the preparation of synthesis gas. Ammonia Casale's process was commercialized in 1928 and became available for licensing in 1929.

No less significant was the synthesis of urea (NH_2CONH_2), first synthesized by Friedrich Wöhler in 1828. The laboratory preparation from ammonium carbamate was reported in 1868 by the Russian chemist Alexander Basaroff, assistant of Hermann Kolbe at Leipzig. The carbamate was made by reacting ammonia with carbon dioxide. Around 1919, commercial manufacture, by a different route, acid hydrolysis of calcium cyanamide, was taken up by American Cyanamid, and the process remained in use until the 1940s. In 1922, BASF introduced a process based on Basaroff's work. It used the high-pressure reaction between liquid ammonia and liquid carbon dioxide, and heating the ammonium carbamate so formed under pressure. As with methanol, this attracted the interest of operators of high-pressure ammonia processes. In the early 1930s, Du Pont and ICI commercialized their ammonia-to-urea processes using the same high-pressure reaction between ammonia and carbon dioxide. Corrosion of equipment and the production of carbon dioxide of sufficient purity posed several challenges. The corrosion problem was originally overcome by using silver-lined equipment, and later by introducing special alloys. The theoretical basis of the urea reaction was published by M. Frejacques in 1948, and the mechanism of the commercial reaction was described by S. Kawasumi during 1951 and 1954.

Urea is especially important in the manufacture of novel plastics, notably those based on melamine. Melamine, a fine white crystalline powder, is a triazine (2,4,6-triamino-1,3,5-triazine), first isolated and named by Justus Liebig in 1834. Its resins are used for molding purposes, surface coatings, laminates, fire retardants, in wood-based panels, and as wood adhesives. Melamine is also used for strengthening of paper for maps, wallpapers, and banknotes, for bowls, plates, mugs and other utensils, for hard-wearing kitchen and shop counter tops, as, for example, decorative laminates, often called formica. The main urea-based melamine process was developed during the 1950s by Johnstone S. Mackay at Pittsburgh Chemical Company. In the more modern production process, urea is converted to melamine non-catalytically and at a high pressure. For this reason, and because a stream of ammonia-CO_2 off gases is reprocessed in the production of urea, a melamine plant is almost always placed adjacent to a urea plant. From the late 1940s urea was manufactured for use as a fertilizer and as an animal feed supplement.

10.3 Commercial and Political Aspects: Italy and Beyond

The 1920s and 1930s were a particularly tumultuous time in the political history of the world. Authoritarianism in its various manifestations was on the rise, notably in the weaker democracies of Europe. Chemical technologies were closely tied to political objectives. In some cases they were indispensable. In Germany instability during 1919–1923 had given way to a period of relative calm, known as the Golden Twenties, which came to an abrupt end in 1929 with the global economic crisis. This foreshadowed the rise to power of National Socialism. The outcome was that in the 1930s, IG Farben's expensive synthetic gasoline program took on a new relevance in the push for autarky. In the Soviet Union, the beginning of Five Year Plans in 1928 prioritized the development of heavy industry, including the production of chemicals for agriculture. This followed the introduction of ammonia technology from Ammonia Casale (Chap. 5). Ammonia and related technologies were imported from Italy, the US, and Germany.

In Italy the development of synthetic ammonia processes coincided almost exactly with the emergence of the authoritarian regime of Benito Mussolini. Strikes in the summer of 1920 precipitated political and economic volatility. In October 1922, Mussolini began his ascent to power. In 1925, his fascist dictatorship was established. Mussolini's government inherited an economy in a state of severe recession. The weak lira had increased the price of essential imports, including wheat and coal, as well as industrial machinery. Immediately, self-sufficiency programs were put in place. Leading chemists, including Giuseppe Bruni, were at the forefront of academic-industrial collaborations. Montecatini and other corporations were closely involved in the national planning. Working with the new regime provided opportunities for creating powerful industrial monopolies. This included in synthetic ammonia, as well as calcium cyanamide, in order to achieve agricultural self-sufficiency. This is why the Italian situation is of special interest here. Moreover, it was in many ways a forerunner of the Green Revolutions in Mexico during the 1940s and India during the 1960s [2, 3].

The introduction of protective tariffs on imported grain stimulated major expansion in agriculture, particularly of wheat. The outcome of this policy encouraged increased production of nitrogen fertilizers. For their purchase, the state provided farmers with generous subsidies. However, this did not favor expansion of Casale's ammonia process in Italy. As mentioned earlier, this was because Montecatini had become committed to the rival Fauser ammonia process and, through the corporation's control of the bulk of the nitrogen and fertilizer markets, restricted the application of the Casale ammonia process in Italy, in part by placing a cap on production [4].

Major state-sponsored infrastructure projects that favored Montecatini included electricity generation, such as the Marlengo hydroelectric facility that provided power to Montecatini's Merano synthetic ammonia factory. Merano, which relied on electrolytic hydrogen, was the largest ammonia factory in Italy. The fact that in the early 1930s six Italian factories used the Fauser process, two (small factories) the Claude process, and one the Casale process, made clear the extent to which

Montecatini controlled the nitrogen industry. Five of the six Fauser-Montecatini factories converted ammonia into the fertilizer ammonium sulfate. Later in the decade, Giamarco-Vetrocoke of Venice installed Casale ammonia units, with a total rated capacity of 100-mtpd.

As in Italy, from the mid-1920s growth elsewhere of the synthetic ammonia industry had been driven by the expectation that ammonium sulfate fertilizer made from synthetic ammonia would be widely adopted by farmers, and, in several countries, that ammonia would enable self-sufficiency in munitions production. There were also the export markets, especially the substantial Asian markets. Germany and the UK, effectively IG Farben and ICI, respectively, the two leading exporters of ammonium sulfate, competed and then collaborated in this business. At the end of the decade the great increase in annual production of synthetic ammonia, as taken up mainly in European countries, led to eventual overcapacity, and downward trends in the price of fertilizer. By 1930, outputs, and prices, were dictated by an international cartel, dominated by Germany and the UK. It was, however, in the authoritarian states, Germany, Italy, the Soviet Union, and Japan, that growth in manufacture and consumption of ammonium sulfate fertilizer from home-based synthetic ammonia took place. In these nations prior to the late 1930s ammonia production was strongly encouraged for its role in maintaining national security. Thus in the Soviet Union, the State Nitrogen Institute (GIA) was inaugurated in Moscow in 1931. It became a center for studies in high-pressure reactions and related chemical engineering developments. In 1943, it was renamed the State Research, Development and Design Institute on Nitrogen and Organic Chemicals (GIAP).

Because of the mandated, and expanding, home production and consumption of nitrogen fertilizers, the Italian synthetic ammonia industry suffered less than in most other European countries. From 1934, and especially after November 1935, when the League of Nations imposed sanctions on Italy in response to its invasion of Ethiopia, chemical autarky programs, notably in nitrogen products, were increasingly planned and executed by Montecatini.

10.4 Strategic Materials and War Planning

New directions and expansions in the chemical industry from 1930 were directed by political pressures that encouraged industrial planning for national defense, far more so than in the 1920s, and expansion in production of war related products. At Montecatini, Fauser, who in 1933 had established the firm's first central research laboratory, invented a process for the oxidation of ammonia into liquid nitrogen tetroxide, which was then converted into concentrated nitric acid, suited to the manufacture of explosives. This was developed by Bamag-Meguin and used in many countries. In 1933, the Terni company was taken over by the newly established state public holding agency, the Istituto per la Ricostruzione Industriale, and moved towards production of strategic materials. Two years later, Ammonia Casale became sole owner of the SIRI research facility at Terni.

In 1933, following the development in Germany of coal hydrogenation for the manufacture of synthetic fuels, Mussolini's government expressed interest in this branch of high-pressure chemistry. However, this was far less promising in Italy because of the lack of coal. Hydrogenation of imported crude oil began in 1938, under the auspices of the government's Azienda Nazional Idrogenazione Combustibili, or ANIC.

SIRI took a different approach. Uncertainty over the supply of hydrocarbon fuels required for internal combustion engines prompted the SIRI research laboratory to design, patent and put together an ammonia-powered motor car engine in 1935. This early design for the use of ammonia as a fuel was developed mainly by Mario Zavka, director of the chemical section from 1934 (Fig. 10.1). The ammonia was cracked into hydrogen and nitrogen by the heat of the exhaust gases. The carburreter was modified to suit the mixture of cracked gas and air. Road trials were made with Zavka's own Fiat 509 car, and then with the same engine installed in the newly introduced Fiat 527 Ardita 2500 truck, which was demonstrated before government officials. However, though a portent of clean fuels of current interest, the ammonia-powered engine was not a success. Later, the ammonia fuel was replaced by synthetic methanol [5]. Ammonia Casale's research into synthetic fuels led to the discovery of routes to higher alcohols and other useful products. SIRI was well equipped to undertake investigations into catalytic high-pressure hydrogenations. One process, developed in 1935, followed extensive scientific and technical research into partial hydrogenation of the hydrocarbon naphthalene to give the product tetralin (1,2,3,4-tetrahydronaphthalene), a colorless liquid, used as a fuel additive when mixed with benzene and alcohol, and as a solvent, and decahydronaphthalene, also a fuel additive. The process was developed for Rumianca. The tetralin process went on stream in 1937.

During 1934–1935, for the first time, Ammonia Casale revamped ammonia converters, an activity which was increased during the following years [6]. SIRI undertook experiments on the cracking of hydrocarbons, as well as the reforming of natural gas for production of hydrogen. Lubricants were also developed at Terni.

In the UK, as part of industrial mobilization, government ammonia plants managed by ICI were erected. Following concerns over the vulnerability of Billingham to bombing by the Luftwaffe, new ammonia and synthetic fuel (coal-to-oil) facilities were located well away from the northeast coast. The first ICI ammonia factory to be built for the Ministry of Supply opened at Mossend, in Lanarkshire, Scotland, in September 1939. In Lancashire, ICI jointly with Royal Dutch-Shell and Trinidad Leaseholders Ltd erected a large synthetic fuel factory at Heysham, near Morecambe (Heysham Aviation Fuel Works) for the Air Ministry. The Ministry of Supply funded the erection of a chemical plant at the site. It produced ammonia, nitric acid, and ammonium nitrate fertilizer from 1941, and later became an ICI ammonia facility. Ammonia was also produced at the government works operated by ICI at Dowlais, Merthyr Tydfil, south Wales, from 1940, and Prudhoe, Northumberland, from 1941. ICI's steam reforming process developed in the 1930s was from around 1940 used in Canada, the US, and Australia. The introduction of steam reforming and developments in ammonia catalysts at Billingham led in 1941

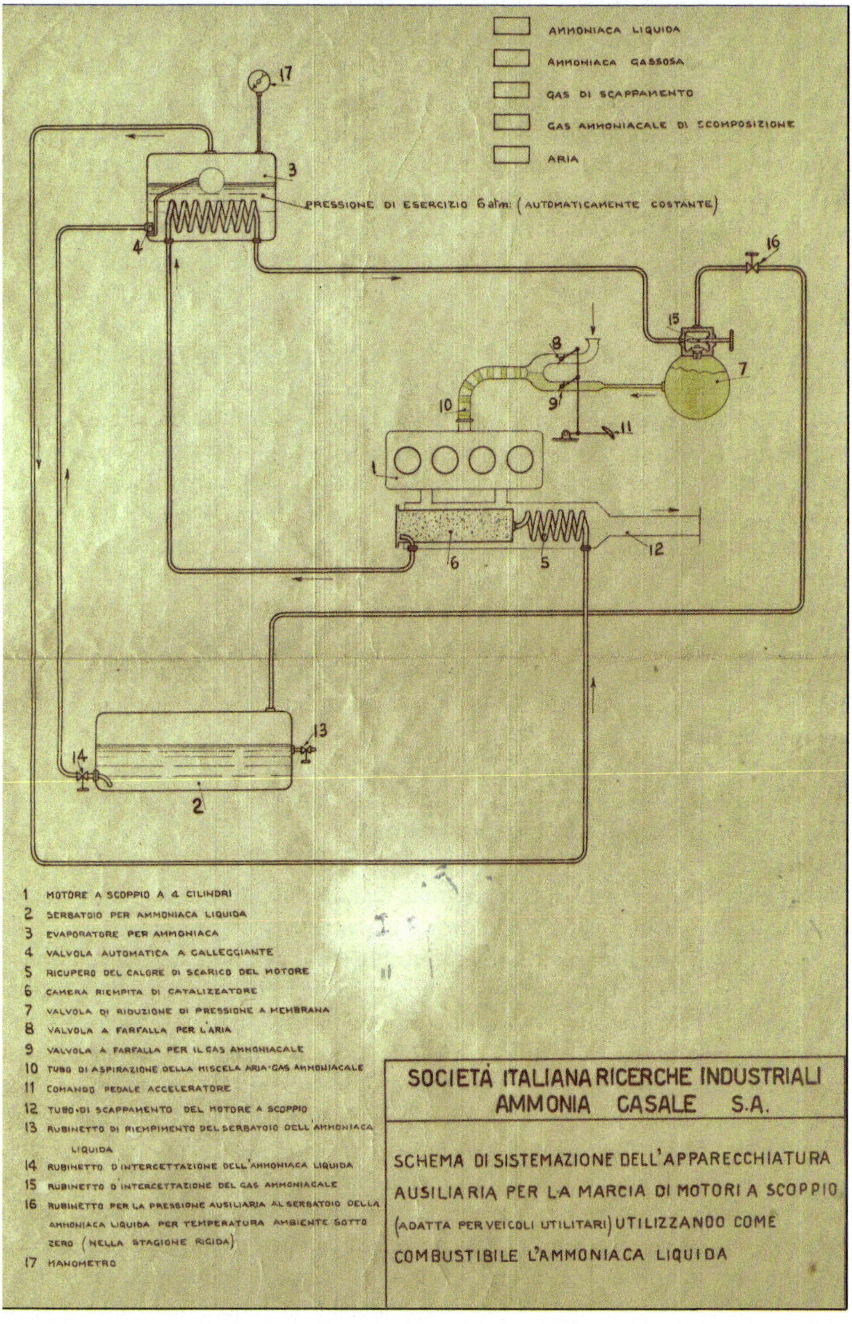

Fig. 10.1 Design for system of powering an internal combustion engine with liquid ammonia fuel, as developed at SIRI (Casale SA. With permission)

to the establishment of an ICI catalyst manufacturing unit located at Clitheroe, in Lancashire. There, nickel-based catalysts for steam reforming were developed.

In the Soviet Union, the development of high-pressure chemistry followed the introduction of Casale, Chemico, and Fauser ammonia technologies around 1930. After Germany attacked Russia in June 1941, great quantities of chemical manufacturing equipment were shipped east by train to the Urals. This included from the Lysychansk Nitrogen Fertilizer Plant in the Donbas, established in 1934 (today part of Sievierodentesk Azot Association/Ostchem group). At Metafrax Chemicals, of Gubakha, established in 1941, construction work was considerably held up when the staff were required elsewhere for assisting the war effort. The situation changed in 1945 with the arrival of trains loaded with coal-to-oil hydrogenation machinery, including converters and compressors, taken from Germany as reparations. This enabled Gubakha to commence regular production, particularly of methanol. Similar German hydrogenation machinery contributed to the startup of ammonia production in 1949 at Industrial Complex No. 18, the forerunner of Salavat-nefteorgsintez (today Gazprom neftakim Salavat), at the then new town of Salavat, on the Balaya (White) River, and the hub of the Volga-Urals oil field.

In Spain, where Sociedad Ibérica del Nitrógeno (SIN) had manufactured ammonia by the Claude process at La Felguera, Asturias, since the 1920s, and where the Casale process was worked in Aragon, production ceased during the years of the Civil War (1936–1939). Under the government of Francisco Franco, SIN built a new ammonia factory in the early 1940s (Nitratos de Castilla). The National Institute of Industry, from its foundation in 1942, participated in support of the ammonia industry.

10.5 Reforming of Hydrocarbons

During the 1930s, cracking, hydrogenation, and reforming were developed for the petroleum processing industry. Following the widespread introduction of cracking, in which long-chain hydrocarbons are broken down into smaller molecules, originally using heat alone, and from around 1940 in the presence of catalysts, high-pressure hydrogenation processes were applied to the conversion of heavy oils into useful fuels. In particular, hydrocracking of crude oil produced aviation and motor fuel; the products contained little or no unsaturation. From these developments, the technology of petroleum reforming developed rapidly.

In the US, Standard Oil Development worked on a process for producing aviation naphthas by high-pressure hydrogenation at around 1000 atm. The hydrogen was produced by the two-stage steam reforming of natural gas, not unlike the ICI process (see next section) [7]. By-product hydrogen from petroleum refining processes became available in the US in 1941. As with the use of natural gas, the feedstock at six of the US wartime ammonia factories, this had long-term implications for the nitrogen industry.

10.5.1 Steam Reforming: Hydrogen from Natural Gas

Processes for producing pure hydrogen are closely associated with a second phase in the development of the synthetic ammonia industry. This included the development of technologies based on coke oven gas during the second half of the 1920s, and studies on hydrocarbons in the 1930s. These fossil fuel processes, including the earlier water gas processes, involved production of synthesis gas, followed by the shift reaction for converting carbon monoxide to carbon dioxide, carbon dioxide removal, and final cleanup for removal of remaining carbon dioxide and carbon monoxide.

It was the emergence of steam reforming of hydrocarbons that drove the transition of the ammonia industry away from coal as feedstock to oil and hydrocarbon gases. Through several process improvements in steam reforming, methane became, and remains, the most important hydrocarbon used in the production of hydrogen.

From 1928, ICI at Billingham undertook studies on steam reforming of gaseous hydrocarbons. Special catalysts were developed to overcome poisoning with sulfur. A viable steam reforming process using off gases from coal hydrogenation was introduced at Billingham in 1936. A similar process was started up at Heysham in 1942.

Hydrogen production for the ammonia synthesis, as developed from the ICI process, involves two stages, primary reforming, which takes place in a furnace, and secondary reforming (Figs. 10.2 and 10.3). Typically, the primary reformer furnace incorporates vertically placed high-temperature alloy steel tubes containing the catalyst. The reaction is endothermic; heat was originally supplied by the external combustion of gas or a suitable fuel, and later by burners in the furnace. Steam and desulfurized natural gas pass down the tubes in which the reforming reaction, the breakdown of the natural gas, occurs. The furnace has a high-temperature refractory lining. Primary reforming furnaces became available in one of four configurations: side fired, e.g. Haldor Topsøe; down or top fired, e.g. ICI and Uhde; bottom- or up fired; and inclined terrace wall fired, as introduced by Foster Wheeler (Fig. 11.5).

The gas mixture, often called "process gas," is then passed to the secondary reformer, where it enters near the top at around 1000 °C. Air is piped in, also near the top; in this way the secondary reformer introduces nitrogen from the atmosphere for the ammonia synthesis. This is an autothermic reformer in which the heat required to maintain the reaction is derived from the exothermic reaction itself, through heat exchange. In part because it is not a furnace, the secondary reformer is of simpler construction than the primary reformer. A catalyst bed is held in a refractory-lined chamber. Mixing and combustion take place above the catalyst bed. Under the air-blown conditions in the presence of an extremely active nickel catalyst the remaining hydrocarbon from the primary reformer is now oxidized. The reformed gas mixture is cooled through a waste heat boiler, and then subjected to the shift reaction. Originally, this was followed by removal of the carbon dioxide by water scrubbing, and then of residual carbon monoxide by

Fig. 10.2 Primary and secondary reforming of natural gas as carried out from 1942–1943 at the Ozark Ordnance Works, Eldorado, Arkansas. *Top left* shows primary reformers and exterior of primary reformer buildings, where steam was raised. *Top right* is the interior of a primary reformer building. The middle row of photographs shows, *from left to right*, an air preheater, a secondary reformer, and the carbon monoxide (shift) converter. The secondary reformer is an unfired vessel. The bottom row shows a waste heat boiler, gas and steam interchangers, and the control panel board. The latter includes instruments for measuring the rate of flow of gases, including in the air line which feeds air into the secondary reformer, and the temperature. After World War II, the facility was leased to the Lion Oil Company. (From Mitchell GS (1946) Ammonia. Its production from natural gas. Petroleum Refiner 25(June):245–258/97–110, on p. 248/100) (Hydrocarbon Processing. With permission)

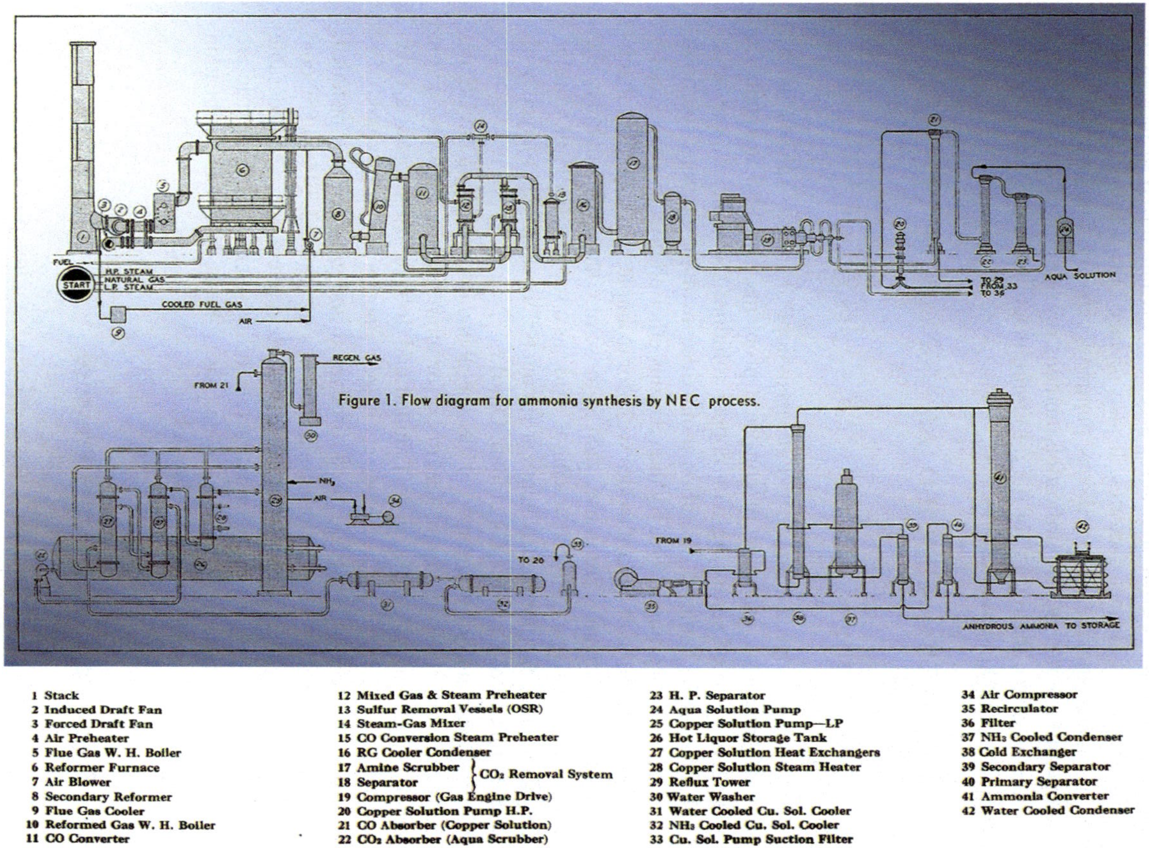

Fig. 10.3 Chemico (NEC) process for synthesis of ammonia based on two-step reforming of natural gas. The reforming section is at upper left. (From Anon (1953) Ammonia and methanol. Petroleum Refiner 32(9):186 (Hydrocarbon Processing. With permission)

scrubbing in copper liquor. The reforming was carried out at atmospheric pressure [8].

New processes for removal of carbon oxides, prior to the synthesis of ammonia, replaced the copper liquor scrubbing. Purification of synthesis gas by scrubbing with monoethanolamine (MEA) (Girbotol process) was introduced in the US in 1943 at the wartime ordnance factories. It also replaced the water scrubbing for removal of carbon dioxide, required less pumping capacity, and operated at a lower temperature. Other novel purification methods for removal of carbon oxides included methanation, the reverse of the reforming reaction, which takes place in the presence of a nickel-based catalyst.

$$CO + 3H_2 \leftrightharpoons CH_4 + H_2O \quad \Delta H = -206.1\,\text{kJ mol}^{-1}$$

$$CO_2 + 4H_2 \leftrightharpoons CH_4 + 2H_2O \quad \Delta H = -164.9\,\text{kJ mol}^{-1}$$

Methanation, introduced around 1950 at Mississippi Chemical, Yahoo City, by the Girdler Corporation, was free of corrosion problems, and also maintenance free, and again replaced scrubbing with copper solutions. Nonetheless, the original BASF process for removal of carbon monoxide, involving cuprous solutions, continued in use until around 1970. By then various other proprietary processes for more efficient absorption of carbon oxides with, for example, a hot solution of activated potassium carbonate, were in use, such as those of Giamarco-Vetrocoke, and Benfield. Vetrocoke also introduced arsenious oxide as a sorbent. (In 1959, Girdler, then the Chemetron Corporation, Chemical Products Division, was merged with Chemical & Industrial Corp., of Cincinnati, Ohio).

After 1945, in countries where natural gas was abundant, notably in the US and Italy, it became the most important feedstock for ammonia. In 1959 production of ammonia from natural gas began in Trinidad and Tobago. This was at a facility owned by W.R. Grace & Co. However, it took another eighteen years, after new deposits off the east coast had been discovered and gas pipelines installed, before other firms began to set up ammonia facilities in Trinidad. Elsewhere, coal based processes remained dominant. In Germany, where there was a considerable demand for methane for use in the manufacture of acetylene by the electric arc process, it was available from the coking ovens gases. In the UK coke gasification for methane made it expensive as a feedstock.

10.5.2 Steam Reforming Under Pressure

Until the early 1950s steam reforming was conducted at a low pressure. A significant advance was achieved in 1953, when reforming was carried out at around 30 atmospheres. To achieve this, the engineering contractor M.W. Kellogg took advantage of the pressure at which the natural gas was available from deep wells. This was used at the 165-tpd synthetic ammonia facility of Shell at Ventura, California (1952–1953), and the 300-tpd synthetic ammonia facility of Standard Oil Company of Ohio, near Lima, Ohio. The higher compression employed in reforming in the 1950s meant that less compression was required in reaching the ammonia synthesis pressure. The downside, at first, was that pressure-resistant equipment was required, as well as a higher operating temperature. However, and not insignificant, was the fact that by the 1950s, the overall cost of the four to six catalysts required in the total ammonia synthesis, including for the production of hydrogen by steam reforming, had fallen considerably [9]. Moreover, several compression stages had been eliminated by this time. Compression was required only prior to steam reforming and prior to synthesis. Steam reforming under pressure was undoubtedly the main innovation in ammonia synthesis in the 1950s. Long term it would have a major impact on the highly energy intensive ammonia industry.

10.5.3 Reforming of Naphtha

ICI continued to play a leading role in large-scale steam reforming even though natural gas was not available in the UK. In 1938, just two years after achieving steam reforming of off gases from coal hydrogenation, ICI established that reforming of sulfur-free liquid hydrocarbon distillates was viable. Reforming of naphtha, a mixture of light distillate hydrocarbons, was investigated, and ICI invented a synthesis gas process involving steam reforming of liquid hydrocarbons. Compared with the coal-based process there was a substantial reduction in the emission of noxious gases.

In 1954, the ICI process was introduced at its Heysham ammonia facility. Three years later, the first reforming unit incorporating a top fired furnace was adapted to a low-sulfur naphtha, which was readily available, and inexpensive. The gas stream, following the shift reaction and carbon dioxide removal, gave almost pure hydrogen. The ICI development team was headed by T. J. J. Pearce. In 1959, using ICI technology, M.W. Kellogg completed a second unit for the reforming of naphtha at Heysham [10].

The next step at ICI involved pressurized steam reforming of the naphtha. This process was introduced on a commercial scale during 1959–1960 at Heysham, in 1962 at Severnside, and in 1963 at Billingham. This was the first large-scale use of pressurized reforming of liquid hydrocarbons. Special catalysts were developed to overcome poisoning with sulfur compounds, and the buildup of carbon in the primary reformer tubes. The tubes, as many as 240, placed in four rows, were made from newly introduced nickel-chrome steels, enabling the use of a higher temperature. The primary reforming took place in the presence of superheated steam, with a nickel catalyst supported on an inert material, at around 700–800 °C. An improved secondary reforming stage was developed for the oxidation of unconverted naphtha [11, 12]. As in earlier secondary reformers, the burning of air provided nitrogen for the ammonia synthesis. The gas mixture leaving the secondary reformer was cooled by passage through a heat exchanger, and then under the conditions of the shift reaction more hydrogen, and carbon dioxide, were formed.

The nitrogen–hydrogen composition was now made up in the proportions required for the ammonia synthesis. Following removal of carbon oxides, and compression of the mixture, it was fed into the converter. Recovered carbon dioxide was used in the manufacture of urea by reaction with ammonia, or converted to dry ice, or sold to the beverage industry (and later also to the meat industry and food processors). Because the synthesis of ammonia from natural gas is accompanied with production of carbon dioxide, it was logical for larger ammonia plants to be matched with urea plants. Likewise, as already mentioned, urea units were matched with melamine production units.

By the 1960s, increases in the steam reforming pressure, mainly as a result of catalyst developments, enabled use of fewer tubes, which were fabricated from expensive alloys, and smaller furnaces. The higher pressures facilitated heat recovery for use in other sections of the ammonia unit [13]. ICI, the largest producer of ammonia, operated five ammonia facilities: Billingham, in County Durham; Heysham; Wilton (opened in 1949; ammonia production began in 1952); Severnside (1963); and Immingham (1966).

Steam reforming of naphtha for the production of synthetic ammonia became important in countries that lacked coal and natural gas. By the end of the 1960s, the ICI process, conducted at 30 atm, had been licensed to six engineering contractors (Humphreys & Glasgow; M.W. Kellogg; Power-Gas Corporation; Selas Corporation of America; Friedrich Uhde; and Foster Wheeler) [14, 15]. Humphreys & Glasgow, founded in 1892, was appointed joint licensor worldwide for ICI's steam reforming technology when used in the manufacture of town gas and hydrogen, and in most countries for various syntheses [16]. In 1965, drawing on naphtha as feedstock, the Fluor Corporation of Los Angeles, California, began work on the design of two identical fertilizer plants for South Korea (the Republic of Korea), incorporating 310-tpd anhydrous ammonia units.

By 1960, over 80% of ammonia produced in the US was already made from natural gas, mainly by two-stage reforming. In Italy, as another example, by the early 1970s, almost 70% of hydrogen was prepared by the reforming of gas fractions produced from distilled crude oil or from natural gas. Half was consumed in the synthesis of ammonia [17].

Reforming of North Sea natural gas at Billingham from the 1970s brought to an end the use of both coal and naphtha as feedstocks in the UK. Usefully, the naphtha two-stage steam reforming process had also been designed for use with methane as feedstock. It is not without interest that in the 1970s during a period of natural gas shortages in the US considerable emphasis was placed on a long term move to naphtha as feedstock for hydrogen [18].

Preliminary purification of hydrocarbon feedstock is essential prior to reforming. In the case of natural gas, it is first compressed and preheated, and the catalyst poison sulfur is reacted with hydrogen over a suitable catalyst. The resulting hydrogen sulfide is then reacted with zinc oxide, as absorbent, to give zinc sulfide, which is readily removed [19].

As a result of these novel technologies, a typical ammonia process based on natural gas involves five steps, of which four involve preparation of gases: two-step reforming of hydrocarbons, for hydrogen; the shift reaction, for more hydrogen and converting carbon monoxide to carbon dioxide; carbon dioxide removal with MEA; and methanation of remaining carbon oxides. The fifth step is the ammonia synthesis.

10.5.4 Partial Oxidation: The Texaco Process

Another route to hydrogen from hydrocarbon sources involved partial oxidation (combustion), especially of heavy fuel oil. This non-catalytic process was investigated at BASF during 1936–1939. It was perfected in the US by the development division of Texaco (Texas Oil Company) and Hydrocarbon Research for heavy hydrocarbons and natural gas, by 1953, and by Shell Development for natural gas in 1956. The flexibility in the source of hydrocarbon, liquid, gas, or solid, was an important feature [20]. Nitrogen, as a separate input, was produced economically by burning air in hydrogen in a special furnace, or by cryogenic air separation.

The process takes place at 1200–1500 °C, considerably higher than used in steam reforming. A hydrocarbon-air mixture, enriched with oxygen, if required, is made up. (The oxygen is obtained from an air-liquefaction unit, which also provides nitrogen.) It is the lack of oxygen in this process which prevents formation of carbon dioxide. The reaction is complex, but can be represented by:

$$CH_4 + {}^1\!/_2O_2 \rightleftharpoons CO + 2H_2 \quad \Delta H = -36\,\text{kJ mol}^{-1}$$

Under the usual shift conditions, the monoxide is then reacted with steam.

The process incorporates a partial oxidation reactor, a shift reactor, and equipment for purification of hydrogen. It does not require a heat exchanger. As a result the plant, which can achieve up to 80% efficiency, is more compact than that required in steam reforming.

10.6 Integrating Texaco, Casale, and Pechiney Processes for Urea

From around 1950, the American engineering contractor Foster Wheeler was the main agent for Ammonia Casale outside Italy. During the 1950s, Foster Wheeler promoted the Texaco process in combination with the Casale process for ammonia, and the French Pechiney process for converting ammonia into urea, as modified at the engineering firm W.R. Grace & Co. (Figs. 10.4 and 10.5). In the ammonia section, nitrogen was added to hydrogen obtained by the Texaco partial oxidation process and the gaseous mixture was then directly introduced via the ejector into the Casale converter, which was still operated at around 800 atmospheres. Process advantages, according to Foster Wheeler, included: the use of the ejector; a "simplified heat exchanger built integrally in the reaction chamber, [which] eliminates outside heat exchanger and simplifies piping"; the high operating pressure which "permits production of liquid anhydrous ammonia at normal cooling water temperature"; and the "great tolerance for carbon monoxide and oxygen which may be contaminants in the synthesis gas." By 1955, the Casale ammonia process was "now used in more than 40 plants with four more being constructed [by Foster Wheeler]," while the "Pechiney process has been designated for two plants under construction" [21].

Of fifteen Casale ammonia plants installed by Foster Wheeler until 1960, nine used the Texaco process for production of hydrogen. (Foster Wheeler represented Ammonia Casale abroad until 1970).

The Texaco partial oxidation process was taken up at ICI Billingham, but its introduction in 1959 was delayed as the result of a serious explosion at the oxygen plant [22]. This problem occurred elsewhere, and was the main disadvantage of the process. At Billingham heavy fuel oil and later naphtha were the feedstocks. By the time that the process at Billingham had been improved, it was about to be superseded by pressurized steam reforming of naphtha. However, as a low cost route to pure synthesis gas, partial oxidation technology was adopted in a number of countries, and on a very large scale in India and China.

Improvements to the partial oxidation process included application of pressure to increase the yield of hydrogen. However, in the end, partial oxidation proved

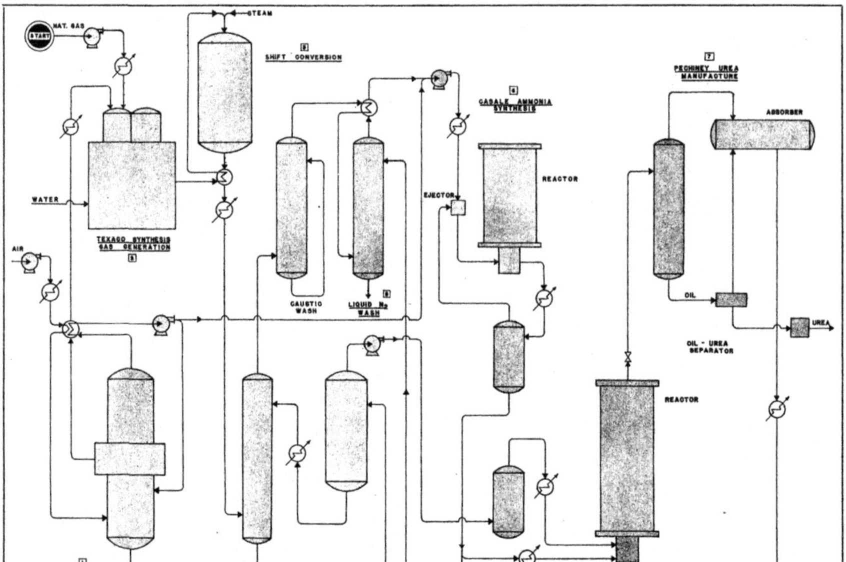

Fig. 10.4 Flowchart for synthesis of urea by the integrated Texaco partial oxidation, Casale ammonia, and Pechiney urea processes. (From Anon (1953) Ammonia and urea. Petroleum Refiner 32(September):178–182, on 178) (Hydrocarbon Processing. With permission)

Fig. 10.5 Foster Wheeler advertisement, 1954, featuring the ammonia (Casale)-urea (Pechiney) facility of the Grand River Chemical Division of Deere & Company, Pryor, Oklahoma. The seven-step manufacture of urea involved: partial oxidation of methane to produce hydrogen by the Texaco process; synthesis of ammonia by the Casale process; and conversion of ammonia into urea by the modified Pechiney process. The ammonia unit was scheduled for completion in July 1954 and the urea unit in September 1954. (Petroleum Refiner 33(December):insert) (Wood Group plc. With permission)

to be no match for steam reforming under pressure: “At one time, it appeared that partial oxidation would supersede steam reforming, but further advancements in the technology, i.e., the advent of steam reforming under pressure, restored steam reforming to its dominant position” [23]. Thus, for example, in 1963 steam reforming, rather than partial oxidation, was adapted to an ammonia unit based on Casale technology, with a rated daily capacity of 400 tons, for Nitrin, Inc., near Cordova, Illinois. The site, owned by Northern Natural Gas Company and International Minerals & Chemical Corporation, also produced ammonium nitrate and urea.

10.7 Electrolysis

In the 1970s, production of hydrogen by electrolysis still saw limited application, mainly at the two ammonia facilities of Norsk Hydro, which had operated ammonia units since 1928. Even then, half the hydrogen required was produced from natural gas. In the late 1970s, Nangal, in the Punjab, India, changed over from electrolytic hydrogen to steam reforming of naphtha. The only other important location where hydrogen for ammonia relied on electrolyzers was at Kima-Aswan of Egyptian Chemical Industries (KIMA), in Egypt. KIMA was established in 1956, and opened the Kima-Aswan works in 1961, utilizing electrolytic hydrogen until 2019. The Hûngnam fertilizer factory in North Korea, which had used the Casale ammonia process, ceased electrolytic production of hydrogen following its destruction during the Korean War. It was subsequently rebuilt with assistance from the German Democratic Republic, and in the 1970s was upgraded with help from China, partly using lignite as feedstock. There were three other, though quite small, electrolytic facilities, one at Reykjavik, Iceland, one in Zimbabwe, and another at Cusco, Peru. In 2021, only the Cachimayo Nitrogen Fertilizer Factory (now Enaex) at Cusco, which opened in 1962, remained committed to electrolytic hydrogen [24]. Today, the revival of electrolytic hydrogen in order to reduce consumption of fossil fuels is a leading research topic (see Chap. 14).

10.8 The Fertilizers

In the 1930s, particularly in Europe, there was a gradual changeover from ammonium sulfate to ammonium nitrate as the main fertilizer, in part because of the higher nitrogen content of the latter (33%). This favored production of synthetic ammonia, at the expense of calcium cyanamide and coking oven ammonia products, which were not readily converted into nitrate. France is a good example of this trend, enabled considerably by the manufacture of synthetic ammonia at the Toulouse government factory and in the private sector.

Extensive fertilizer trials in farmers’ fields supplemented those of the well-established agriculture research stations. The impact of nitrogen on the rice crop in different agro-climatic regions in Asia received special attention [25]. In high

Fig. 10.6 Foster Wheeler 1959 advertisement for its modified Pechiney urea process developed in collaboration with W.R. Grace & Co. At left is the prilling tower where the urea melt turns to droplets that fall, cool down, and solidify into a granular (prilled) material in a countercurrent flow of air. This is necessary for the production of a useful fertilizer. (Petroleum Refiner 58(July):33. With permission.)

income countries between 1949 and the end of the 1960s nitrogen consumption more than trebled. "Spectacular increases" were registered by the Food and Agriculture Organization of the United Nations in the Soviet Union, the US, and Canada. "Chemical fertilizers…have also given a significant advantage to the large farm," because they could be "transported anywhere and…spread by machines so that crops grown under prairie conditions receive supplementary plant nutrients" [26].

These events were paralleled by changes in the scale of production of ammonia and the distribution of nitrogen fertilizers, particularly in North America after 1945, where bulk-blending, marketing and distribution was increasingly carried out by producers and large cooperatives. There was a surge of activity as farmers undertook spring planting in the month or so prior to the end of the fertilizer year (30 June). This imposed pressure on railway delivery services, and sometimes led to a shortage of hopper car space.

Three products dominate the modern ammonia-based fertilizer industry: granular ammonium nitrate (with names such as Nitrochalk, Nitrolime, and ICI's Nitram), urea, and liquid anhydrous ammonia ("Direct N"). Ammonium nitrate is extremely reactive with combustible materials, and has the potential to become explosive. At the end of the 1940s, Leonard A. Stengel at Commercial Solvents developed a safe (and a less costly) manufacturing process in which the vapor phase reaction between nitric acid and ammonia takes place in a stainless steel reactor. The dry ammonium nitrate is sprayed in a tall tower and concentrated. This creates small lumps, called prills. Prilled ammonium nitrate, known as High Density Ammonium Nitrate (HDAN), in the form of granules (as well as crystals) is used as fertilizer. The solid Low-Density Ammonium Nitrate (LDAN) is used in the manufacture of explosives, usually for the mining industry.

During the 1950s and 1960s, the white, water soluble urea (NH_2CONH_2), with its high nitrogen content, became an increasingly important fertilizer. It was manufactured for this purpose in the US by Du Pont and the Solvay Process Division of Allied Chemical [27]. Manufacturing processes were also introduced by Chemico, Inventa (Switzerland), Lonza (Switzerland), Montecatini, Pechiney, Snamprogetti (Italy), Stamicarbon (Holland), and Toyo Engineering (Japan) (Fig. 10.6). Urea's use as a fertilizer, particularly in rice growing regions in hot climates, relied on the favorable economics of production and ease of transport and use. Urea is stable and easy to transport, neutral, and well suited to most soils and plants. Thus it was found that the use of urea on the alluvial soils in southern India was 45% more efficient than application of ammonium sulfate [28]. During the 1960s, China, India, Pakistan and Indonesia, major cultivators of tropical rice, moved increasingly towards urea as fertilizer. More recently, urea has been introduced in cooler climates for cereals.

The application of anhydrous (liquid) ammonia as fertilizer was important in North America from around 1950. However its introduction elsewhere, including in the UK during the 1960s, was a mixed success. The use of liquid ammonia declined in the 1990s, and has been replaced by milky-looking urea-ammonium nitrate fluids. Nitrogen products are used in compound fertilizers, originally developed in the late 1920s, in admixture with compounds of phosphorus and potassium (NPK fertilizers). For example, a 10–10–20 NPK fertilizer consists of 10% nitrogen ("0.5% nitric nitrogen" and "9.5% ammoniacal nitrogen"); 10% phosphorus; and 20% potassium.

During and after the 1950s, several governments provided subsidies to farmers (and occasionally to manufacturers), in order to encourage use of nitrogen fertilizers and other agricultural chemicals. In the European Union excessive agricultural subsidies until the 1990s led to massive overuse of nitrogen fertilizer. Leakages arising from this situation and inappropriate agronomic practices elsewhere caused water borne "reactive" nitrogen to damage aquatic systems leading to eutrophication of lakes and coastal waters. From the late 1980s, efforts were made to use nitrogen fertilizers more efficiently through surveys based on satellite remote sensing, in order to improve application, including its timing.

10.9 Post-war Recovery and Fertilizers

After 1945, heavily damaged ammonia factories on mainland Europe were restored and brought back into production to help alleviate the severe food crisis following the ravages of World War II. This at first did not include Germany, where the western Allies intended to prevent the resurgence of its chemical industry, particularly of those sectors connected with production of armaments, including ammonia, which was one of the "prohibited" industries. In 1946, if not earlier, this did not prevent the IG Farben Oppau works, in contravention of the orders of the occupying military authorities, from starting up its ammonia units for the production of fertilizer. The US Civilian Production Administration was in any case by then contemplating the reopening of German nitrogen and fertilizer factories. This conflicted with the deliberate removal of German chemical and other machinery as restitution for use in former non-enemy countries. In 1949, "organized resistance," and ugly confrontations, followed attempts to dismantle chemical plant at Fischer–Tropsch works in the Ruhr and elsewhere. The Germans argued that the plants were for peace-time rather than war purposes [29]. The same difficulties applied in Japan. At the request of the United States Department of Defence, Frederick Pope, recently retired president of Chemico, drew up recommendations for revival of the chemical industries in both countries, following a visit to Germany in late 1948, and to Japan early in 1949.

The post-war reconstruction of the strategically important ammonia and other chemical factories that had been destroyed or damaged by bombing, created plenty of urgent work for engineering contractors, their subcontractors and manufacturers of machinery. Allied corporations had access to the latest developments in German chemical technology through commissions of enquiry, notably the US Field Intelligence Agency, Technical (FIAT), and the British Intelligence Objectives Sub-committee (BIOS). Both included civilian consultants, many connected with leading industrial firms, chemical and otherwise. Of particular interest were

German developments in the manufacture of synthetic gasoline by the Bergius and Fisher-Tropsch processes. In the ammonia industry, the investigators were impressed by developments in the continuous Winkler gas generator process.

As was the case after 1918, synthetic ammonia was required for fertilizers to make up for wartime depletion of soil, and to enable much needed food production. Nitrogen fertilizer shortages were severe in Japan, Korea, other parts of Asia, and in Africa. Drought in Europe followed by a difficult winter in 1946–1947, led the US to send 16.5 million tons of food, mainly wheat, to Europe and Japan as emergency relief. The shortages of nitrogen fertilizers globally, and the threat of starvation, brought about the rebuilding of the ammonia factories in both Germany and Japan.

European national targets for recovery included: the French Monnet plan (1946), for modernization of the nation's agricultural production, with more than doubling of fertilizer use compared with prewar levels; and the UK plan (1947), which was aimed at increasing agricultural production by 50% above the prewar level. In both countries there was a steep rise in wheat yields during the 1950s. In contrast, political difficulties in Italy caused the wheat yield to remain almost static until the early 1960s [30]. The UK was the largest recipient of the American Marshall Plan, aimed at the rebuilding of European economies, which was implemented from June 1947 and lasted until 1951. The plan was rejected by the Eastern Bloc states, the outcome of which was its contribution to the emerging Cold War. This was one reason why after 1947 the Western zone of occupied Germany was allowed to repair and restart war damaged factories, work that included upgrading of ammonia plants. The Federal Republic of Germany was created in May 1949.

New and restored factories for nitrogen products as fertilizers were prominent features of national self-sufficiency plans, in addition to preventing hunger. In Norway, production at Norsk Hydro of calcium nitrate and ammonium nitrate, equivalent to 93,000 metric tons of nitrogen annually, was restored in 1948. Away from Europe, anti-colonial and nationalist movements were reshaping large tracts of the globe. The UK partitioned India in 1947, Israel achieved independence in 1948, the People's Republic of China was declared by Mao Zedong in October 1949, and armed conflict began on the Korean Peninsula in 1950. The demand for US ammonia was not only driven by the food needs of war ravaged countries but also from 1950 by renewed production of munitions.

10.10 Ammonia Casale in the 1950s

From the mid-1930s, Ammonia Casale had fallen behind. The research at Terni into Ersatz materials, the dictates of wartime, in particular the break with clients in the Allied nations, the occupation by the German Army from 1943, and the struggle to recover after 1945 had left little time for ammonia research. However, the tremendous post-war need for fertilizers and the simplicity and ready availability of its technology created opportunities for Ammonia Casale to achieve some success in reviving its former worldwide activities in synthetic ammonia. This included, around 1950, the design and construction of a 40-mtpd ammonia plant, with gas preparation based on charcoal, for Fabricaciones Militares Sociedad del Estado, at Rio Tercero in Argentina. This organization, founded in 1941 for manufacture of military equipment, became widely diversified including into railway rolling stock, and, with the ammonia unit, into fertilizers. The Rio Tercero facility is of note because it was the last one to be designed and engineered by Ammonia Casale prior to its changing hands in the 1970s [31]. Thereafter Ammonia Casale became mainly engaged in licensing its technology, notably in Italy to Snamprogetti, the contracting arm of the state-owned ENI (Ente Nazionale Idrocarburi).[1] Elsewhere, Ammonia Casale technology was, as already mentioned, licensed through Foster Wheeler, which in the mid-1950s erected a 100-tpd anhydrous ammonia unit in Brazil for Cia Braziliera des Estirinio Petrobas, at Cubatao, and in the early 1960s an ammonia unit at the PUSRI Palembang (PT Pupuk Sriwidjaja Palembang) facility in South Sumatra, Indonesia. PUSRI, or Persero, was founded in 1959 after the government of Indonesia drew up plans for home-based fertilizer manufacture in order to reach self-sufficiency in grain production. The ammonia unit began operating in 1963. The facility later also manufactured urea.

There were a few exceptions to the licensing arrangements made between Ammonia Casale and Foster Wheeler. Around 1960, Uhde and Casale were taken on as joint contractors for a 300-mtpd ammonia facility built for the Greek government at Ptolemaeus, and scheduled to be completed in 1963. In 1966, Casale ammonia technology enabled production of 450-mtpd at the fertilizer complex of Gujarat State Fertilizers at Bajwa, near Baroda, India. In Italy, Rumianca installed Ammonia Casale units for production of fertilizer. The feedstock, natural gas, was processed in the single-stage Shapleigh reformer unit licensed from Hercules Powder, and erected by Société Chimique de la Grande Paroisse (Grande Paroisse).

Notwithstanding certain modifications and process improvements—including at Ammonia Casale, and Air Liquide (owner of the Claude process, and associate of Grande Paroisse), and the important transition from coal and coke processes to liquid hydrocarbons and natural gas for hydrogen—there had been few other notable advances in ammonia technologies since around 1930. Moreover, until the late 1950s, in contrast to the US, coke oven gas was still widely used as the source of hydrogen in France, Belgium and Germany [32].

There was little incentive for change. One reason was that ammonia synthesis was considered a mature technology. More exciting developments were taking place in polymers and resins. Another reason was the wartime expansion, especially in the US, which led to a doubling of ammonia capacity. In 1949, the total annual capacity of the ten US government-backed plants was 529,000 tons, and of the prewar commercial plants 552,000 tons.

Ammonia Casale continued to promote its existing processes without facing a great deal of competition, and benefited from the Italian economic miracle of

[1] ENI from 1953 was parent company of AGIP, Azienda Generale Italiana Petroli, founded in 1926.

the 1950s and 1960s. During the period 1948–1958 the Italian growth rate per annum in industrial production was second only to Germany, and during 1958 to 1972 was almost equal to the first place, the Netherlands. Snamprogetti used the Casale ammonia process from the late 1950s at the new government owned fertilizer plant of ANIC, near Ravenna (ANIC Ravenna). The rated annual capacity was 165,000 metric tons of equivalent nitrogen. The hydrogen, as elsewhere in Italy, was obtained from natural gas. The main products were ammonium sulfate and prilled ammonium nitrate [33]. Casale technology was later used at the fertilizer factory of Chimica Meridionale, at Potenza, in southern Italy, with a capacity of 150-mtpd. ENI's involvement with ammonia using Casale technology contributed towards the corporation's break up of Montecatini's long-held monopoly on nitrogen fertilizers.

One notable innovation at Montecatini was its total-recycle process for urea, introduced during 1954–1958. In this process, recycle of unreacted ammonia and carbon dioxide as a water solution was achieved through a multi-stage system operating at different pressures. Montecatini overcame equipment corrosion problems through maintaining a high concentration of ammonia in the reaction, and employing relatively mild conditions. Total recycle processes introduced by other firms used a common process principle.

They included Snamprogetti, which in 1966 designed a 300-mtpd ammonia and urea facility for ANIC Gela, at Gela, in Sicily. The urea unit was based on a new energy-saving process, carried out under pressure at 100–150 atm. Subcontractors at Gela included Foster Wheeler, Air Liquide, Stone & Webster, and M.W. Kellogg. The ENI subsidiary Nuovo Pignone (until 1954 Pignone), since the 1930s a specialist in plant machinery, supplied most process equipment. Bellini of Montova manufactured the Gela plant's distillation towers.

10.11 Regional Expansions in the 1960s: The Middle East, Asia, and the Eastern Bloc

Egypt was the first country in the Middle East to take up manufacture of synthetic ammonia, as well as nitric acid and calcium nitrate, in this case at Suez. Construction work, undertaken by Chemico, commenced at the end of the 1940s. The plant was built for the Société Egyptienne (El Nasr) d'Engrais et d'Industries Chimiques (SAE). However, the hostilities between Egypt and Israel, and reduction in the supply of refinery gas, led to closure in 1969. Subsequently most operations were moved to Talkha, in the Nile Delta, north of Cairo. Egyptian Chemical Industries (KIMA) from 2019 manufactured urea and liquid ammonia from natural gas.

The Casale ammonia process was used in Iran at the Shiraz fertilizer plant, which manufactured urea by the Montecatini process, and ammonium nitrate. Manufacture at Shiraz commenced towards the end of 1963, and full production was achieved two years later. Plans were drawn up for additional facilities closer to the sea shore and oilfields [34]. Around the same time, the Casale process was used by the Kuwait Chemical Fertilizer Co., at one of its plants near Shuaibah.

Foster Wheeler was the main contractor. Apart from Ammonia Casale, licensors were Stamicarbon, for urea and ammonium sulfate, and Simon-Carves Ltd, of Stockport, Cheshire, England, for sulfuric acid [35]. Stamicarbon NV, a subsidiary of the state-owned Dutch mining company NV Nederlandse Staatsmijnen (Dutch State Mines, DSM), was founded in 1947, and based at Geleen. Snamprogetti constructed a 150-mtpd ammonia plant for the Syrian government.

In 1965, Saudi Arabia created the Saudi Arabian Fertilizer Company (SAFCO), the nation's first petrochemical company, based in Dammam, on the Persian Gulf. Occidental Petroleum Corporation was taken on to supervise design, construction, recruitment and startup of a fertilizer plant for SAFCO. The intended capacity was 600-mtpd of anhydrous ammonia, for conversion into 1,020 tons of urea (with recovery of 50 tons of sulfur). The feedstock was natural gas, previously flared off to waste, supplied by Petronim, the General Petroleum and Minerals Organization of Saudi Arabia. Construction work, involving Chemico as prime contractor, began in 1966, but, as a result of major technical and staffing problems, it took just over a decade (and after the contract with Occidental was terminated), in 1971, before the facility was operating in a satisfactory manner. Dammam serves as an important example of how the large-scale manufacture of synthetic ammonia required suitably qualified and trained management and technical personnel [36, 37]. SAFCO became an affiliate of the Saudi government corporation, Saudi Arabian Basic Industry Company (SABIC), following its foundation in 1976.

In Burma (today Myanmar), the government awarded Hitachi-Zosen of Japan a contract for a 40,000-mtpy ammonia unit; the engineering work was entrusted to Toyo Engineering Corp. At Ulusan, Korea, Toyo built what was claimed to be the Orient's largest fertilizer complex (Korea Fertilizer Co., Ltd), with a capacity of 195,000-mtpy of ammonia and 330,000-mtpy of urea.

European firms received large orders from the Eastern Bloc. In 1958, Nikita Khrushchev placed great emphasis on the development of the Soviet chemical industry, and three years later, around the time of rapprochement with European nations, focused on fertilizers. This led to contracts with European engineering contractors, including Snamprogetti, for construction and supply of equipment. In the mid-1960s, using Casale technology, Snamprogetti constructed a 430-mtpd ammonia plant, to the order of Techmashimport, the Soviet import agency, at Tula, an industrial city about 190 km south of Moscow. Snamprogetti also installed a 650-mtpd methanol rectification unit at Tula.

British firms were important suppliers to the Eastern Bloc, often bringing in subcontractors to help in the design and construction of plant, as well as to supply manufacturing equipment. In 1963, Power-Gas Corporation supplied a urea plant to Poland [38].[2] Three years later, Simon-Carves exported four ammonia converters to eastern Europe, three to Poland, and one to Hungary (the latter was 63 ft long, and weighed 91 tons), all manufactured by English Steel Forge & Engineering Corporation (of English Steel Corporation, Sheffield) [39]. For Humphreys &

[2] Power-gas was founded in 1901 to manufacture Mond producer gas from coal.

Glasgow Ltd, as main contractor, John Thompson & Co., of Wolverhampton, manufactured an ammonia converter weighing 141-tons and other high-pressure vessels for delivery to the Schwedt-on-Oder fertilizer facility in the German Democratic Republic. To the order of Simon-Carves, Thompson supplied an ammonia absorption tower for Hungary [40]. The main ammonia producer in Hungary, from the mid-1960s, was the Tisza Chemical Works (Tiszai Vegyi Kombinát), which drew on natural gas piped in from Romania. The Tisza ammonia process was based on Soviet designs and equipment, which included the reforming stage. The original capacity was 450-tpd; by 1970 it had been expanded to 600-tpd. Tisza produced ammonium nitrate and urea, the latter also using a Soviet-designed process. Chemoprojekt, a Czechoslovakian national corporation founded in 1950, signed a long-term technical exchange agreement with Humphreys & Glasgow [41]. Among other arrangements, this gave Chemoprojekt access to ICI reforming technology.

These examples reflect the growing international scope, and scale, of the nitrogen industry in the early to mid-1960s. This was also a time when a new situation in the affairs of the world would lead to a major change in the technology of ammonia production. The outcome was that global ammonia capacity would more than double in the decade and a half after 1965.

References

1. Kastens ML, Dudley JF, Troeltzsch J (1950) Synthetic methanol production. Modern chemical processes: a series of articles describing chemical manufacturing plants. Reinhold, New York, pp 72–82
2. Sollai M (2024) The fascist green revolution. Plants, People, Planet 6:1094–1103
3. Amatori F (2002) The fascist regime and big business: the Fiat and Montecatini cases. In: James H, Tanner J (eds) Enterprise in the period of fascism in Europe. Routledge, London, pp 55–66
4. Francisci L (2023) Una vittoria di Pirro. Cronaca della vertenza legale del 1928 tra la Società Terni e la Montecatini per il predominio nel mercato dei fertilizzanti azotati. Memoria Storica XXXI 61:103–130
5. Covino R, Francisci L (2020) Il processo Casale durante l'autarchia. Tra successi industriali e sperimentazioni. Ricerche Storiche XLX (January–April):41–57
6. Annual reports to shareholders, Ammonia Casale, 22 June 1935, and 13 June 1936 (Casale SA)
7. Murphree EV, Brown CL, Gohr EJ (1939) Production of aviation fuels by high pressure hydrogenation. Papers presented before the Petroleum Division of the American Chemical Society, part 1, Baltimore April 3 to 8. American Chemical Society, Washington DC, pp I 1–I20
8. Yeandle WW, Klein GF (1952) Purification of ammonia synthesis gas. Chem Eng Prog 48(July):349–352
9. Habermehl R (1965) Nitrogen fertilizer plants: improved catalysts reduce costs. Chem Eng Prog 61(January):57–61
10. Borgars DJ (1968) The naphtha reforming process used by Imperial Chemical Industries Limited for the production of ammonia. In: Fertilizer production, technology and use. Papers presented at the United Nations Interregional Seminar on the production of fertilizers, Kiev, Ukrainian Soviet Socialist Republic, 24 August–11 September 1965. United Nations, New York, pp 169–175, on 170
11. Brightling J, Pach J (2013) Ammonia production: a century of safety, health & environmental improvements. Ammonia Technical Manual, pp 231–245, on 234–235

12. Fowles M, Carlsson M (2021) Steam reforming of hydrocarbons for synthesis gas production. Top Catal (31/08/2021). https://doi.org/10.1007/s11244-021-01496-z (accessed 20 November 2024)
13. Habermehl R (1969) Optimizing industrial catalysts. Chem Eng Progr 65(October):55–58
14. Rostrup-Nielson J (2004) Steam reforming of hydrocarbons: a historical perspective. In: Bao X, Xu V (eds) Natural gas conversion VII. Studies in surface science and catalysis, vol 147. Elsevier, Amsterdam, pp 121–126
15. Murkin C, Brightling J (2016) Eighty years of steam reforming: the technological milestones of catalysts, reformer design and operation in Billingham since 1936. Johns Matthey Technol Rev 60(4):263–269
16. Anon (1962) Chemical and Process Engineering. July, column 43, p 349
17. Anon (1957) Quaderni dell'ingegnere chimico italiano 11(Oct):155–160
18. Nakles DV (1972) The natural gas shortage, summary of lecture by James A. Finneran. A chemical engineering commentary. Chem Eng Prog 68(July):107–108
19. Travis AS (2022) History of chemical industry: technologies and products versus corporate history. In: Giunta C, Seeman J (eds) Centenary issue. Bull Hist Chem 47(1):50–61
20. Oele AP (1968) Recent trends in the technology and economics of ammonia production. In: Fertilizer production, technology and use. Papers presented at the United Nations Interregional Seminar on the production of fertilizers, Kiev, Ukrainian Soviet Socialist Republic, 24 August–11 September 1965. United Nations, New York, pp 153–164
21. Anon (1953) Ammonia and urea. Pet Refin 32(Sept):178–182; 32(Nov):170–171
22. Holroyd, Sir R (1966) Presidential address. Chem Ind, 29 Aug, pp 1430–1436
23. Kenard, Jr RJ (1968) Natural gas and naphtha reforming. In: Fertilizer production, technology and use. Papers presented at the United Nations Interregional Seminar on the production of fertilizers, Kiev, Ukrainian Soviet Socialist Republic, 24 August–11 September 1965. United Nations, New York, pp 176–182, on 176
24. Rouwenhorst KHR, Travis AS, Lefferts L (2022) 1921–2021: a century of renewable ammonia synthesis. Sustain Chem 3:149–171. https://doi.org/10.3390/suschem3020011. Accessed 25 Nov 2024
25. Statistics of Crop Responses to Fertilizers (1966) Food and Agriculture Organization of the United Nations, Rome
26. Yates PM (1972) Agricultural adjustment in developed countries. Prepared by FAO in cooperation with the Economic Commission for Europe under the direction of Mr. Paul Lamartine Yates, Consultant. Food and Agriculture Organization of the United Nations, Rome, pp 53–54
27. Anon (1949) Urea triples in two years. Chemical Industries 63(4):559–560
28. Anon (1966) Statistics of crop responses to fertilizers. Food and Agriculture Organization of the United Nations, Rome, pp 9–12
29. Anon (1949) Dismantling delayed: organized Ruhr resistance. The Times (London), 9 June
30. Fauri F (2000) "Economic miracle" and Italy's chemical industry, 1950–1965: a missed opportunity. Enterp Soc 1(2):279–314
31. Hepner IL (1963) Chemical plant in the common market. Italy. Chem Process Eng 44(Sept):520–527
32. Harding AJ (1959) Ammonia: manufacture and use. Oxford University Press/ICI, London, p 5
33. Anon (1959) Petrochemicals multiply in Italy. C&EN 37(5 Jan):86–89
34. Salimi H (1968) The fertilizer industry in Iran: current status and plans for future development. In: Fertilizer production, technology and use. Papers presented at the United Nations Interregional Seminar on the production of fertilizers, Kiev, Ukrainian Soviet Socialist Republic, 24 August–11 September 1965. United Nations, New York, pp 93–95
35. Anon (1964) Kuwait Chemical Fertilizer Co. Hydrocarb Process 43(Nov):64
36. Anon (1966) Chemico: fertiliser contract for Saudi Arabia. Chem Ind, 19 Nov, p 1934
37. Quidwai A (1968) Operation and management of a nitrogen complex under Middle-Eastern conditions. In: Fertilizer production, technology and use. Papers presented at the United

Nations Interregional Seminar on the production of fertilizers, Kiev, Ukrainian Soviet Socialist Republic, 24 August–11 September 1965. United Nations, New York, pp 365–382
38. Power-Gas advertisement (1963) Chemical Process Engineering 44(Mar):149
39. Anon (1966) Humphreys & Glasgow Ltd Ammonia synthesis converters for Hungary and Poland. Chem Ind, 20 Aug, p 1423
40. Anon (1966) Absorber tower for Hungarian ammonia plant. Chem Ind, 19 Nov, p 1934
41. Anon (1966) Humphreys & Glasgow technical collaboration with Czechoslovakian National Corporation. Chem Ind, 24 Dec, p 2138

11 The Green Revolution, 1000-tpd Single-Train Ammonia Synthesis

Abstract

In the 1960s, American engineering contractors made massive contributions to the design and construction of ammonia and nitrogen fertilizer plants. Innovations at M.W. Kellogg led to a shift in the scale of production of individual ammonia units from around 250–300-tpd to 1000-tpd by the mid-1960s. Kellogg's single-train energy-integrated units, powered by centrifugal compressors, not only tripled the maximum rated capacity of individual synthesis loops, or trains, but brought about tremendous savings in energy and reduced the cost of fertilizer to the consumer. American and Japanese engineering firms, often in collaboration with European contractors and chemical firms, in particular Britain's Imperial Chemical Industries (ICI), dominated the synthetic ammonia industry. New markets appeared beyond the Western hemisphere, brought about by the need to confront growing famine and political changes. Synthetic ammonia was essential to the success of the Green Revolution, with its need for self-sufficiency, particularly in India. The global map of the ammonia industry was about to be redrawn.

11.1 Introduction

During the second half of the 1960s, public interest in, and even admiration for, technology was dominated by the Space Race and especially the US Apollo program. However, for many of the world's population daily living was connected to the less heralded innovations in synthetic ammonia technologies. These innovations were mainly due to the combined efforts of research and development teams at the American engineering contractor M.W. Kellogg of New York. Kellogg pioneered single-train energy-integrated units powered by centrifugal compressors that enabled production of 1000-tpd, and more, of ammonia. This not only tripled the maximum rated capacity of individual synthesis loops, or trains, but brought

A. S. Travis, *The Global Synthetic Ammonia Industry*, Perspectives on the History of Chemistry, https://doi.org/10.1007/978-3-031-87978-4_11

about tremendous savings in energy and reduced the cost of fertilizer to the consumer. The new single-train units enabled the emergence of the Green Revolution and with it the redrawing of the global map of the synthetic ammonia industry. The large-scale manufacture of nitrogen fertilizers began to move east to new sites, at first, and notably, to India.

By the early 1970s, the boom in the West resulting from the new ammonia-related technical innovations, especially among engineering contractors, was over. The subsequent general global downturn in business, at first the outcome of increases in the price of oil and natural gas, in addition to rising capital costs, held back new investment in plants. Because of the depressed economic conditions, certain newly developed energy-saving designs incorporating lower pressures and novel catalysts, such as those developed at Britain's Imperial Chemical Industries (ICI), were not taken up immediately (see Chap. 13). Nevertheless, the demand for ammonia continued to grow in those economies based on central planning. Between 1952 and 1971, the Soviet Union, Romania, Bulgaria, and Yugoslavia had already doubled agricultural output. During the same period, tractors and combine harvesters brought mechanized agricultural practices in Western Europe to the level earlier achieved in the US. In this chapter, we embark on a review of the main changes in the global ammonia industry—invariably driven by the need for increased agricultural output—which created the circumstances that were so well suited to new and far-reaching technologies. Of particular relevance are: the aforementioned Kellogg single-train energy-integrated synthetic ammonia technology, including a novel bottleneck converter; ICI's synthetic methanol process; and the tremendous markets created by expansion of ammonia production in India. The next chapter continues the story by describing the impact, both political and technological, during the 1970s in the People's Republic of China and the Soviet Union.

11.2 Engineering Contractors and New Ammonia Technologies

In the 1960s, cutting edge research into novel ammonia technologies was something that could be carried out only by major chemical corporations and the large engineering contractors that had diversified during and after World War II, particularly firms in the US with abundant research and development resources, infrastructure, expertise, and generous research budgets [1].

In order to capture new markets and satisfy new needs, processes were developed embodying far-reaching process design changes that would have a profound impact on the capacities, economics, and environmental performances of ammonia units. In part this was enabled by a number of peripherals, including improved bulk distribution systems for ammonia, especially the new long-distance pipelines that reduced the cost of shipping, refrigerated storage, and an increase in new plants around 1960 that employed steam reforming of natural gas. The outcome was that rated ammonia production capacities of individual units were increased from around a maximum of 350- to 400-tpd—though more usually 250-tpd—the norm until the early 1960s, to 1000-tpd, and more, from 1966. This followed the introduction of novel centrifugal compressors and the major redesign of several production stages, from which emerged the single-train energy-integrated ammonia units suited to large-scale, energy-saving production [2, 3].

Until the mid-1960s, the ammonia synthesis loops were driven by reciprocating compressors, which, though efficient, accommodated relatively small volumes of gas, and required frequent maintenance. The inherent limitations of these compressors made it necessary for large ammonia factories to operate several trains, or loops, in tandem. Moreover, the reciprocating compressors were noisy in operation and required skill in operation. With the switchover to centrifugal compressors, not only were the disadvantages overcome, but they provided all pressure requirements for each ammonia unit. High-capacity ammonia units became possible. The technology is mainly associated with the entry into the ammonia sector of M.W. Kellogg, an engineering contractor that until the early 1940s was not involved in the design and erection of plants for manufacture of ammonia. However, Kellogg in the 1930s had developed important chemical processing know-how through research into a moving-bed catalytic process for the petroleum industry and subsequently became a member of a consortium called Catalytic Research Associates. By 1938, Kellogg had invented methods for converting natural gas (methane) into synthesis gas (hydrogen and carbon monoxide), intended for production of organic chemicals. During the 1940s, Kellogg was active in research into synthetic gasoline, also from natural gas, its Synthol process, a variant of the German Synthine process [4] (Kellogg, through a subsidiary, also participated in the Manhattan Project.)

This type of experience was common to several other American engineering contractors that entered into process engineering. They included the Bechtel Corporation. Bechtel developed expertise in large-scale project management while engaged in a range of New Deal projects from the mid-1930s, including the Hoover Dam, and, during the war years, while working, invariably to tight deadlines, the Liberty ships. American firms were also engaged in the building of large and complex petrochemical complexes, mainly at various locations on the Gulf and Pacific coasts, but also in Pennsylvania, West Virginia, Ohio, and elsewhere. In general, they designed and erected, but did not fabricate, machinery and equipment. One firm might act as the main, or prime, contractor, taking on all responsibilities for procurement, engineering, and supervision of construction. The lead contractor engaged the specialist services of other contractors or licensed their processes. This style of organization and management, suited to expertise in both petroleum and natural gas technologies, was something quite new to Europe with its often ongoing and heavy reliance on coal-based chemical manufacture.

Kellogg's know-how in ammonia production grew out of its wartime government contracts for the construction of two facilities in the US. As a result of this work, Kellogg by the late 1940s had designed a converter in which the catalyst was held in beds contained in as many as ten baskets (cartridges), with cooling (quenching) gas flowing between the baskets, enabling control of the heat removed from the reaction zone (Fig. 11.1). This quench converter, "with provision for injecting quench between beds," and with "virtually point-by-point temperature control throughout," was incorporated in new construction projects by around 1950 [5]. Also contributing to developments were more active catalysts that enabled an increase in reforming pressures, often developed by specialists, such as Girdler, of National Cylinder & Gas, and Catalysts and Chemicals Inc., of Louisville, Kentucky.

In novel technical innovations, there were several notable successes by firms outside the US. ICI (UK), Uhde (Germany), Toyo Engineering, Chiyoda Chemical Engineering & Construction, Mitsui Toatsu, and Hitachi Zosen (all of Japan), Haldor Topsøe (at Lyngby, Copenhagen, Denmark), Stamicarbon (Holland), and Montecatini and Snamprogetti (both of Italy), were among the community of chemical and engineering firms that followed Kellogg's lead with advanced technology systems. Like the Kellogg converter, the Montecatini-Fauser converter of around 1950 used point-to-point temperature control; this enabled heat removal at each point in the converter. No less significant was the fact that firms such as Uhde, of Dortmund, Germany, began to follow American practice in project management.[1] However, it is noteworthy that the merger of Montecatini and Edison in 1966, to form Montedison, presaged a period of decline in innovative activity throughout the Italian chemical industry.

Here we need to draw attention to certain international events and review the leading technical developments that are closely connected with the mid-1960s and beyond. First, we take a brief look at the very special economic and political situations that in many ways encouraged engineering firms to devote vast resources into fixed nitrogen technologies in order to produce greater yields, at lower operating pressures and costs, and at increased efficiencies.

11.3 Towards the "Green Revolution"

Wartime disruptions had compounded the volatility in the production, supply, and price of food. The post-war economic and geopolitical issues, notably agrarian unrest—particularly among small farmers and in low-income countries—the decline of colonial empires, and the emergence of independence movements, impacted on the widening post-war interests of Western countries, and the US in particular [6]. America had a history of providing foreign food aid and agricultural supplies, notably from 1917 until the early 1920s, with a program funded by Congress and supervised by Herbert Hoover. After 1945, the US was the main contributor to the United Nations Relief and Rehabilitation Administration (UNRRA) and the UN Food and Agricultural Organization (FAO). In the 1960s, food aid became an integral part of US government policy and diplomacy.

[1] Uhde later became Krupp Uhde, then ThyssenKrupp Uhde, and from 2014 ThyssenKrupp Industrial Solutions.

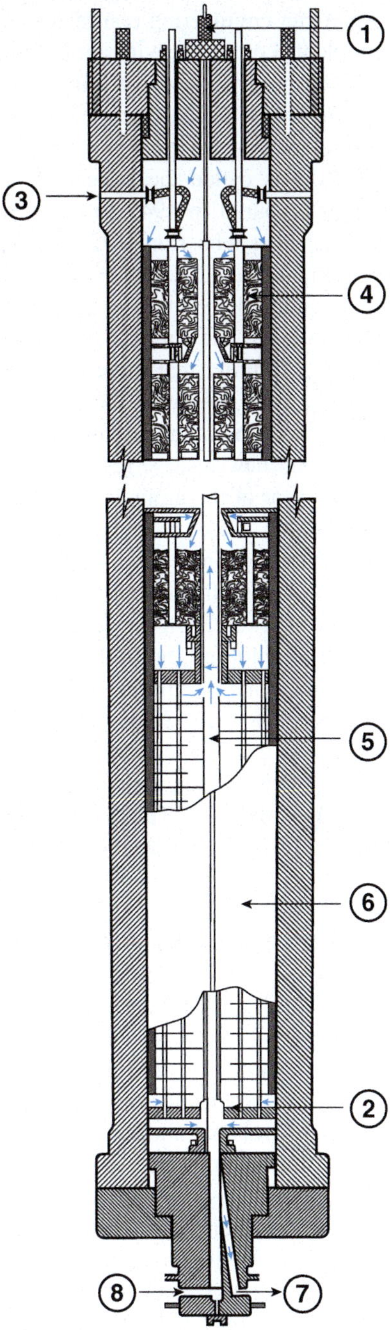

Fig. 11.1 M.W. Kellogg ammonia converter, late 1940s, operated at around 350 atm, with a capacity of 100-tpd. The catalyst was divided among two or three (or more) baskets, or cartridges, with the inter-basket temperature controlled by injecting quench gas (unheated synthesis gas) between the baskets. This enabled control of the heat removed from the reaction zone. In the figure, the gas mixture is charged to the converter through the inlet nozzle at the top (1) and then flows down the annular region between the shell and the internals to the bottom of the heat exchanger (2). It then moves up the shell side of the exchanger (6), is directed through the central gas transfer pipe (5), and then down within the baskets over the catalyst beds (4). Following the synthesis reaction, the gas mixture passes through the heat-exchanger tubes and leaves the converter at the outlet (7). Quenching gas for control of temperature enters at inlets (3). Control of preheat of the catalyst bed is through the bypassing of a portion of the inlet gas around the heat exchanger. The inlet for the by-pass gas is at (8). Redrawn from Hein LB (1952) Chem Eng Progress 48(August):412–418, on p. 414

American engineering contractors provided answers to some of the most pressing problems connected with agricultural needs, and strategies for emulation, through transforming the scale of production of nitrogen and other fertilizers.

The major incentive for expansion in the production of synthetic ammonia for fertilizers was the emergence of the new markets created by what, from the late 1960s, was called the "Green Revolution." This development emerged out of a report prepared in 1949 for UNESCO by Sir E. John Russell, a leading British soil scientist who from 1912 to 1943 directed the agricultural experiment station at Rothamsted. He emphasized the overriding importance of fertilizers based on nitrogen in alleviating hunger, famine, and malnutrition in many parts of the world. During 1951, the statistician Frank Yates, then associated with Rothamsted, produced a formula that quite accurately predicted future requirements in nitrogen. Closely associated with the early period of the Green Revolution was the American scientist Norman Borlaug, who in the 1940s, as a result of research in Mexico, developed new disease-resistant high-yield varieties of wheat. Mexico, until then an importer of the grain, became self-sufficient and by the 1960s was even exporting surplus wheat. Likewise, the US—which imported half of its wheat before 1950—through new strains of wheat, as well as increased ammonia production, became self-sufficient in the 1950s and then an exporter by the next decade [7].

The Rockefeller Foundation, the Ford Foundation (as major sponsors of the Green Revolution), and other major funding organizations adopted Russell's recommendations, promoting both nitrogen fertilizers and research into new strains of wheat and rice. Studies by Borlaug, associated with the Rockefeller Foundation's Mexican Agricultural Program and the Ford Foundation, led to a new variety of rice, IR8, which, following improvements from around 1970, produced more grain per plant [8]. This, and enhanced access to groundwater for irrigation, was especially significant to the situation in India.

11.4 Synthetic Ammonia in India

Under the British Raj, which ended in August 1947, the Indian synthetic ammonia industry had taken off slowly, beginning with the plant of Mysore Chemicals and Fertilisers Ltd, at Belagola, Mysore State (now Mandya District, Karnataka State), for the manufacture of ammonia at 5-tpd and ammonium sulfate at 20-tpd. Hydrogen was obtained by electrolysis of water.

An important consideration, as perceived by the British Commonwealth Relations Office, was in promoting Indian self-sufficiency in order to reduce the strain on the UK's sterling reserves. There was also the need to combat political unrest, brought on by the growth of the independence movement in India [9]. In November 1944, a British technical mission consisting of George S. Gowing and J. Rigg, both of ICI, and T. H. Riley, of the Association of British Chemical Manufacturers, recommended the building of a large fertilizer factory to produce annually 350,000 t of ammonium sulfate. The technical mission recommended Sindri, in southern Bihar, northeast India (today Sindri is in Jharkhand State).

Independently, in 1946–1947 the first ammonia factory of Fertilisers and Allied Chemicals, Travancore, Ltd (FACT, later Fertilizers and Chemicals Travancore) was erected in the small state of Travancore, on the southwestern coast of India. It used wood gasification for producing hydrogen. The rated capacity was 40-tpd of ammonia, which was converted into ammonium sulfate. During 1950–1956, in what was now postcolonial India, Travancore was part of Travancore, Cochin, which, with the addition of administrative areas, from 1956 was renamed Kerala. In 1960, FACT became a public sector company. Two years later, the Indian government became the main shareholder.

In December 1951, following the earlier recommendation of the British technical mission, Sindri Fertilisers & Chemicals Limited, at Sindri, was established; it was the first public sector company owned and managed by the Indian government. Construction was carried out by a consortium of Anglo-American contractors, which included the UK firms ICI and Power-Gas Corporation Ltd of Stockton-on-Tees. Sindri Fertilisers produced hydrogen from coke oven gas. Gypsum was the raw material used for ammonium sulfate production. The factory was inaugurated in March 1952 by the Prime Minister of India, Jawaharlal Nehru, who, anxious to bridge the gap between tradition and modernity, declared that Sindri was (along with new dams and steelworks) one of the temples of modern India. During 1954–1956, decisions were made to expand both Sindri and FACT, and build three new ammonia plants. Two of them, at Nangal, in Punjab, and Rourkela, in Orissa, produced calcium ammonium nitrate fertilizer. Nangal and Rourkela used, respectively, electrolytic hydrogen and coking oven hydrogen [10–12]. The electrolytic plant at Nangal supplied heavy water for the Indian nuclear program. Rourkela was part of a complex that included an iron and steel works. Around 1960, engineers at Sindri, with some assistance from Stamicarbon NV, the subsidiary of DSM (Dutch State Mines), designed and constructed, as part of a turn-key project, nitric acid, ammonium nitrate, and nitrolime facilities at Rourkela.

The third plant, that of Neyveli Fertilizers Ltd (of Neyveli Lignite Corporation Private Ltd), was in the southeast at Neyveli, Madras State (from 1969, Tamil Nadu). It converted ammonia into urea, the import and use of which had grown tremendously since the mid-1950s. The integrated ammonia-urea plant, with a rated capacity of 300-tpd of ammonia and 500-tpd of urea was ordered in 1959 on behalf of the Indian government and constructed by Montecatini, and Ansaldo, of Genoa. Ammonia was produced by the process of engineering contractor Fauser-Montecatini. At the time when the contract was signed, the urea unit was claimed to be "the world's largest." Pintsch Bamag and Linde provided technology for gasification of lignite and fractional distillation of air [13]. The complex was completed in 1966. The need to import nitrogen products (mainly ammonium sulfate and urea), and phosphate rock, muriate of potash (potassium chloride), and sulfur for producing mixed fertilizers, favored coastal locations for most Indian fertilizer plants (Fig. 11.2).

After experts from the Ford Foundation visited India and subsequently recommended the introduction of an emergency program to bolster agricultural production, plans were drawn up in the early 1960s for two new ammonia plants

Fig. 11.2 Map of India showing locations of ammonia and ammonia-phosphate fertilizer facilities in 1965 (and other sites mentioned in this chapter). Included are facilities for which high-capacity single-train units were proposed during 1965–1966: 600-tpd at Durgapur (the first to be approved), in Kerala, and at Goa, and Trombay (expansion); 750-tpd at Madras (Channai); and 1000-tpd at Haldia and Kanpur. The feedstock for hydrogen in all cases was naphtha. Redrawn from reference [10], p. 90

using Casale technology. One was at Ennore, near Madras (today Chennai), in Madras State (today Tamil Nadu), of East India Distillers (E.I.D.-Parry). The main contractor was Simon-Carves, and the feedstock was petroleum fractions [14]. The other plant was for Gujarat State Fertilizer & Chemicals (GSFC), which was established in 1962, and immediately set about construction of a fertilizer complex at Bajwa, near Baroda (the Vadodara plant), on the west coast. GSFC awarded the construction contract to the Japanese shipbuilding and engineering firm Hitachi Zosen, of Osaka, which worked closely with Toyo Engineering [15–18]. The ammonia section included the introduction into India of ICI naphtha steam reforming technology for hydrogen, provided by licensee Humphreys & Glasgow of London. This facility included a Toyo Engineering urea unit. Power-Gas Corporation was also involved in the project. Early expansion included Chemico's ammonium sulfate phosphate process. The Vadodara plant at Bajwa began production with a daily capacity of 450-mtpd of anhydrous ammonia from two Casale units. In 1967, two more Casale units were installed, with a total rated capacity of 500-mtpd. This project represented the main entry point of Ammonia Casale and Japanese engineering contractors into the Indian nitrogen industry.

In 1964, Japanese experts reported that refinery by-products, "wasted in many cases," could be used "in making ammonia, methanol, and other chemicals" [19]. The refinery by-products included light petroleum naphtha. Over the following years, as the Indian ammonia industry expanded, extensive use was made of ICI's naphtha steam reforming process, as used at Bajwa. However, there was a significant falling off of the UK's involvement in the Indian chemical industry.[2]

In 1965, with the failure of monsoon rains over northern India, and a food aid program dating from 1954 with the US coming to an end, the wheat specialist Chidambaram Subramaniam (1910–2000), head of the Indian Ministry of Food and Agriculture, emphasized the urgent need to introduce ammonia-based fertilizers and novel seeds. It was six decades since a monsoon failure had occurred, and India was facing widespread hunger and starvation. Hasty negotiations between Subramaniam and Lester R. Brown of the US Agency for International Development led to a diplomatic agreement that brought about immediate relief: the US delivered six-hundred shiploads of wheat to India. However, in return, and as part of the agreement, President Lyndon B. Johnson insisted on major reforms in agriculture. At the end of 1965, the US and the World Bank encouraged India, especially, as well as Pakistan, to accept assistance from Western multinational engineering contractors in developing large-scale nitrogen fertilizer factories [20, 21]. The World Bank together with mainly American and European banks agreed to underwrite the construction of major fertilizer projects. The plant installation

[2] Construction work on ammonia plants, including expansions, undertaken during the country's third five year plan (1961–1966) included, in 1963: Ennore (East India Distillers, licensors Shell/Casale); Alwaye, Kerala State (FACT, licensors Texaco/Power-Gas/British Oxygen/Dorr-Oliver); Trombay (Hindustan Chemical and Fertilisers Corporation of India, licensor Chemico); and Rourkela (Hindustan Steel, licensor, Uhde). Urea plants were under construction at Trombay (Hindustan Chemical and Fertilisers); Neyveli (Neyveli Lignite); and Nahorkatiya, Assam (Fertilizer Corporation of India). In 1961, Hindustan Chemical and Fertilisers became part of the Fertilizer Corporation of India. In 1965, the construction situation, just prior to the introduction of high-capacity single-train units, was: Alwaye (FACT, licensors Texaco/Power-Gas/British Oxygen/Dorr-Oliver/ICI), 154-tpd; Bajwa (Gujarat State Fertilizer, licensors Humphreys & Glasgow/Casale/Power-Gas/Chemico/Vetrocoke), 450-tpd; Trombay (Fertilizer Corporation of India, engineering Chemico), 390-tpd; and Visakhapatnam (Coromandel Fertilisers). Urea plants were under construction at Alwaye (FACT); Bajwa (Gujarat State Fertilizer); Neyveli (Neyveli Lignite); Visakhapatnam (Coromandel Fertilisers), and Trombay, Namrup, Gorakhpur, and Korba (Fertilizer Corporation of India). Sources: Construction Boxscore sections, Hydrocarb Process 42(January 1963):198, and Hydrocarb Process 44(September 1965):70H. The foundation stone for Korba was laid down in 1973 by Prime Minister Indira Gandhi, but the facility faced severe technical difficulties and was closed shortly after. Between 1990 and 2002, the factories of the Fertilizer Corporation of India were closed down. Sindri closed in 2002.

costs, covered in large part by the foreign financial aid, were high. However, it was emphasized that home-produced ammonia and nitrogen fertilizers would no longer make necessary the need for valuable foreign currency to pay for imports. For India, it was none too soon.

The situation became critical again in 1966 as the result of a second monsoon failure, during July to September, and severe drought. Bihar State was threatened with famine. Urgent intervention was called for. The US Agency for International Development provided loans to India and Pakistan enabling them to cover the cost of fertilizer and grain imports. The Indian government took control of the situation and immediately imported 18,000 tons of the seed varieties developed by Borlaug and his team.

Fortunately, the construction of seven fertilizer plants by multinational Engineering firms had already been approved by the government of India. This was during the final months of India's Third Five-Year Plan, which came to a close in March 1966 [10].[3]

All seven plants were based on single-train high-capacity ammonia units as introduced by M.W. Kellogg (see next section). Three were for the Fertilizer Corporation of India, a government corporation established in 1959 and which two years later created a consolidation of several companies, including Sindri and Hindustan Chemical and Fertilisers Ltd. Fertilizer Corporation's new Trombay plant, near Bombay (now Mumbai), was financed by the US Development Loan Fund. (Toyo Engineering's urea technology was introduced at Fertilizer Corporation's Gorakhpur facility.) Montecatini-Edison (merged as Montedison in 1966) and Ansaldo were commissioned to erect two nitrogen plants, both to produce 600-mtpd of ammonia and 100-mtpd of urea, for FACT, at Alwaye, Cochin, in Kerala state, and Durgapur, Western Bengal [22]. The synthesis gas plants were designed and built by the UK's Power-Gas Corporation. Among project engineering companies set up by the Indian government in the 1960s, two were assigned to fertilizers: FACT Engineering and Design Organization (FEDO), a unit of FACT, which had responsibility for ammonia and phosphatic fertilizers, and Projects and Development India Ltd (PDIL), a unit of Fertilizer Corporation of India, which handled urea and nitrogeneous fertilizers [23].

During 1966–1967, the Lummus Company of New York constructed for Coromandel Fertilisers what was claimed to be India's largest fertilizer manufacturing facility, located near Visakhapatnam, Andhra Pradesh State, in the southeast. For Zuari Agro Chemicals Ltd, established in 1967 in Goa, Power-Gas Corporation, in collaboration with H.K. Ferguson Co., of Cleveland, Ohio, worked with the main

[3] New ammonia plants approved in India during 1965–1966: Fertilizer Corporation of India, Trombay, 90,000-tpa (expected completion 1965); Neyveli Fertilizers, Neyveli, 70,000-tpa (1965); Fertilizer Corporation of India, Gorakhpur, 80,000-tpa (1967–1968); Gujarat State Fertilizers, Baroda, 95,000-tpa (1967–1968); Fertilizer Corporation of India, Namrup, Assam, 45,000-tpa (1966–1967); Coromandel Fertilisers Pr. Ltd, Vizakhapatnam, 80,000-tpa (1967–1968); Fertilisers and Allied Chemicals, Travancore Ltd (FACT) Alwaye (Cochin), 40,000-tpa (1965–1966). Reference [11], p. 304.

contractor, Toyo Engineering, on a fertilizer project with a rated daily capacity of 1100-metric tons of urea. Toyo installed its urea technology and licensed the ammonia technology from Kellogg. Investors included the US Steel Corporation, the International Finance Corporation, and the Bank of America. The complex began operating in 1973. Power-Gas and Ferguson acted as joint consultants for a similar complex erected for the East Pakistan Development Corporation [24].

The high-capacity ammonia plants started the main trend in the moving of production sites from Western countries and Japan to new locations close to areas of consumption [25]. In the case of India, it reflected the changes in the organization of science and technology initiated by Indira Gandhi, who had been appointed prime minister in January 1966 and was an enthusiast for large-scale technological development [26].

In 1969, ICI naphtha reforming technology was adopted at two fertilizer facilities at Kanpur, Uttar Pradesh State, in northern India. Also in 1969, Japan's Chiyoda Chemical Engineering & Construction received a contract to build an ammonia-urea complex at Kota, Rajasthan State, in the northwest (Shriram Fertilisers and Chemicals, SFC). The feedstock at Kota was also naphtha. The ammonia section licensors included Shell (USA), for naphtha desulfurization, Haldor Topsøe, for the reforming and synthesis sections, and the Benfield Corporation (of Pittsburgh), for carbon dioxide capture for use in the urea process. For the latter, Stamicarbon provided its total recycle process technology. The complex was completed in the mid-1970s [27].

By the mid-1970s, with the support of home-produced nitrogen fertilizer, the increase in food grain yield was tremendous, as was the knowledge transfer to Indian specialists from engineering firms across the globe, including Denmark, Holland, Italy, Japan, the US, and the UK. Moreover, as a non-aligned nation, India also sought and received assistance from the Soviet Union.

Nitrogen fertilizers and the new strains of wheat and rice were located at the very heart of the Green Revolution. Also significant, particularly in India, were the increasing reliance on tubewell access to water for irrigation and improved distribution networks. Reflecting the changing geography of the nitrogen industry, as it increasingly served and was installed in Asian countries, national programs involving ammonia production for fertilizers were responsible for the most significant growth. More than any other factor, this was enabled by the novel Kellogg technology. It represented a turning point in large-scale ammonia production, and at much reduced cost.

11.5 Kellogg's Ammonia Technology

This section describes the major technical developments that contributed to the ammonia industry during and after the 1960s, in part arising from the symbiosis between synthetic ammonia technology and the Green Revolution. Another important factor was the growing reliance on hydrocarbons for production of hydrogen, which came about at the very same time when the chemical industry in the organic sector accelerated away from coal-based processes towards petrochemicals and natural gas. It provided the opportunity for M.W. Kellogg, with a strong background in petroleum plant operations, to become a major player in the ammonia sector.[4] Kellogg in the 1960s made the first major breakthroughs in the design and development of high output (600- and then 1000-tpd) single-train energy-integrated synthetic ammonia units.

Kellogg's involvement began in the early 1950s, when its Chemical Process Division was working on contracts for erection of facilities for ammonia, methanol, and nitric acid, in addition to production of ammonium nitrate, ammonium sulfate, and urea (the latter by the new Montecatini process, which had been licensed to Kellogg). At that time, the largest American ammonia factory produced 1000 t of ammonia daily, based on a battery of three synthesis trains. The daily average installed capacity from a single train at some sixty US ammonia plants was just under 200 t [28].

Kellogg, following on from its studies conducted in the 1950s, undertook pilot and semi-commercial plant studies on pressure reforming of natural gas at up to around 60 psig (4.14 bar). This led to the design of a reforming unit for Shell's anhydrous ammonia facility at Ventura in California [29]. The primary reformer was fed with natural gas as it was usually available under a relatively high pressure. This reduced by about 25–35% of the compression required for the ammonia synthesis.

Improved pressurized steam reforming drew in part on ICI's work [30]. ICI developed its more robust high-capacity pressurized steam reforming process, initially for naphtha. Kellogg recognized how the ICI process could be the basis of a proprietary pressurized reforming process. As a result of this and other innovations, Kellogg would maintain important advantages over its competitors. In 1966, Kellogg filed a pressure reforming patent in the names of James A. Finneran, Hays C. Mayo, Robert H. Multhaup, and Ronald B. Smith [31].

While the move to hydrocarbons as sources of hydrogen until the mid-1960s represented one of the main transformations in the ammonia industry, a no less important development involved the redesign of the converter. Most innovations in the design of converters were concerned with the flow of gas through the catalyst, temperature control of the catalyst bed, and heat recovery. These derivatives

[4] Kellogg was founded by Morris W. Kellogg in New York in 1901. Four years later, in Jersey City, Kellogg started the manufacture of piping for electricity power generation plants. During the 1930s, the company's headquarters were moved to New York City. In 1942, Kellogg designed and built the first fluid catalytic cracking unit, for a refinery in Baton Rouge, Louisiana. In 1944 Kellogg was taken over by Pullman Inc. After the war, Kellogg perfected the important BWR (Benedict-Webb-Rubin) equation of state, used to describe the behavior of mixtures of light hydrocarbons. This led to Kellogg's strength in cryogenic separation technology, particularly in the field of liquefied natural gas (LNG). In 1970, the headquarters were moved to Houston. In 1976, the company changed its name to Pullman Kellogg (the Pullman Kellogg Division of Pullman Inc.) and in 1988, after a series of changes in ownership, Kellogg became the main engineering division of Dresser Industries. In 1998, Kellogg merged with Brown and Root Engineering and Construction (Brown & Root) to create KBR.

of first-generation tubular converters, with technologies dating from the 1920s to 1940s, retained the axial flow of gas, that is, the mixed gases flowed longitudinally along the axis of the reactor tubes that surrounded the catalyst. To mitigate the inherent pressure losses of this type of reactor, the catalyst particles had to be of a certain size, which limited their active surface area and thus overall activity. Typically, until the 1950s, most converters were of the tube-cooled type, in which cooling tubes carried away the reaction heat and maintained near-isothermal conditions throughout the catalyst mass. This included the TVA and Chemico converters. A second type of converter design, as introduced by Kellogg and mentioned earlier, did not use cooling tubes in the catalyst bed. This was the multi-bed converter, in which the catalyst was divided among baskets, or cartridges, with the inter-basket temperature controlled by injecting quench gas (unheated synthesis gas) between the baskets (and later by internal gas–gas heat exchangers) (Fig. 1.1). Kellogg's converter design of around 1953–1954 for a single-train 300-tpd anhydrous ammonia unit for Standard Oil Company of Ohio, at Ventura, was one of the first examples of this quench-type converter working at a capacity approaching the maximum possible with existing technology [32, 33]. This configuration became the standard for Kellogg converters until the mid-1960s.

When during the early 1960s, Kellogg moved into large-scale ammonia synthesis with the novel energy-integrated process, its new multi-bed converter, still designed for axial flow, was distinguished by a feed/exhaust gas heat exchanger located above the catalyst chamber, creating the "bottleneck" reactor profile (Fig. 11.3). This heat exchanger could be removed as a unit for servicing or repair without disturbing the catalyst; the converter was now operating under essentially adiabatic conditions. The adiabatic multi-bed synthesis reactors were designed to handle much larger gas flows than the tube-cooled isothermal type. Simple and reliable in operation it was highly successful. This and other innovations permitted the synthesis loop to run at lower pressures and enabled high-capacity ammonia facilities to operate with a single train instead of multiple trains as previously [34, 35]. Later designs incorporated enlarged openings in the pressure shell which allowed the catalyst cartridges to be easily lifted out of the reactor.

According to Kellogg engineers in 1965, "Acceleration of this development began to take shape in the period 1953–1956 with the introduction of pressure reforming, the direct use of energy in the process to satisfy plant energy requirements, and the first use of the single-train centrifugal machinery for synthesis-gas compression in plants conceived, designed and built by Kellogg for Shell Chemical [Ventura] and Standard Oil of Ohio (United States of America). Concomitant with and necessary to the introduction of these features was the recognition of the ammonia plant as a whole—a single design and operating entity, as compared with a group of separated plants: one for the generation of synthesis gas; one in which it was purified; another in which it was compressed; and a separate synthesis plant." As for the design of reformer furnaces: "Availability of more data covering a broader range of operating conditions now (1965) permits designs to be made for economic [reformer] furnaces for operation at significantly higher pressure levels. This improvement reflects in lowered compression costs and better heat

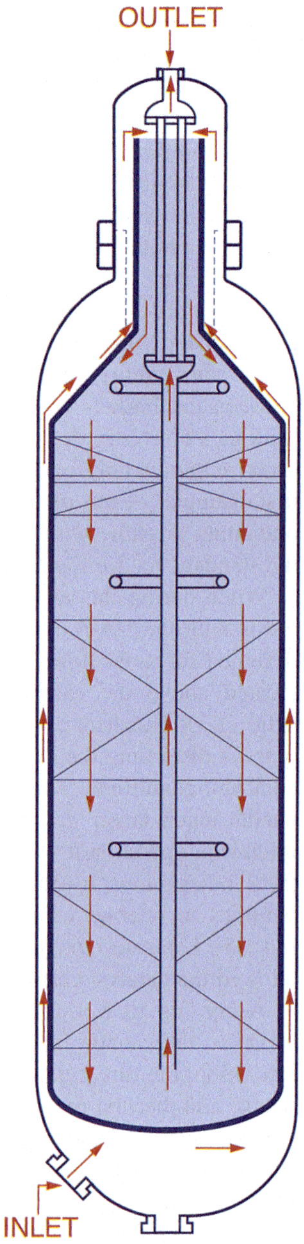

Fig. 11.3 M.W. Kellogg "bottleneck" quench-type converter for a high-capacity single-train ammonia plant. In the original design, the pressure shell contained an internal cartridge (basket) with three or four catalyst beds and three quenches. Redrawn from reference [35], p. 262

economy" [36]. Higher pressure reforming reduced the lower downstream pressure requirements, which became significant once low synthesis pressures became the norm in high-capacity units. Kellogg's design culture was born from its involvement in the design and erection of integrated single-train processing units for the petrochemical industry [37].

11.5.1 600- to 1000-tpd

In December 1963, Kellogg completed the first ever full-scale single-train ammonia unit, with a capacity of 600-stpd, twice that previously in general use, for the American Oil Company (Amoco), at Texas City, on the Gulf Coast [38]. One year earlier, Kellogg had competed with, and outbid, Chemico and Foster Wheeler for the contract. Amoco had specified that the unit had to be single train, apart from reciprocating compressors for the final gas compression before entry into the converter. As a result, the Kellogg design, based on a study conducted by its electrical engineer Hays Mayo, used a combination of centrifugal and reciprocating compressors. A complete unit was originally operated at Tuloma Gas Product, an affiliate of Amoco, perhaps for development purposes, including establishing proof of concept on an industrial scale. The success of the Texas City Amoco unit, described as "the largest 'single-train' plant in the nation," was followed by a similar 600-stpd unit for Monsanto, at Luling, Texas. By this time, Kellogg personnel were figuring out how to improve manufacturing efficiencies by matching reforming and ammonia synthesis pressures to the requirements of a single centrifugal compressor, and, moreover, to enhance the single-train capacity to 600-tpd. (The backgrounds to the centrifugal compressor and its role in high-capacity units are told in the next section.)

The impact of centrifugal compressors on the scale of production in the mid-1960s was rapid. In 1966, for Olin Mathieson Chemical Corporation (the 1954 merger of Olin Industry, Inc., and Mathieson Chemical Corporation), the Bechtel Corporation completed what was claimed to be the "world's largest synthetic ammonia plant," at Lake Charles, Louisiana, with a capacity of around 1400-stpd (100,000-tpy). However, though this was not from a single-train unit, the facility was notable for incorporating fifteen Cooper-Bessemer centrifugal compressors, of which the compression for processing came from seven of them—two for syn gas, two for air, two for ammonia refrigeration, and one acting as a gas booster [39]. The facility was associated with a 400-stpd Stamicarbon urea unit, built in collaboration with the engineering contractor Arthur G. McKee & Co., of Cleveland, and included the world's second-largest steam reforming terrace wall furnace for ammonia, supplied by Foster Wheeler. The largest furnace for an ammonia unit was for the Standard Oil Company of Kentucky (Fig. 11.4; see also Sect. 11.6) [40].

Fig. 11.4 Foster Wheeler 1966 advertisement for its terrace wall steam-hydrocarbon reformer furnace at the Olin Mathieson Chemical Corporation's 1400-tpd ammonia facility at Lake Charles, Louisiana. This was the world's second-largest Foster Wheeler reformer furnace for ammonia works. The largest such furnace was for the 1000-tpd single-train ammonia unit for Standard Oil of Kentucky (Hydrocarbon Processing 45(December):20. Wood Group plc. With permission)

11.5.2 Compressor Developments and Energy Integration

Until the mid-1960s, the synthetic ammonia industry was characterized by high investment costs and high energy consumption. There were limitations on the technology. Thus, several compression stages in series were necessary in order to achieve the required pressure for ammonia synthesis. It was the need to compress the synthesis gas to the pressure of the synthesis section which made the plants expensive to operate. In order to produce 200- to 300-tpd from a single train, three reciprocating compressors driven by gas engines or electric motors were needed. Another downside was the low recovery of waste heat from the process, which made the plants inefficient and very expensive to run. Not only was much energy wasted, but carbon dioxide was released to the atmosphere, unless required for manufacture of urea. In the mid-1950s, the largest reciprocating compressors were of 5000 horsepower. Two, each carrying half the load, were required for a typical 300-tpd ammonia plant. For larger outputs of ammonia, this represented a major challenge to designers of high-capacity plants.

Inputs towards change were to come from wide-ranging technological improvements made during and after World War II, in particular the development of materials suitable for use at high pressures and temperatures. Rapid progress in avionics provided a wealth of experience that was applied to gas turbines and compressors [41]. In 1953, centrifugal compressors were first used in synthesis gas compression. This encouraged Kellogg to change its approach to the design process by incorporating the new equipment and technologies that had recently become available. Engineers previously engaged in design of aircraft joined chemical engineers to tackle improvements in high-pressure chemistry. Kellogg proposed the use of a centrifugal compressor, driven by a steam turbine, to perform at least part of the compression of the synthesis gas. Because centrifugal compressors could not achieve pressures as high as reciprocating pressures, the synthesis pressure had to be reduced. This enabled a major reduction in energy consumption. It also led to the Kellogg all-centrifugal single-train synthetic ammonia unit.

Development took place over three years, between 1962 and 1965, during participation in four tenders for the construction of ammonia plants, the first two for the already mentioned Amoco and Monsanto facilities, and the other two for ICI in the UK, one at Severnside, and the other at a new facility, for supply of ammonia to Fisons Fertilizers Ltd, at Immingham in northeast Lincolnshire (opened in 1966). The ammonia storage sphere at Immingham had a capacity of 400 t.

A major advantage of the new design incorporating centrifugal compressors was that the capital cost of the plant, on the basis of output, was reduced by around 30%. Further, increases in reformer pressures enabled greater recovery of heat, which improved the energy balance of the ammonia process, and enabled considerable reductions in the compression power requirements.

For each of the three tenders, Kellogg engineers carried out numerous studies to improve various aspects of the process. Unexpected problems were discovered and needed time to put right. The ICI case is of special interest, because it would transform ammonia production at the large Billingham works.

In 1963, ICI introduced the world's first single-stream ammonia unit with a capacity of 360-mtpd, for urea and fertilizer production, at its then new Severnside facility. This unit incorporated parallel reciprocating compressors for gas compression, the ICI pressurized steam naphtha reforming process, and carbon monoxide removal with hot potassium carbonate [42, 43]. ICI was interested in introducing a second high-capacity unit at Severnside. Significantly, when around 1964, Foster Wheeler erected a 400-tpd ammonia unit based on Casale technology for Nitrin, Inc. near Cordova, Illinois, this was approaching the limit for reciprocating compressors. In 1964, Foster Wheeler advertised a 1000-tpd ammonia unit, again using Casale technology, with centrifugal synthesis gas compression (Fig. 11.5). No doubt the design was based on the bid for the Amoco 600-tpd facility.

Kellogg, Chemico and Ammonia Casale were aware that at capacities of 600-stpd (544-mtpd) and more it was practical and economic to incorporate a turbine-driven multi-stage centrifugal synthesis gas compressor. Thus Kellogg's ammonia unit for Amoco was equipped with a tandem-drive centrifugal compressor system, a train of four compressor cases, driven by a single gas turbine, and reciprocating compressors just for bringing gas to the synthesis pressure. Significantly, 600-stpd of ammonia was the lowest capacity suited to centrifugal compressors, which operated at a very high speed and were ideally driven by a steam turbine, using waste heat available from plant operations. Such a compressor had, for example, been developed by Nuovo Pignone in Italy; a license for its manufacture was granted to C.A. Parsons & Company, of Newcastle upon Tyne, the leading UK manufacturer of turbines. In the new ammonia synthesis design, high-pressure steam was raised from process waste heat in the synthesis gas plant and the synthesis loop. Kellogg calculated that by matching the sources of waste heat in the process to the thermal requirements of the complete process, it was possible to vastly improve the overall energy efficiency and thus the economics of the process; all the main machinery of an ammonia unit could now be driven by steam, with a minimum consumption of electrical energy. In sum: "The high-capacity, single-train ammonia plant design has been made possible by the bringing to fruition of two basic concepts: (a) an integration of the energy available within the process with that required to operate the plant; and (b) selection of process conditions to make most efficient and economic use of proved machinery and equipment" [36, p. 165]. This concept was also found useful in the design of other high-tonnage processes, such as for methanol and urea.

Kellogg submitted to ICI a design for a single-train unit with a capacity of around 600-stpd. It was arranged so as to completely avoid the use of reciprocating compressors. This was made possible by the decision to operate the ammonia synthesis at only 150 atm, around half the pressure previously used, and also the maximum pressure allowed with centrifugal compressors. The lower synthesis pressure design, however, needed much more catalyst and operation of two reactors (converters) in parallel (or a larger reactor). Kellogg improved the heat recovery system beyond that achieved in the Amoco plant and increased the pressure of steam generation. Lacking sufficient data to investigate the kinetics of the synthesis reaction for the lower synthesis pressure, Kellogg asked Haldor Topsøe of

Fig. 11.5 Foster Wheeler 1964 advertisement for a 1000-tpd ammonia unit based on Ammonia Casale technology. It features a photograph of a 400-tpd unit, probably for Nitrin, Inc., at Cordova, Illinois. This was at a time when Foster Wheeler, Chemico and Kellogg were competing in the design of a high-capacity unit for Amoco (Hydrocarbon Processing 43(August). Wood Group plc. With permission)

Copenhagen, Denmark, to handle this part of the project. Because Haldor Topsøe had recently supplied catalysts to Collier Carbon and Chemical for a Mont Cenis ammonia unit at Brea, California, it had to hand all the data for a full size plant of the type that Kellogg had designed. Haldor Topsøe and Kellogg engineering staff worked together in Copenhagen to calculate the volumes of catalyst required for Kellogg's low-pressure synthesis. Kellogg had access to the computers in Copenhagen. The results dictated the size of the converter [40]. However, developmental delays, particularly the time-consuming, complex calculations, prevented completion of the new design before ICI's deadline. The contract for Severnside was assigned to Bechtel [44]. This second Severnside unit, like its predecessor, used reciprocating processors.

Centrifugal compressors were, however, introduced very soon after at ICI's main ammonia factory. In January 1964, ICI awarded Kellogg tenders for two 1000-stpd (972-short-tons-a-day) single-train ammonia units for installation at Billingham. A third, and identical, unit was ordered very shortly after. The annual capacity of each unit was 300,000-standard tons. By the time of the third ICI tender, Kellogg had developed a unique synthesis unit which, in spite of the low pressure, not only had a rated capacity of as much as 1000-stpd of ammonia, but incorporated full energy integration and standardization. As noted above, because the synthesis was conducted at only 150 atm, enlarged converters were necessary. However, there were delays in completion, which was originally scheduled for 1966. These were caused by difficulties in procuring equipment, due to a "surge of ordering from all other chemical companies" [45]. A spokesman for ICI's Agricultural Division claimed that the delays were caused by "Shoddy subcontracted engineering" [46]. There were also teething problems when the ammonia units were started up. Named Ammonia I, II, and III, the ICI units went on stream during 1967, and replaced the older units. But for the delays, they would have been the very first 1000-stpd units to be placed in operation. Credit for that went to the unit installed by Kellogg during 1965–1966 for Coastal Chemical Corporation, at Yazoo City, Mississippi (a subsidiary of Mississippi Chemical Corporation; they merged in 1972). The converter, with 5-inch-thick steel walls, was ninety feet long. It was, as the advertisements trumpeted "A New Era for Ammonia" [47, 48]. Shortly after, CF Industries (until 1970, Central Farmers Industries) at Donaldsonville, Louisiana, began operating a similar unit, adjacent to the ammonia plant of First Nitrogen Corporation [49] (First Nitrogen Corporation was jointly owned by First Mississippi Corporation, of Jackson, Mississippi, and FS Services Incorporated, of Bloomington, Indiana.)

Still, Billingham, was certainly the "Largest Ammonia Project in the World!" [50]. Once in full operation, the three single-train, energy-integrated units contributed towards Billingham's annual rated capacity of around one million metric tons of ammonia. A 350,000-mtpy urea plant at Billingham followed soon after.

Through efficient recovery and use of waste heat, no separate power station was required, as had been the case at Billingham since the 1930s. The front end of the ICI process produced hydrogen by steam reforming of the naphtha feedstock at an operating pressure of over 400 psig (28 bar, 27 atm), double the pressure when the process was first introduced at ICI's Heysham site (where ammonia manufacture ceased in 1977). The naphtha was produced from low sulfur North African crude oil available from the UK's first oil refinery in the northeast—built on land adjacent to Billingham—which also supplied the nearby ICI factory at Wilton. Opened in September 1966, the oil refinery was a joint project of Phillips Petroleum and ICI (Phillips-Imperial Petroleum). To support the expanding ammonia business, and catalyst sales, during the mid-1960s, ICI's Agricultural Division invested two million pounds sterling on extending its catalyst manufacturing facility at Clitheroe in Lancashire.

In summary, the single-train high-capacity process designed by Kellogg incorporated certain innovative concepts whose main contributions were twofold: to reduce energy consumption, and therefore the cost of production; and to enable a net increase in the capacity obtainable in a single unit, and therefore a considerable economy of scale. The process, as was typical of natural gas-based ammonia processes, was separated into a reform section and a synthesis section, each based on several unit operations. The Kellogg design, featuring centrifugal compressors in ammonia units with capacities of between 600- and 1000-stpd, high-pressure steam systems, and high-pressure reforming, was recognized as superior to that of its competitors. This was borne out by its rapid spread. Kellogg made tremendous inroads into the ammonia business. It received orders for seven units in 1964 (the three for Billingham, and four in the US), and seventeen by mid-1965 [51, 52]; and as many as twenty (eight of which were outside the US, including from India), even before the first of the 1000-stpd units was completed in 1966 [53, 54]. One special feature of these facilities, as mentioned earlier, was that they no longer looked like traditional walled in and covered factories, since most processing equipment was placed outdoors, as was typical of oil refineries and petrochemical complexes.

11.6 The Single-Train Boom

In the spring of 1966, demand for ammonia-based fertilizers by American farmers was especially strong. However, their demands could not be met. Engineering contractors were unable to maintain schedules for installing single-train units mainly as the result of a shortage of engineers. According to *Chemical & Engineering News* of 15 August 1966, eleven 1000-stpd, or more, units were originally scheduled to be completed during 1966 and 1967.[5] Of these, only three, including the unit at Yazoo City, had come into production by late summer 1966 [55].

[5] The following companies and locations were listed, along with intended (if not actual) opening dates: for Allied Chemical (Geismar, La., and Hopewell, Va.), first half of 1967; American Cyanamid (Fortier, La.), third quarter 1966; Borden Chemical (Geismar, La), fourth quarter 1966; Chevron Chemical (Pascagoula, Miss.), April 1967; Commercial Solvents (Sterlington, La.), mid-1967; Continental Oil (Blytheville, Arkansas), May 1966; First Nitrogen (Donaldsonville, La.), September 1966; Coastal Chemical Corporation/Mississippi Chemical (Yazoo City, Miss.), March

In October, Kellogg publicity announced that four facilities based on its new single-train technology were in operation [56]. The construction delays, in the end, mattered little. The single-train process had more than proved its worth, including outside the US. By the end of 1966, Kellogg International Corporation, with offices in London, England, was building for the Dutch firm Ammoniak Unie, N.V., a joint venture of BASF and MEKOG, a 1000-tpd ammonia plant at Pernis, a major petrochemical complex near Rotterdam. Independently, Chemico also introduced a single-train energy-integrated process and contributed its process know-how towards a 1000-mtpd anhydrous, liquid ammonia unit for the Algerian government. Advances in refrigeration made possible efficient cooling of ammonia into its anhydrous, liquid form, which became the norm. Distribution of ammonia was carried out in pressurized railcars, barge and truck containers, and pipelines. Pipelines were used for distribution of natural gas and, increasingly, hydrogen, as well as ammonia.

The last months of 1966 were marked with high-powered advertising by equipment manufacturers and chemical engineering contractors in *Hydrocarbon Processing*, featuring for example the Bechtel-designed 1400-stpd ammonia facility for Olin Mathieson at Lake Charles, Louisiana [57, 58];[6] Chemico's one-hundredth ammonia plant, incorporating a 1000-tpd unit, for Borden Chemical at Geismar, Louisiana [59]; and the world's largest terrace wall reformer furnace for production of hydrogen, as already mentioned, at the 1500-tpd ammonia unit of Standard Oil Company of Kentucky, by Foster Wheeler [40].

The seventh Kellogg-designed 1000-stpd single-train unit to commence production in the US, in early 1967, was that of First Nitrogen Corporation, at Donaldsonville Louisiana [60]. This was followed by similar units at American Cyanamid's Fortier Complex, at Avondale, Louisiana [61, 62], and, at the end of 1968, Hill Chemicals Inc., at Borger, Texas, located 48 km north of Amarillo [63, 64].[7] In keeping with the increasing demand for urea as fertilizer, two large urea units, one for American Cyanamid (Fortier) and the other for Borden Chemical (Geismar, Louisiana), were added to their ammonia facilities [65]. Chemico built what in 1968 was claimed to be the "world's largest single-train methanol plant," with a rated capacity of 800-tpd, for Borden [66].

In 1967, Kellogg received the *Chemical Engineering* Kirkpatrick Chemical Engineering Achievement Award for the design and development of high-capacity, single-train ammonia units. Warren L. Smith, president of Kellogg, gave credit for the achievement to twelve of his colleagues in engineering and research groups: Len Axelrod, Jack Dwyer, Gunther Eschenbrenner, James (Jim) A. Finneran, Ben Jesser, Hays C. Mayo, Robert (Bob) H. Malthaup, Fritz Peterson, Larry Quartully, Herman Rickerman, Ronald (Ron) B. Smith, and Joe Yarze. At the same

1966; Olin Mathieson (Lake Charles, La.), March 1966; and US Steel (Clairton, Pa.), first half of 1967.

[6] The associated urea unit, with the "highest urea prill tower in existence," was engineered by Arthur G. McKee & Co.

[7] Hill Chemicals, with headquarters in Dallas, was incorporated in November 1965.

time, Smith emphasized that design of an ammonia plant was now a multi-divisional operation, integrating a range of skills, including chemical engineering, design engineering, catalysts, and pilot plant groups. There were also important contributions from suppliers, subcontractors, and client groups [67, 68].

By the end of the year, Kellogg had completed thirty-three single-train units with capacities of 600- to 1000-stpd, all based on high-pressure catalytic steam reforming of hydrocarbons [69]. This included at Allied Chemical's Hopewell synthetic ammonia factory, which had been producing 1000-stpd using technology dating back to 1928. The new single-train unit, introduced in 1967, had the same capacity. It occupied one-fifth the space of the existing 850,000-square-foot synthetic ammonia installation. Around the same time, Du Pont at Belle, West Virginia, completed an upgrade of ammonia production with a Kellogg unit, also of 1000-stpd capacity. Kellogg received additional orders, including from Gulf Oil (Donaldsonville, Louisiana), for 1000-stpd single-train units. In most projects, Kellogg acted as the main contractor, with responsibility for procurement, engineering, and supervision. Its subcontractors included not only suppliers and designers of equipment, but also firms with databases that could be applied to specific problems. This was the case when Kellogg asked Haldor Topsøe to undertake critical calculations concerning catalysts based on the latter's computing facilities and accumulated data. Contractors also joined forces with chemical companies to improve, patent, and market novel versions of existing processes. Here a good example is Uhde's joint development with ICI of low-pressure processes for ammonia and methanol, and thus suited to the new single-train technology. Humphreys & Glasgow designed a 1000-mtpd methanol unit based on this low-pressure technology for use at Billingham (see also later) [70]. By late 1967, Humphreys & Glasgow was working on the fourteenth ammonia plant in collaboration with ICI, in this case for Industriali Import Romania, of Bucharest. Humphreys & Glasgow was involved in the design and engineering of an ammonia plant for Norsk Hydro, at Herøya, near Porsgrunn, Norway. Norsk Hydro designed the ammonia loop. Construction of the plant was completed in 1967. Also in 1967, Uhde introduced a top-fired furnace for generation of synthesis gas that employed high-temperature cracking of hydrocarbons with steam. In 1968, Uhde, jointly with Chemie Linz of Austria developed an 889-mtpd ammonia converter with indirect heat exchange between catalyst beds, claiming advantages over conventional quench-type converters, as used in most large single-train plants. Other firms that joined the single-train boom included McKee, Snamprogetti, Stamicarbon, Toyo Engineering, and, for a short time, Ammonia Casale. (Ammonia Casale received orders for seven high-capacity units, ranging from 650- to 1000-mtpd: in 1964, a 650-mtpd unit for installation in France; in 1965, a 600-mtpd unit for Yugoslavia; in 1966, and 1967, two 1000-mtpd units destined for France; in 1969 a 1000-mtpd unit for Romania; in 1971, a 810-mtpd unit for W.R. Grace in Memphis, Tennessee; and in 1972 a 650-mtpd unit for France. See also Sect. 14.2)

Kellogg designed a 600-stpd ammonia unit installed at Terra Chemicals International, Inc., of Port Neal, Iowa (south of Sioux City), a site which included integrated urea fertilizer and ammonium nitrate units. Kellogg was also responsible for a 600-tpd ammonia unit supplied to African Explosives & Chemical Industries (AECI), an ICI affiliate, situated at Umbogintwini, south of Durban, South Africa [71].[8] For hydrogen, AECI used Koppers-Totzek low-pressure, high-temperature coal gasification technology. The main product was ammonium nitrate. The steam reformer was based on both ICI and Kellogg designs [72].

Also of 600-stpd capacity was the anhydrous ammonia plant of Farmland Industries, Inc., at Fort Dodge, Kansas. Opened by April 1967, it was "one of five Kellogg-built ammonia plants with capacities up to 1000 t per day to go on stream in the U.S. in the past 18 months. Around the world, four more Kellogg-designed 1000-t-per-day plants are now in startup—three are already producing ammonia—and 16 other large-capacity units are in various stages of engineering or construction" [73]. A second 600-tpd Kellogg unit for Farmland Industries went on stream in 1968. Farmland Industries, Central Farmers (CF Industries), and Mississippi Chemical were the principal manufacturers of ammonia for the US farm cooperatives. Farmland Industries was the largest agricultural cooperative in the US. In 1997 with Mississippi Chemical, it set up a joint venture to manufacture ammonia in Trinidad, taking advantage of cheap natural gas.

Considerable expansion also took place in Canada. In 1965–1966, Canadian Kellogg Co. Ltd constructed for Canadian Industries Limited (an ICI affiliate) an ammonia facility at Courtright, near Sarnia, Ontario, with total rated capacity of 1000-stpd [74]. The ammonia was used for both fertilizers and explosives. It was the last major ammonia facility built in the east. In 1969, Esso Chemical (Exxon) erected an ammonia unit at Redwater, Alberta, which with its access to cheap natural gas became the center of the Canadian ammonia industry. In 1979, the capacity of Redwater was 680-mtpd. By then, fourteen ammonia plants existed in Canada, of which just four were in the east, in Ontario, one of which was idle. The others were in the gas-rich provinces of Alberta and British Columbia [75]. CF Industries held a 49% interest in an ammonia-urea facility operated by Canadian Fertilizers Ltd at Medicine Hat, Alberta. On 17 June 1976, the first of two ammonia-urea complexes at Medicine Hat came on stream. Canadian Kellogg was the design contractor (Medicine Hat was acquired by CF Industries in 2010).

Cominco (later Agrium, and from 2018 Nutrien) operated nitrogen plants near Calgary (the former World War II plant, Alberta Nitrogen Products, Ltd), Alberta, and at Trail, British Columbia [75, pp. 112–113}.[9] It supplied nitrogen fertilizers to farmers in western Canada and the upper Midwest of the US (the Corn Belt). However, the Canadians could not match the cost advantage enjoyed by US manufacturers of ammonia, which by then were mainly concentrated in the Gulf region,

[8] AECI had used the original ICI process at Modderfontein since 1931–1932. In the mid-1960s the capacity was 135,000-tpy. Ammonia production at AECI ceased in 1999. Thereafter, Sasol (South African Synthetic Oil Limited, established in 1950) was the only producer of ammonia in South Africa, at Secunda and Sasolburg.

[9] Another operator in Calgary was, from 1965, Western Co-operative Fertilizers (Westco), which previously purchased ammonia from Cominco. Producers of ammonia in Alberta included Alberta Ammonia Co., of Calgary, and Sherritt, at Fort Saskatchewan.

with ready access to natural gas, and networks of pipelines linked to both suppliers and consumers. Almost 30% of rated US ammonia capacity was in Louisiana, and 10% in Texas. This was one reason why Cominco was also active in nitrogen production in the US, including 50% ownership in Hill Chemicals. The new Mid-America Pipe Line Co. carried anhydrous ammonia from Texas to Iowa, Kansas, and Nebraska. In 1977, at Carseland, Alberta, Cominco began large-scale production of ammonia and urea. By then, the government of Alberta had placed limits on the use of natural gas and brought the price of gas in line with the price in eastern Canada.

11.7 Ammonia in Japan and Southeast Asia

The much increased scale of production had a considerable impact in Japan, where engineering contractors, such as Chiyoda, Hitachi Zosen, Mitsui Toatsu, and Toyo Engineering—which in the 1950s had diversified away from major construction projects quite unrelated to chemicals, including shipbuilding—achieved major successes in ammonia, urea, and related processes. They were well-placed to bring about improvements and expansion of the country's nitrogen industry, and later also of several Asian countries. Japan, the world's fourth-largest producer of ammonia—after the US, the Soviet Union, and the Federal Republic of Germany—had at the close of 1966, twenty-two firms operating 27 ammonia plants. Three new high-capacity single-train units were about to come on stream, the largest, for Sumitomo Chemical, with a daily capacity of 750-mtpd. The two others were for Showa Denko (500-tpd) and Ube Industries (600-tpd) [76]. A year later, as a result of developments at Kellogg, Japanese firms were planning coordinated construction of five single-train units, each of 1000-tpd capacity, and closure of smaller, older, and less efficient plants.

11.7.1 Self-sufficiency in Southeast Asia

Kellogg's single-train design, by tripling of the maximum attainable capacity in a single unit, reduced by half the production cost of synthetic ammonia. The widespread introduction of these units in Western countries was in part predicated on expectations of large export orders for nitrogen fertilizers from the temperate regions. The demand was there, sometimes as the result of natural disasters. In both 1965 and 1966, harvests were not only disastrous in India but in most of Southeast Asia. This contributed towards the growing output of anhydrous synthetic ammonia in the US, which increased from 8.86 million tons in 1965 to 13.82 million tons in 1970 [77]. However, the demand from abroad soon began to level off, not because of the absence of need, but because of the absence of foreign exchange that prevented poorer countries from covering the cost of imports. Asian countries responded with self-sufficiency programs, not unlike those of the 1960s, using external assistance. In the early 1970s, the US Agency for International

Development and the United Kingdom Overseas Development Association contributed towards a 120-million dollar ammonia fertilizer complex at Kalol, Gujarat State. The Kalol site was the second in India based on Kellogg technology and natural gas as feedstock. It was inaugurated by Prime Minister Indira Gandhi on 8 November 1974 [78]. (Previously the large Indian facilities were based on naphtha.) The complex belonged to the Indian Farmers Fertiliser Cooperative Limited (IFFCO), a 30,000-village cooperative. The main products were urea and NPK fertilizer. Cooperative Fertilizers International, a not-for-profit US organization, provided managerial assistance. As with Western designed nitrogen plants erected in China and Russia around this time, local engineers and organizations were closely involved in the construction phase.

In the 1960s, Chemico erected ammonia-urea plants in Pakistan, where nitrogen fertilizers were needed as much as in India. In 1975, Kellogg undertook construction of a 1000-stpd ammonia plant for Pakistan's Pakarab Fertilizers Ltd, financed by the International Bank for Reconstruction and Development [79]. Most of the ammonia manufactured in Pakistan was converted into urea, which was also imported. The situation was similar in Bangladesh, where at the end of the 1970s urea was produced at Ghorasai and Fenchigani.

The Kellogg single-train ammonia unit and ICI's steam reforming technology, particularly as applied to naphtha as feedstock, had contributed immensely to the Green Revolution. High-capacity nitrogen plants became a front line against hunger and malnutrition. Food importing countries reduced the growth of their imports. Indian food production increased from 50 million metric tons in 1950–1951 to 152.4 million metric tons in 1984–1985 [80]. Today most of India's ammonia production is concentrated in Gujarat, Uttar Pradesh, and Andhra Pradesh. Production capacity is around 35 million metric tons per year, and for agricultural purposes, the ammonia is mostly converted into urea.

There was no question that the several changes in ammonia technology introduced during the first half of the 1960s represented the greatest transformation since the 1920s [29, pp. 426–437]. Process efficiencies introduced in the US included a 25% reduction in use of natural gas as a result of the introduction of centrifugal compressors [81]. This, and high-capacity single-train urea units, would bring about global benefits following the discovery of new gas deposits in Asia and elsewhere [82].

Quite unexpected, however, were events in the 1970s that would upset the world energy balance and at the same time encourage the US to trade its technology with the communist powers.

References

1. Freeman C (1968) Chemical process plant: innovation and the world market. National Institute Economic Review, no. 45, pp 29–57
2. Labrow S (1968) Influence of ammonia synthesis on compressor design. Chem Process Eng 48(Jan):55–57
3. Begg GAJ (1968) Turbo-compressors. Chem Process Eng 48(Jan):58–61

4. Weil BH, Lane JC (1948) Synthetic petroleum from the synthine process. Remsen Press Division, Chemical Publishing Co., New York, pp 185–186, 190, 199–200
5. Kellogg advertisement (1954) Pet Refin 33(Dec):303
6. Jacoby EH (1949) Agrarian unrest in Southeast Asia. Columbia University Press, New York
7. Baranski MR (2022) The globalisation of wheat: a critical history of the Green Revolution. University of Pittsburgh Press, Pittsburgh
8. Borlaug NE (1970) Seed for the future. In: International conference on mechanized dryland farming: proceedings of an international conference on mechanized dryland farming sponsored by Deere & Company under the auspices of the Food and Agriculture Organization of the United Nations, August 11–15, 1969. Deere & Company, Moline, Illinois, pp 70–88
9. Page A (2016) "The greatest victory which the chemist has won in the fight (…) against Nature": nitrogenous fertilizers and the British Empire, 1910s–1950s. Hist Sci 54:383–398, on 396–397
10. Mukherjee SK (1968) Plans for chemical fertilizers in India: retrospects and prospects. In: Fertilizer production, technology and use. Papers presented at the United Nations Interregional Seminar on the production of fertilizers, Kiev, Ukrainian Soviet Socialist Republic, 24 August–11 September 1965. United Nations, New York, pp 86–92
11. Dandapani SM (1968) Fertilizer marketing in a developing country. In: Fertilizer production, technology and use. Papers presented at the United Nations Interregional Seminar on the production of fertilizers, Kiev, Ukrainian Soviet Socialist Republic, 24 Aug–11 Sept 1965. United Nations, New York, pp 302–306
12. Saha M (2013) Food for soil, food for people: research on food crops, fertilizers, and the making of 'Modern' Indian agriculture. Technol Cult 54(2)(April):289–316
13. Anon (1959) Orders and contracts. Chem Process Eng 40(Nov):40
14. Anon (1963) Indian compound fertiliser plant. Chem Ind, 4 May, p 747
15. Anon (1964) Hydrocarb Process 43(June):214 (Gujarat State Fertiliser Co., Ltd)
16. Anon (1966) Construction boxscore section. Hydrocarb Process 45(Sept):91-K
17. Anon (1966) Hydrocarb Process 45(Oct):72-J (Gujarat State Fertilizers Co. Ltd)
18. Anon (1966) Chem Ind, 26 Nov, p 1980 (Humphreys & Glasgow Ltd)
19. Anon (1965) India must build bigger, say experts. Hydrocarb Process 44(Sept):71F
20. Brown LR (1970) Seeds of change: the green revolution and developments in the 1970s. Praeger Publishers, New York, pp 56–59
21. Brown LR (2013) Breaking new ground: a personal history. WW Norton, New York
22. Anon (1966) The Indian government. Hydrocarb Process 45(Oct):72-B
23. Mashelkar RA, Rajan JV (1989) Chemical engineering developments in India. In: Peppas NA (ed) One hundred years of chemical engineering: from Lewis M. Norton (M.I.T. 1888) to present. Kluwer, Dordrecht, pp 153–221
24. Power-Gas Corporation Limited advertisement (1968) Hydrocarb Process 47(Nov):20
25. Kane GP (1965) Chemical industry growth in India. Chem Eng Progr 61(June):107–112
26. Phalkey J, Wang Z (2016) Planning for science and technology in China and India. Br J Hist Sci Themes 1:83–113, on 92–101
27. Chandran NB (1984) Setting up and operating ammonia plants. The Indian scenario today. Econ Polit Wkly 19(50)(15 Dec):A21, A23–A24
28. Smil V (2001) Enriching the earth: Fritz Haber, Carl Bosch, and the transformation of world food production. The MIT Press, Cambridge, Mass, p 125
29. Spitz PH (1988) Petrochemicals: the rise of an industry. Wiley, New York, p 434
30. Anon (1954) Kellogg ammonia process. Pet Refin 33(Mar):150–154
31. Finneran JA, Mayo HC, Multhaup RH, Smith RB (1966) US patent 3441393 (filed 19 Jan 1966, approved 29 April 1969). Process for the production of hydrogen-rich gas (assigned to Pullman Incorporated)
32. Kellogg advertisement (1953) Pet Refin 32(Mar):232–233
33. Kellogg advertisement (1954) Pet Refin 33(Mar):244
34. Quartulli OJ, Fleming JB, Finneran JA (1968) Best pressure for NH_3 plants. Hydrocarb Process 47(Nov):153–161

35. Eschenbrenner GP, Wagner GA III (1972) A new high capacity ammonia converter. Chem Eng Pract 68(Jan):62–66
36. Axelrod LC, O'Hare TE (1968) Aspects of large-scale ammonia production. In: Fertilizer production, technology and use. Papers presented at the United Nations Interregional Seminar on the production of fertilizers, Kiev, Ukrainian Soviet Socialist Republic, 24 August–11 September 1965. United Nations, New York, pp 165–168, on 165.
37. Pattabathula V (2019) Ammonia. In: Kirk-Othmer encyclopedia of chemical technology. https://doi.org/10.1002/0471238961.011313150326211 6.a01.pub4
38. Kellogg advertisement (1964) Hydrocarb Process 43(July):14
39. Cooper-Bessemer advertisement (1966) Hydrocarb Process 45(Sept):63
40. Foster Wheeler advertisement (1966) Hydrocarb Process 45(Nov):140
41. Constant EW II (1980) The origins of the turbojet revolution. Johns Hopkins University Press, Baltimore
42. Brightling J, Pach J (2013) Ammonia production: a century of safety, health & environmental improvements. In: Ammonia Technical Manual, pp 231–245, on 234–235
43. Fowles M, Carlsson M (2021) Steam reforming of hydrocarbons for synthesis gas production. Top Catal. https://doi.org/10.1007/s11244-021-01496-z
44. Pattabathula V, Richardson J (2016) Introduction to ammonia production. Chem Eng Prog 112(Sept):69–75
45. Anon (1966) Hidden profits for ICI. Chem Ind, 19 Mar, p 474
46. Anon (1967) Startup delays gripes ICI. Hydrocarb Process 46(July):53
47. Kellogg advertisements (1966) Hydrocarb Process 45(July):16, and 45(Nov):33
48. Anon (1967) Ammonia plant. Chem Ind, 1 April, p 521
49. Prichard JG (1988) Making things grow: the story of Mississippi Chemical Corporation. University Press of Mississippi, Jackson, Mississippi
50. Kellogg advertisement (1965) Hydrocarb Process 44(April):285
51. Kellogg advertisement (1965) Hydrocarb Process 44 (Feb):2
52. Kellogg advertisement (1965) Chem Eng Progr 61(July):113
53. Anon (1966) Kellogg's single train large ammonia plants achieve lowest costs. C&EN, 20 Nov, p 112
54. Lyon SD (1975) Development of the ammonia industry. Chem Ind, 6 Sept, pp 731–739
55. Anon (1966) Ammonia market in America. Chem Ind 29 Aug, pp. 1423
56. Kellogg advertisement (1966) Hydrocarb Process 45(Oct):286
57. Bechtel advertisement (1966) Hydrocarb Process 45(Oct):57
58. Anon (1966) Hydrocarb Process 45(Dec):67 (Olin Mathieson Chemical Corp.)
59. Chemico advertisement (1966) Hydrocarb Process 45(Nov):333
60. Kellogg advertisement (1967) Chem Eng Progr 63(June):9
61. Anon (1967) American Cyanamid ammonia plant. Chem Ind, 13 May, p 765
62. Kellogg advertisement (1967) Hydrocarb Process 46(Aug):2
63. Anon (1967) Texas fertiliser project. Chem Ind, 3 June, p 887
64. Anon (1967) Hill Chemicals ammonia fertiliser complex. Chem Ind, 14 Oct, p 1711–1712
65. Anon (1965) Chemico, DSM develop new urea processes. Hydrocarb Process 44(Oct):47
66. Anon (1968) Hydrocarb Process 47(Nov):34 (Borden Chemical Co.)
67. Anon (1967) Kellogg's single-train large ammonia plants achieve lowest costs. Chem Eng 20(Nov):112–117
68. Ostuni R (2021) Kellogg technology. In: Cariati V (ed) Ammonia for life: the synthesis of ammonia and the birth of modern chemical industry. Casale SA, Lugano, pp 95–110, on 109
69. Anon (1967) Ammonia: The M.W. Kellogg Company. Hydrocarb Process 46(Nov):148
70. Woolcock JW (1968) Low-pressure methanol process. Chem Process Eng 47(June):69–72 (see also Humphreys & Glasgow advertisement, on p 73)
71. Anon (1985) Ammonia Production at AECI. The South African Chemical Foundation: Information centre on the chemical industry, Bulletin no. 17, June, pp 9–10
72. Kellogg advertisement (1967) Hydrocarb Process 46(Jan):18
73. Kellogg advertisement (1967) Chem Eng Progr 63(April):93

74. Kellogg advertisement (1965) Hydrocarb Process 44(July):14
75. Foran WJ (1979) Fertilisers. Chem Ind, 17 Feb, pp 111–116
76. Ponder TC (1966) Keep an eye on the Japanese HPI (hydrocarbon processing industry). Hydrocarb Process 45(Dec):176C–176P, on 176K
77. Greenberg E, Christopher T, Hill CT, Newburger DJ (1979) Regulation, market prices, and process innovation: the case of the ammonia industry. Westview Press (reprinted: Routledge, London, 2019), p 4
78. Prime Minister Indira Gandhi dedicates major new Indian fertilizer complex. M.W. Kellogg press release, 8 November 1974. M.W. Kellogg documents located in Box 219, folder "Pullman-Kellogg (1)" of the U.S.-China Business Council Records, at the Gerald R. Ford Presidential Library. https://www.fordlibrarymuseum.gov/library/document/0029/7490136. Accessed 20 Nov 2022
79. £250,000 Pakistan export order for APV-Paramount (1975) Chem Ind, 3 May, p 361 (APV-Paramount supplied 350 tube assemblies)
80. Randhawa NS (1987) Agricultural research in India: an overview of its organization, management and operations. Food and Agriculture Organization of the United Nations, Rome, p 208
81. Anon (1980) Chem Ind, 21 June, pp 491–499, on 493
82. Cook LH (1966) Use these design criteria for large complete recycle urea plants. Hydrocarb Process 45(Feb):129–136

12 The Energy Crises, the People's Republic of China, the Soviet Union, and the 1970s

Abstract

The boom resulting from the introduction of Kellogg and other high-capacity single-train ammonia units lasted, in the West, until the energy crises of the 1970s. Fluctuations in the prices of energy and ammonia militated against new investments in the industry. However, the demand for ammonia continued to grow in those economies based on central planning. This chapter reviews the major changes in the global ammonia industry—invariably driven by the need for increased agricultural output. Ammonia technologies became tools in the complex world of international diplomacy; they brought about better relations between the US and the People's Republic of China and the Soviet Union. Transfer of technology was brought about through the engineering contractors, cross-licensing of technologies, and international financing. ICI's new low-pressure methanol process, borrowed from ammonia technology, was introduced into the Soviet Union and elsewhere. Asian and Middle East countries were emerging as major producers of ammonia.

12.1 The Energy Crises

Increases in energy costs from October 1973 as a result of the Yom Kippur War in the Middle East adversely affected the manufacture of most kinds of heavy chemicals, particularly of ammonia and fertilizers, in Western countries. Arab oil producers belonging to OPEC (the Organization of Petroleum Exporting Countries) imposed an embargo on exports to the US and other countries. A fourfold rise in the price of petroleum precipitated an economic crisis, particularly in Europe, which imported over 60% of its energy requirements, mostly as petroleum, from countries in the Middle East. This was followed by an even greater shock to oil consumers during 1978–1979, a tenfold increase in the price of crude oil. With

A. S. Travis, *The Global Synthetic Ammonia Industry*, Perspectives on the History of Chemistry, https://doi.org/10.1007/978-3-031-87978-4_12

the cost of energy responsible for 80% or more of the cost of ammonia manufacture, there was a substantial decline in profitability in the fertilizer industry [1]. This had a major impact in the US where at the end of the 1970s there were over one-hundred ammonia plants with an aggregate rated capacity of 20 million tons [2]. Less efficient ammonia factories were closed, including Du Pont's Belle facility, which ceased producing ammonia in 1978. Within a few years, Gulf Oil was running down the ammonia facilities at its former Spencer Chemicals sites.

Though the situation improved for a short while around 1980, the crisis persisted until well into the middle of the decade, by which time the price of oil had fallen (diluting the purchasing power of OPEC nations). In the West, one major outcome of the fluctuations in supply and price was, both in the mid-1970s and early 1980s, and with long-term security in mind, an interest in seeking new sources of energy. This included improved pressurized gasification of coal for synthesis gas production through the Koppers-Totzek, Lurgi, Texaco, Winkler, and other processes [3, 4]. Significantly, the dynamics of the relationship between coal and petroleum as sources of gases would continue well into the 1980s. This was nothing new. Natural gas was another matter. In the mid-1950s, there had been predictions that the use of natural gas in the US as a chemical feedstock would decline because it was a premium domestic fuel. This was one of the reasons why the Texaco Development Company's oxidation process for converting natural gas to synthesis gas was modified for use with heavy fuel as feedstock [5]. Later, even the potential of nuclear power as a source of hydrogen in the synthesis of ammonia was considered. However, petroleum and, especially, natural gas remained the most important feedstocks in the ammonia industry (the exceptions, relying on coal, were mainly the People's Republic of China, South Africa, India, Pakistan, and until the 1990s, the former German Democratic Republic).

Notwithstanding concerns over the costs and availabilities of feedstocks, there were opportunities for promoting the new reformer and single-train ammonia technologies once the US opened up to business with the main communist powers, China and the Soviet Union. The technologies of nitrogen products, and also of methanol, became tools in the complex world of international diplomacy; their global transfers were brought about through engineering contractors, cross-licensing of technologies, and international financing. The greatest impact was in the 1970s when the People's Republic of China, the Soviet Union, and the Middle East countries were emerging as major producers of ammonia. Significantly, China had devoted more land to cultivation of wheat than any other nation and adopted policies based on agrarian self-sufficiency. Introduction of the new ammonia technologies and demand for nitrogen fertilizer contributed to the doubling of global ammonia capacity, from 50 million tons in 1970 to around 100 million tons in 1980.

12.2 The People's Republic of China

While many Western chemical and engineering contractors were badly hit by the economic impacts and competitive pressures of the 1970s, they found some relief from the growing markets in Asian and Eastern Bloc countries. The People's Republic of China, already an importer of ammonia technologies, mainly from Europe, is an important case in point. China was of particular significance to American contractors.

During China's first five-year plan (1953–1957) the country's two prewar ammonia factories were reactivated (Dalian, Mont Cenis/Uhde process; and Yongli, Nanjing, Nitrogen Engineering Corporation process) and, with assistance from the Soviet Union, three medium-sized plants, with a total rated capacity of 50,000-tpy were erected at Jilin, in the northeast, Lanzhou (China National Petroleum Corporation), in the northwest, and Taiyuan, in the north (Taiyuan Chemical Industry Group). During 1956–1958, the Design Institute of the Ministry of Chemical Industry developed a synthetic ammonia unit with a rated capacity of 50,000-tpy, installed first at the Sichuan Chemical Complex of the Sichuan Chemical Plant, and subsequently at the Zhejiang Synthetic Ammonia Plant, the Shanghai Wujing Chemical Plant, and the Guangzhou Nitrogen Fertilizer Plant [6]. Around the same time, a facility for manufacture of catalysts was opened at Nanjing to serve the sulfuric acid and ammonia industries. In an attempt at gaining self-sufficiency, some 1500 small and not very efficient ammonia units were installed at widely dispersed locations. Capacities ranged from 15- to 60-tpd [7, 8]. Many of them relied on anthracite and converted the ammonia into ammonium bicarbonate, the use of which as a fertilizer was almost a chance discovery. From 1958, bicarbonate became the main fertilizer in China and in the early 1970s accounted for half of Chinese nitrogen fertilizer use.

Between 1963 and 1965, following the great famine of 1959–1961, in part the result of Mao Zedong's Great Leap Forward, five ammonia units were ordered from the UK (and others from Italy), in addition to a Stamicarbon urea unit from the Netherlands. The transportation of ever larger ammonia converters from Europe to China created logistical problems, for both land and sea transport. For example, in 1968 Humphreys & Glasgow shipped a 130-t converter from Glasgow to Liverpool by rail, whereby special arrangements had to then be made for its loading onto a China-bound ship. This included the challenge of finding suitable lifting equipment [9]. In 1965, ICI had signed a contract with China for supply of 40,000 t of ammonium sulfate. Just under two years later, around the time that the new Kellogg ammonia units came into production at Billingham, a second contract was signed with China for nitrogenous fertilizers to the value of one million pounds sterling [10].

Despite the ravages of the Cultural Revolution (1966–1976), peasants were encouraged to bring about improvements in agricultural production by technically educated individuals sent from the cities to the countryside. Great emphasis was placed on the scientific application and use of fertilizers. In 1971, at the mid-point of the revolution China became interested in trade with the US.

President Nixon was then dealing with problems in the US economy, notably unemployment and inflation, in addition to a battered dollar. Wage and price controls were introduced in the summer of 1971. Tariffs were imposed, for a short time, to give a jump start to American industry, which was encouraged to increase exports in order to support the dollar. This included expansion of the agricultural sector, which led to massive production of nitrogen fertilizer to satisfy home use. There was also the war in Vietnam. The American administration believed that by opening up to the People's Republic of China, it might be possible to take advantage of frictions between China and the Soviet Union, then the main backer of North Vietnam, in order to exert pressure to bear on the Russians, which could in turn be used to help bring about an end to the war. The first contact for many decades between the US and China took place in April 1971, when nine members of the US table tennis team made a trip to the communist country. On their travels, they were greeted by Premier Zhou Enlai. In July, US National Security Advisor Henry Kissinger traveled in great secrecy from Islamabad, Pakistan, to Beijing, where he met with the Chinese leadership to discuss trade and defense issues. This period of so-called ping-pong diplomacy led to a thaw in relations between the US and the People's Republic of China. In trade affairs, the US lifted its embargoes, going back to 1949, against China. Chemical processes were among the main bargaining tools. In December 1971, the US government approved the first export license to the Chinse mainland for the acrylonitrile process of the Standard Oil Company of Ohio. In the same month, war broke out between India and Pakistan (resulting in the establishment of Bangladesh in place of East Pakistan on 16 December).

In February 1972, Nixon famously traveled to China to meet with Mao Zedong and Zhou Enlai. The Chinese government soon after ordered thirteen large ammonia-urea complexes, of which eight were assigned to Kellogg, and five to European and Japanese engineering contractors [11]. The Kellogg orders, all for 1000-mtpd single-train ammonia units, were "the largest dollar volume [order] ever placed by the PRC with a U.S. firm in the industrial sector" [12].

The combined size of the Chinese orders was unprecedented in the history of chemical industry. Early in 1974, Kellogg found it worthwhile to become a member of the National Council for United States-China Trade. One corporation that greatly assisted Kellogg in gaining access to the Chinese market was Farmland Industries, which had substantial overseas interests, and at Enid, Oklahoma, in 1974 placed on stream a 1000-tpd Kellogg ammonia unit. By 1980, all the new high-capacity single-train Chinese ammonia units were in operation or under construction.

The engagement of China with European engineering contractors followed the establishment in 1975 of trade relations between the communist nation and the European Community (and a major trade agreement in 1978). The original contracts with European firms for ammonia and urea plants were based on the use of natural gas as feedstock. However, two plants were subsequently converted to naphtha, and others to coal. These changes away from natural gas may have been the result of difficulties with the transport of gas to the manufacturing sites. For

1978–1979, Chinese urea capacity was seven million metric tons, and ammonia capacity was over four and a half million metric tons.

At least five more contracts for 1000-mtpd capacity single-train ammonia units were awarded to Western firms. One unit was based on coal and two on heavy oil as feedstocks. In 1978, there were fifty-five ammonia facilities in China, mainly of medium size, in addition to the 1500 small factories [7]. Around 1980, the German firm Lurgi Kohle und Mineraloltechnik GmbH obtained a contract from the Chinese government for a coal-based 1000-mtpd ammonia unit, incorporating Lurgi coal and pressure gasification technology and gas purification and conditioning processes. Lurgi also received a contract for a methanol plant [13, 14]. Using technology licensed by the DSM subsidiary Stamicarbon, Kellogg provided technical services for a 1740-mtpd urea unit, constructed by China National Chemical Construction Corporation. Stamicarbon was favored for its know-how in coal technologies, as well as licensor of DSM's urea and nitric acid processes [15, 16]. China National Technical Corporation signed a ten-year agreement with Linde AG for supply of ammonia technologies.

After 1980, as a result of the importation of ammonia technologies, particularly the high-capacity single-train ammonia units, Chinese farmers were enabled to set records in grain production. By 1986, there were 14 ammonia units operating with rated capacities of 1000-mtpd and three more under construction. Around one thousand of the small factories were still in operation, generally based on improved efficiencies. Notwithstanding the existence of medium and large plants, 57% of synthetic ammonia production was still derived from the small factories [7]. The main Chinese products were urea, ammonium nitrate, ammonium bicarbonate, ammonium chloride, and ammonia liquor. With a growing emphasis on urea in agriculture, the use of ammonium bicarbonate declined. Total production of nitrogen fertilizer was over 12 million tons in 1984, compared with just one million tons in 1965.

China, by building world-scale ammonia plants, became the global leader in ammonia production in the early 1990s. It accounts for one-third of international production and is the only nation that undertakes large-scale production of ammonia based on coal, amounting to approximately 80% of capacity. The balance is produced from natural gas. China National Petroleum Company controls petrochemical production in the north of the country; Sinopec International controls the same sector in the south.

Of all the Kellogg ammonia plants erected during the 1970s and 1980s, 80% were located in 23 different countries outside traditional ammonia-producing Western nations, though mainly in the People's Republic of China and the Soviet Union. Developments in the Soviet Union paralleled those in China.

In the 1960s, and in contrast to American contractors, European contractors, including the Schneider group and other French firms, were already open to direct business with the developed centrally planned economies of the Eastern Bloc. ICI's dealings with the Soviet Union followed a meeting between its chairman, Sir Paul Chambers, and Russian statesman Alexei Kosygin in Moscow in May

1964. Following exchanges of technical delegations, a five-year agreement covering cooperation was signed on 17 October 1966 at ICI's head office in London. This would pay off in the 1970s with, for example, the export of ICI's low-pressure methanol technology (see next section).

To obtain a foothold in the Eastern Bloc markets, Kellogg engaged in partnerships, licensing, and other arrangements with European and Japanese contractors. In 1969, Kellogg provided Toyo Engineering with process design for a 1500-stpd ammonia unit ordered by the Soviet import agency Techmashimport for installation at the Nevinnomysskiy [Nevinniyssk] Nitrogen Fertilizer Factory located in the foothills of the Caucasus Mountains [17]. It went on stream in 1973. Today the Nevinnomysskiy complex belongs to the EuroChem Group, a major producer of ammonia in Russia with headquarters in Switzerland.

12.3 The US and the Soviet Union

In May 1972, a few months after his visit to China, President Nixon traveled to the Soviet Union in an effort to improve relations between the two world powers. He returned there two years later to sign an economic agreement and the Threshold Test Ban Treaty [11].

In the meantime, the National Association of Manufacturers had sponsored a US-Soviet trade conference attended by, among others, engineering contractors, including Bechtel. On 12 April 1973, Armand Hammer's Occidental Petroleum Corporation of California signed a "chemical-fertilizer barter contract" with the Soviet government, represented by Leonid Kostandov, minister of the chemical industry, which involved transfer to Russia of American technology for ammonia production, and export of superphosphoric acid to Russia, to be paid for by the Russian-made ammonia, as well as urea and potash [18]. As part of the deal, over $400 million would be spent in the US for the supply of ten ammonia units.

William J. Casey, chairman of Export–Import Bank of the United States (Eximbank), the official export credit agency of the US government, authorized a loan of $180 million to the Soviet Union in order to facilitate a so-called fertilizer *détente* [19]. As a result, in June 1974, Chemico (then a division of Aerojet-General Corporation, a subsidiary of General Tire & Rubber Company; later GenCorp, Inc.) received a $200 million contract for construction of four ammonia plants, the first one at Togliatti, a city on the Volga River [20]. This joint project involving Occidental Petroleum and the Soviet government was the forerunner of the TogliattiAzot corporation. Integral to its success was the 2471 km Togliatti-Odessa ammonia pipeline, the world's longest. Construction began in 1975, and in 1981 the pipeline began pumping ammonia from western Russia to ports in southern and western Ukraine for export.

Certainly, American engineering contractors had become better placed to supply the Soviet Union with chemical technology. The large Eastern Bloc market was a factor in Kellogg's acquisition in January 1973 of 51% of the Dutch firm Continental Engineering from Verenigde Machinefabriken (VMF). Continental Engineering had been founded in 1962 in part to serve the Eastern Bloc, and in particular the Soviet Union, where it built large urea factories. In the mid-1960s, Continental Engineering supervised construction of a 250-tpd urea plant at Hûngnam, North Korea. Continental Engineering, renamed Kellogg Continental BV, opened up a more direct market for Kellogg technology in the Soviet Union [16]. Thus, following on from the 1969 agreement between Kellogg and Toyo Engineering, Kellogg Continental licensed design technology to Toyo, which together with Russian engineers erected four more ammonia plants in Russia, one each at Severodonetsk (Ukraine) and Novgorod, in the Novgorod Oblast, and two at Novomoskovosk, in the Tula Oblast. Their rated capacities were 1500-mptd. The final two to be completed, at Novgorod, 150 km southeast of Leningrad (today St Petersburg), and Novomoskovosk, 220 km south of Moscow, came on stream in the summer of 1975 [21]. In August 1975, another four plants were commissioned, for Cheropovets (Vologda Oblast), Dneprodzershinsk (Ukraine), Dorogobuzh (Smolensk Oblast, western Russia), and Novgorod [22, 23]. Kellogg Continental also supplied design services to Creusot-Loire Enterprises which erected four 1500-mtpd ammonia units, two at Gorlovka, in Ukraine, and two at Kemerovo, a coal mining center in south-central Russia. As with the Occidental deal, Italy and France received deliveries of Soviet-made ammonia in compensation for ammonia plant and other industrial equipment supplied by Italian and French contractors.

The *Wall Street Journal* in December 1974, under the heading "No slump here: Builders of ammonia plants benefit from global demand," trumpeted the fact that American engineering contractors were involved in twelve ammonia projects in the Soviet Union, nine with Kellogg and three with Fluor Corporation, in addition to eight projects in China [24]. President Nixon's *rapprochement* with the communist powers had given a much-needed boost to at least one sector of the American economy. However, Nixon had resigned in August 1974 as a result of the Watergate scandal. His successor, Gerald Ford, inherited the worst economy since the Great Recession, though it was alleviated to some extent by his continuance with the policy of *détente*. The orders from the Soviet Union continued to arrive, and Kellogg through its subsidiary Kellogg Continental built up a substantial business with the communist regime. Kellogg's signing in late 1976 of a contract for supply of a 1500-tpd synthetic ammonia unit to the Soviet Union brought to eighteen the number of Kellogg designed ammonia units operating in, or under contract with, Russians enterprises [25].

In early January 1980, however, geopolitical issues intervened when US President Jimmy Carter responded to the Soviet Union's invasion of Afghanistan in the previous December by introducing various sanctions, especially an agricultural embargo. This prevented the shipping of seventeen million tons of wheat and corn sold under contract to the Soviet Union prior to October 1980. The International Longshoremen's Association, which controlled Atlantic and Gulf Coast docks, immediately boycotted all Soviet ships and cargo. However, the International Longshoremen's and Warehousemen's Union, which controlled docks on the Pacific coast, where one-quarter of US imports of ammonia from the Soviet Union were normally unloaded, did not introduce a boycott. This difficult situation impacted on US farmers, who, despite the government's assumption of contractual obligations for wheat, lost sales of already grown grain scheduled for export, and also on Armand Hammer's joint ammonia venture with the Soviet government. (The US also led a boycott of the 1980 Olympics in Moscow.) The embargo was ended, after vigorous campaigning by the US farm lobby, by President Ronald Reagan on 24 April 1981. The fertilizer *détente* was restored, and the Soviet Union resumed export of natural gas, ammonia, urea, and potash fertilizers to the US. Also restored were shipments of superphosphoric acid from Florida to Russia, in accord with the USSR-Occidental agreement.

In part, the embargo may have encouraged a move towards self-sufficiency in design and manufacture of high-capacity ammonia units in the Soviet Union. During the 1970s, the state-owned contractor GIAP (State Research, Development and Design Institute of Nitrogen and Organic Chemicals) designed a Kellogg-type ammonia unit based on those licensed to Toyo Engineering. Thirty-seven were installed by the late 1980s. The converter, with a rated capacity of 1360-mtpd, was still of the bottleneck style, but smaller, and the working pressure was higher than in the Kellogg model. These high-capacity ammonia units contributed to the Soviet Union's role as the leading global producer of synthetic ammonia, at least until the late 1980s. Russian engineers also designed a converter that was horizontally placed. It had interchangers between the beds, and the gas crossed the beds from top to bottom, that is, by cross flow, instead of radial flow. Today GIAP is a leading Russian engineering contractor specializing in various sectors of the chemical industry, including ammonia, nitric acid, ammonium nitrate, and methanol.

The Soviet Union's urea technology came from Europe. In 1976, Montedison's engineering division, Tecnimont SpA, received an order for a urea plant of 500,000-tpy rated capacity at Berezniki, a city located on the banks of the Kama River, in the Ural Mountains [26]. Berezniki, today part of UralChem, produced ammonia from 1932, and in 1974 changed over from coke oven gas to natural gas for production of hydrogen. A second urea unit was ordered for Gorlovka. Both were erected by the Soviet government. Most new urea units ordered during the 1970s drew on Snamprogetti and Stamicarbon technology, as licensors. Snamprogetti also constructed ammonia units at Novomoskovsk and Togliatti, while Stamicarbon licensed its urea technology to Chemoprojekt of Prague (founded in 1950), which built urea facilities at Berezniki, Grodno (Belarus), and Salavat (for Salavat-nefteorgsintez). Uhde introduced Stamicarbon urea technology in Odessa. As in the 1930s, and the 1973 arrangement with Occidental Petroleum, Russia repaid loans with commodities and natural resources, originally wheat, and industrial products, including nitrogen fertilizer.

In the late 1970s, Kellogg technology was adopted for a 1360-mtpd ammonia unit at the Leuna works in the German Democratic Republic (East Germany), then called the VEB Leuna-Werke Walter Ulbrecht; Toyo Engineering provided design, engineering, and procurement services. This was the site chosen by BASF for expansion of Haber–Bosch ammonia production during World War I.

An important UK contribution to industrial high-pressure chemistry in the Soviet Union was a new synthetic methanol technology. This was ICI's low-pressure methanol process, based on a 1963 patent, claiming operation at 30 to 120 atm, filed in the names of Phineas Davies and Frederick Snowdon. The process, introduced in late 1967, and in use at Billingham from the mid-1970s, involved novel catalysts and drew on synthesis gas produced by steam reforming of light hydrocarbons or partial oxidation of heavy hydrocarbons [27, 28]. Contracts for two methanol units were drawn up between ICI and Techmashimport in the summer of 1977. Finance came through a buyer credit arranged by Morgan Grenfell Ltd, under the terms of the 1975 Anglo-Soviet Credit Agreement. The engineering firms involved were Davy Powergas and Klockner Ina Industrial Plants.[1] The methanol units, installed at Gubakha (Metafrax Chemicals) in the Urals, and Tomsk, in Siberia, were the largest in the world, and represented the UK's largest ever deal with the Soviet Union. This brought to twenty-six the number of facilities that worked the ICI low-pressure process, which accounted for some 80% of world methanol capacity [29].

12.4 Developments Elsewhere from Around 1970

This section provides some further examples of the spread of the modern nitrogen industry on the global stage. Between 1970 and 1974, according to estimates of Arthur D. Little, Inc., global nitrogen fertilizer consumption increased fivefold. From the mid-1970s, and despite economic downturns, Western firms began to plan for large-scale expansion in fertilizers to meet anticipated high demand. In mid-1974, Humphreys & Glasgow (later Jacobs) signed a contract with ICI to provide engineering services for Billingham's no. 4 production unit, Ammonia IV, which was designed by ICI with a rated capacity of 1000-mtpd [30]. The novel converter, of 15 m in length, with a 150 mm thick shell, was fitted with a lozenge type catalyst cartridge, in which two wire mesh lozenges introduced cooling gas into the continuous bed of large-size catalyst. As required, catalyst was discharged by gravity, for maintenance or recharging of the converter. This single-train ammonia unit, utilizing North Sea natural gas as feedstock, operated with ICI's low-pressure technology at 220 atm. Performance was high, achieving outputs of 1125–1200-mtpd.

Countries with large agricultural sectors continued to expand in, and modernize, production of ammonia and basic fertilizer chemicals, despite, and in many way because of, high energy costs. This included the Republic of Ireland, probably the fastest-growing market for nitrogen fertilizers in Europe. In 1974, EIR-N Chemicals Ltd (jointly owned by Cooperative Agricultural Purchase, Ltd, and N-REN Corp., of Cincinnati, USA) of Cork, awarded Coppée-Rust of Belgium a contract

[1] Davy Powergas was successor to Power-Gas Corporation Co., Ltd; it now belongs to Johnson Matthey.

for a large urea plant, at a cost of $8.5M, to produce 54,000-mtpy. In 1975, Nítrigin Eireann Teoranta (NET)—which had commenced manufacture of ammonia in 1965 at Arklow, County Wicklow—awarded Kellogg a contract for a 1350-mtpd ammonia and 1000-mtpd urea complex, the latter based on Stamicarbon technology, at Marino Point, Cork Harbour. The feedstock was natural gas. This was shortly after Stamicarbon, the subsidiary of DSM, granted Kellogg worldwide rights to its nitrogen and phosphate products [31].

Taiwan's China Petrochemical Development Corporation, a state-owned enterprise established in 1969, awarded a contract to C-E Lummus for a 88,000-tpy methanol plant and 80,000-tpy acetic acid plant at Kaoshiung, for completion in 1982.[2] Kaoshiung drew on natural gas, then recently discovered offshore in the Gulf of Thailand.

Around 1970, Kellogg was commissioned to undertake expansion at the PUSRI Palembang facility in South Sumatra, Indonesia, starting with a 660-mtpd ammonia unit integrated with a 1150-mtpd urea unit. They came on stream in 1974, at which time a second contract was signed, for a 1000-mtpd ammonia unit and a 1725-mtpd urea unit, followed in 1975 by a third, and similar, contract. Toyo Engineering supplied the urea technology [32]. P.T. Kaltim Pasifik Ammonia, at Bontang, Kalimantan, started production of ammonia and urea in 1984 (in 2014 Kaltim Pasifik was acquired by Pupuk Kaltim). These facilities rely on natural gas as feedstock.

Mexico was another beneficiary of the expanding ammonia industry. In the late 1940s, state-owned Petróleos Mexicanos (PEMEX) commissioned Chemico to build a fertilizer plant near Mexico City for production of ammonia, sulfuric acid, and ammonium sulfate. The feedstock was natural gas. In the late 1960s, The Lummus Company of London erected a 1200-mtpd ammonia unit for PEMEX at Minatitlan [33]. The converter, weighing 240-t, was supplied by Babcock & Wilcox (Operations), Ltd, of London and Glasgow [34]. PEMEX remained an important user of American and European technologies. In the early 1970s, PEMEX placed an order with Kellogg for an ammonia unit of 1000-mtpd capacity, to be installed at Cosoleacaque, in Veracruz, southern Mexico. In July 1974, Kellogg received a further order, this time for three ammonia units, two of 1500-mtpd capacity for Cosoleacaque, and one of 1000-mtpd capacity, for Salamanca (Guanajuato). The first unit went on stream early in 1975 [35]. In 1977, PEMEX awarded Kellogg a contract for two 1500-tpd ammonia units at Tabasco, scheduled to go on stream in 1980 [36]. Also for PEMEX, in the early 1980s Uhde designed a single-train methanol unit, with a daily capacity of 2500 metric tons, using the low-pressure process earlier developed by the alliance between ICI and Uhde, for installation in Veracruz. It was scheduled to be in operation by 1984.

Thus far, the international spread of the ammonia industry has been described in terms of loans and grants, licensing arrangements, and the procurement of

[2] From 1966, Lummus was wholly owned by Combustion Engineering (C-E), subsequently C-E Lummus and Engineering Services Group.

equipment, involving engineering contractors, including consortia of firms. In the case of the chemical corporations there were also joint equity ventures. At first, such arrangements were invariably created between multinational American or European firms. However, during the 1960s, joint ventures with corporations and governments in the developing countries were increasingly pursued in order to capture new markets and gain ready access to resources. There were several additional advantages, notably the attraction of foreign and local capital, the opportunities for hands-on training of local managers and of technically qualified staff already acquainted with engineering skills, and the chance to overcome cultural differences. An outstanding example was the Bandar Shahpur Chemical Company, Ltd, an Iranian petrochemical and fertilizer complex in the Persian Gulf jointly owned by the Iranian National Petroleum Company and Allied Chemical. It opened in 1970 and included a 1000-mtpd single-train ammonia unit installed through Kellogg Continental's London office. A second, identical, ammonia unit was commissioned in 1975. Kellogg Continental also supplied urea technology, licensed from Stamicarbon [37, 38]. Though the foreign equity involvement was brought to an end with the Islamic Revolution in 1979, this joint venture laid the foundations of Iran's present-day substantial ammonia, urea, and methanol capacity. (Bandar Shahpur is now Bandar Khomeini.)

In 1981, Kellogg led a five-company consortium which built a fertilizer complex in Nigeria for the National Fertilizer Co. of Nigeria Ltd. The location was Onne, southeast of Lagos. This contribution to the country's Green Revolution included facilities for production of ammonia, urea, and NPK fertilizers. Kellogg engineered the 1000-mtpd ammonia unit and Stamicarbon the 1500-mtpd urea unit [39, 40]. In addition, Kellogg assisted in establishing a marketing and distribution program that brought fertilizers directly to the farmers.

References

1. Greenberg E, Christopher T, Hill CT, Newburger DJ (1979). Regulation, market prices, and process innovation: the case of the ammonia industry. Westview Press (reprinted: Routledge, London, 2019)
2. Anon (1980) Chem Ind, 21 June, pp 491–499, on 493
3. Strelzoff S (1974) Make ammonia from coal. Hydrocarb Process 53(Oct):133–135
4. Buividas LJ (1981) Coal to ammonia: its status. Chem Eng Progr 77(May):44–53
5. Anon (1954) Petrochemicals: tomorrow. Pet Process 9(Sept):1384–1386
6. Zhang F, Zhang W, Ma W et al (eds) (2009) The chemical fertilizer industry in China: a review and its outlook. Translated from the original Chinese version (2007) Chemical Industry Press. International Fertilizer Industry Association, Paris
7. Bookey JB (1986) China's fertilizer industry and the role of the Shanghai Research Institute of Chemical Industry. Chem Ind, 3 Mar, pp 168–174
8. Takeshi M (2015) The small-scale ammonia production of China in the time of Mao Zedong. Paper given at the international workshop on the history of chemistry, Tokyo
9. Clay RH (1968) Transportation of chemical plant. Chem Process Eng 48(Mar):63–66, 81
10. Anon (1967) Imperial Chemical Industries: contract with China. Chem Ind, 4 Feb, pp 167–168, on 168

11. Travis AS (2022) History of chemical industry: technologies and products versus corporate history. In: Giunta C, Seeman J (eds) Centenary issue, Bull Hist Chem 47(1):50–61
12. Chinese technical/commercial mission returns to PRC after long U.S. tour. for your information (Kellogg newsletter), no. 101, February 1974, p 1. MW Kellogg documents located in Box 219, folder "Pullman-Kellogg (1)" of the U.S.-China Business Council Records, at the Gerald R. Ford Presidential Library. https://www.fordlibrarymuseum.gov/library/document/0029/7490136. Accessed 20 Nov 2022
13. Anon (1979) Focus: Lurgi Kohle und mineraloltechnik GmbH. Chem Ind, 20 Jan, p 32
14. Marshall JC (1980). New producers: China. Chem Ind, 5 Jan, pp 38–40
15. Anon (1979) Focus. Chem Ind, 3 Feb, p 64
16. van Rooij A, Homburg E (2002) Building the plant: a history of engineering contracting in the Netherlands. Walburg Pers/Eindhoven, Foundation for the History of Technology, Zutphen, p 77
17. U.S.S.R. activities. For your information, no. 111, March 1975, p 6. MW Kellogg documents located in Box 219, folder "Pullman-Kellogg (1)" of the U.S.-China Business Council Records, at the Gerald R. Ford Presidential Library. https://www.fordlibrarymuseum.gov/library/document/0029/7490136. Accessed 23 Nov 2022
18. Smith H (1973) Soviet and Occidental Oil in multibillion-dollar deal. New York Times, 13 April, p 81
19. Anon (1974) $180-million loan to Soviet Union is made by U.S. New York Times, 22 May, p 9
20. Anon (1974) Chemico to build 4 Soviet plants. New York Times, 21 June, p 49
21. Fourth ammonia plant accepted by U.S.S.R., Kellogg press release 20 June 1975. MW Kellogg documents located in Box 219, folder "Pullman-Kellogg (1)" of the U.S.-China Business Council Records, at the Gerald R. Ford Presidential Library. https://www.fordlibrarymuseum.gov/library/document/0029/7490136. Accessed 23 Nov 2022
22. Anon (1975) Fertilizzanti azotati le URSS. La Chimica e l'Industria 57(Oct):725
23. Four Kellogg fertilizer ammonia plants contracted to be built in the U.S.S.R. Kellogg press release 26 August 1975. MW Kellogg documents located in Box 219, folder "Pullman-Kellogg (1)" of the U.S.-China Business Council Records, at the Gerald R. Ford Presidential Library. https://www.fordlibrarymuseum.gov/library/document/0029/7490136. Accessed 23 Nov 2022
24. Anon (1974) Wall Street J, 26 Dec
25. Anon (1977) Pullman Kellogg Division. Chem Ind, 15 Jan, p 39
26. Anon (1976) Implanto di urea fornito all'URSS. La Chimica e l'Industria 58(Dec):892
27. Royal MJ, Nimmo NM (1969) Why LP methanol costs less. Hydrocarb Process 48(Mar):147–153
28. Anon (1975) 1974 MacRobert Award to ICI for manufacture of high activity catalysts. Chem Ind, 4 Jan, p 4
29. Anon (1977) Russians sign methanol contract. Chem Ind, 4 June, p 409
30. Anon (1974) Humphreys & Glasgow to design ICI ammonia plant. Chem Ind, 13 Aug, p 581
31. Kellogg, Stamicarbon, sign licensing pact, for your information, no. 112, April 1975, p 1. MW Kellogg documents located in Box 219, folder "Pullman-Kellogg (1)" of the U.S.-China Business Council Records, at the Gerald R. Ford Presidential Library. https://www.fordlibrarymuseum.gov/library/document/0029/7490136. Accessed 23 Nov 2022
32. Third major Indonesian contract awarded for Kellogg by PUSRI, for your information, no. 116, Aug 1975, p 1. MW Kellogg documents located in Box 219, folder "Pullman-Kellogg (1)" of the U.S.-China Business Council Records, at the Gerald R. Ford Presidential Library. https://www.fordlibrarymuseum.gov/library/document/0029/7490136. Accessed 23 Nov 2022
33. Lummus advertisement (1967) Hydrocarb Process 46(June):40
34. Babcock & Wilcox advertisement (1968) Chem Process Eng, May
35. Kellogg ammonia plant on stream in Cosoleacaque, for your information, no. 111, March 1975, p 5. MW Kellogg documents located in Box 219, folder "Pullman-Kellogg (1)" of the U.S.-China Business Council Records, at the Gerald R. Ford Presidential Library. https://www.fordlibrarymuseum.gov/library/document/0029/7490136. Accessed 23 Nov 2022

36. Anon (1977) Focus: Pullman Kellogg Division. Chem Ind, 19 Nov, p 881
37. Work continues at Bandar Shahpur. For your information, no. 112, April 1975, p 4. MW Kellogg documents located in Box 219, folder “Pullman-Kellogg (1)” of the U.S.-China Business Council Records, at the Gerald R. Ford Presidential Library. https://www.fordlibrarymuseum.gov/library/document/0029/7490136. Accessed 23 Nov 2022
38. For your information, March 1975, p 2. MW Kellogg documents located in Box 219, folder “Pullman-Kellogg (1)” of the U.S.-China Business Council Records, at the Gerald R. Ford Presidential Library. https://www.fordlibrarymuseum.gov/library/document/0029/7490136. Accessed 23 Nov 2022
39. Anon (1981) Focus. Chem Ind, 3 Oct, p 659
40. Anon (1983) New fertilizer complex set for Nigeria. Hydrocarb Process 62(June):39–40

Process Innovations in the 1970s and 1980s. International Growth

13

Abstract

The 1970s and 1980s saw several major improvements in the technologies of ammonia-related processes as a direct outcome of innovations in catalysis. In steam reforming, ICI's novel nickel catalysts enabled increases in the pressure, permitting use of even fewer furnace tubes, and also smaller furnaces. Iron oxide and copper catalysts were used in the important shift reaction; a nickel catalyst brought about more efficient removal of carbon oxides from synthesis gas by methanation, which replaced the copper liquor process. ICI's high-activity ammonia synthesis catalysts made possible the greatest lowering of the ammonia synthesis pressure. The ICI AMV synthetic ammonia process featured a low-pressure synthesis loop, and a cobalt-promoted high-activity magnetite catalyst. The low pressure enabled cryogenic removal of methane, argon, and excess nitrogen. Other notable developments included Haldor Topsøe's radial-flow converter, and C.F. Braun's novel process for purification of synthesis gas. Changing economic conditions brought about considerable reorganization in the ammonia and chemical industries. US manufacturers suffered from competition in the form of cheap imported ammonia, sometimes made by US technology. High energy costs brought about the decline of the UK synthetic ammonia industry. Elsewhere, however, the industry continued to expand.

13.1 Introduction

Manufacture of synthetic ammonia became highly standardized by the 1960s. Notwithstanding the introduction of several novel features, the ammonia converter remained the one distinguishing feature of all ammonia plants. This included the long-established internal arrangement for tube-cooling the catalyst with the inlet gas mixture and suited converters of rated capacities of up to around 300-mtpd. The main exception was the 1950s M.W. Kellogg design in which the catalyst was

A. S. Travis, *The Global Synthetic Ammonia Industry*, Perspectives on the History of Chemistry, https://doi.org/10.1007/978-3-031-87978-4_13

divided into two or three (or more) layers, or beds, supported in a cartridge, with the interbed temperature controlled by injecting quench gas (unheated synthesis gas) between the beds.

For the high-capacity single-train units with capacities of 1000-mtpd or more, first introduced by Kellogg in 1966, an important design consideration was the fact that the configuration of the earlier tube-cooled converter, referred to as tubular converters, including the Casale converter of the 1920s, the Chemico (NEC) design, and the Tennessee Valley Authority (TVA) converter of the 1940s, with the catalyst embedded in tubes running all the way through the converter, was not viable [1]. Far from it. This is because the number of tubes required in a large high-capacity converter would become excessive and also the tubesheets (to which the tubes are welded) would be overly large and thick. Such a converter with a single bed of catalyst would be too complicated, heavy, and costly to produce. This difficulty was overcome with the adiabatic bed-type arrangement, with separated beds of catalyst and interbed cooling by way of a quench gas—a configuration introduced by Kellogg and in the mid-1960s used in its bottleneck high-capacity converter (see Figs. 11.1 and 11.3).

In the 1960s designs of high-capacity converters incorporating the intercooled catalyst beds the thermal efficiencies were increased threefold, and the energy input was dropped by almost two-thirds. Recovered heat was used to generate high-pressure steam for the turbines that drove the compressors. Energy consumption of compressors was arranged to match energy released in hot parts of the ammonia process.

The lower pressure requirements made necessary larger converters, though the lower pressure enabled reduction of the thickness of the pressure shell, which, overall, needed no more steel than the smaller converters. As a result by around 1980, a typical 1500-mtpd converter had grown to 34 m, with a 2800 mm diameter. For sake of comparison, the largest World War I converter design, of 85-mtpd capacity, was 12 m in length and 800 mm in diameter.

In the epoch of great expansion in ammonia, from 1966 until the mid-1970s, with its emphasis on large scale production, there was less interest in smaller production units. The downside of the large units, however, was when demand declined, as it would periodically, following periods of overproduction. There were some situations when demand in any case was not great. The novel TVA converter, with its capacity limited to around 225-mtpd by 1970, was a tube-cooled vessel, containing a single catalyst bed, with heat exchange below, within the catalyst cartridge. In the late 1970s, a TVA ammonia unit at the National Fertilizer Development Center, Muscle Shoals, by modification of the front end, was demonstrated in connection with Texaco coal gasification technology [2]. It attracted much interest at the May 1979 "Symposium on Ammonia from Coal," organized by the center. The TVA design became useful when there was a revival of interest in smaller plants, especially during the periods of slower growth in the 1980s and 1990s, and in particular in former Eastern Bloc countries, and in some parts of the European Union [3, 4]. Such innovations and new high-capacity units receive the main attention here.

13.2 Innovations at ICI: Active Catalysts and the AMV and LCA Processes

Catalyst improvements from 1940 until the end of the 1960s enabled the introduction of novel and improved reforming processes as well as a lowering of the cost of production of ammonia. In the 1940s, as in previous decades, the chemical companies undertook research into catalysts for their own ammonia processes. This was especially true of BASF and ICI. The ammonia synthesis catalysts continued to be based on iron (magnetite), as used in the first BASF facility from 1913.

The 1970s and 1980s saw several major developments in the technologies of ammonia-related processes as a direct outcome of innovations in catalysis. In steam reforming, ICI's novel nickel catalysts enabled increases in the pressure, permitting use of even fewer furnace tubes, and also smaller furnaces. Iron oxide and copper catalysts were used in the shift reactions. A nickel catalyst brought about more efficient removal of carbon oxides from synthesis gas by the methanation reaction [5–7].

Of the many modifications and refinements introduced from the early 1970s, it was the high-activity ammonia synthesis catalysts of ICI which made possible the greatest lowering of the ammonia synthesis pressure. In October 1975, ICI announced that it intended to invest two million pounds sterling on extending its catalyst manufacturing plant at Clitheroe in Lancashire. Among the catalysts produced there were those suited to the new low-pressure processes. They had been developed at the Billingham catalyst division and included a high-activity catalyst based on iron, promoted with 20% cobalt. For a given conversion, the necessary volume of such catalysts was about 25% less than that of a conventional catalyst. ICI, through its Katalco division, exported catalysts for use in ammonia, methanol, syngas, and town gas production, and in this sector had a strong presence in the US.

Design of a new, lower energy, ICI ammonia unit began in the 1980s, based on concepts developed by Alwyn Pinto. It was called Ammonia V, abbreviated to AMV, and announced by 1983 (Fig. 13.1). The process featured a low-pressure synthesis loop, operating at about 80 to 110 atm. It employed a unique cobalt-promoted high-activity magnetite catalyst (Katalco 74–1), which had been developed specifically for the AMV process. The front end of the process involved primary reforming under pressure at a mild temperature. This was followed by secondary reforming with an excess of air; shift reforming; methanation for removal of residual carbon oxides as methane; and entry of the methanated gas, containing an excess of nitrogen and methane, at a low pressure into the synthesis loop. The low pressure enabled cryogenic removal of methane, argon, and excess nitrogen [4, pp. 277–278; 8]. Though the same basic process was also common to Kellogg and Haldor Topsøe designs, the operating conditions were different.

Unfortunately for ICI, depressed economic conditions, particularly in 1984, were impacting on all sectors of the economy, and no less the synthetic ammonia industry. As a result, the AMV design was not introduced at the Billingham facility. The first prototype, using for the first time the new cobalt-enhanced catalyst,

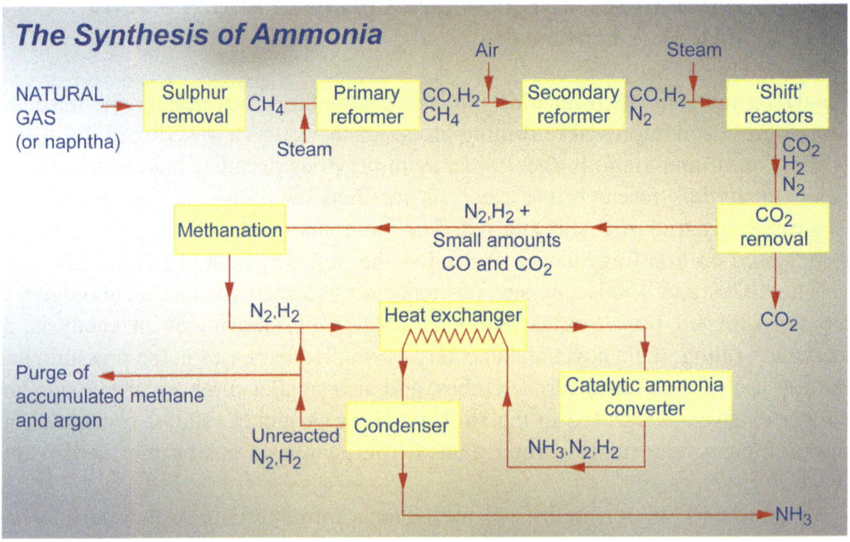

Fig. 13.1 Flowsheet for ICI's AMV process. Redrawn from Cane MCV (1985) Ammonia and protein: why use fertilizers? University of York Science Education Group, York

was put into operation in August 1985 at the former CIL (Canadian Industries, Ltd), factory at Courtright, Ontario, Canada. This location, on the eastern side of the country, was significant because the Canadian nitrogen industry, following considerable expansion during the wartime and post-war years, had lost its competitive edge due to increases in the cost of natural gas from the 1970s. And, as noted earlier, the hub of the ammonia and fertilizer industry had moved west to Alberta and British Columbia.

In the UK, existing ammonia units were operated by ICI at Billingham, nearby at Wilton, and at Immingham and Severnside, and by DSM at Ince Marshes, near Chester, a facility erected during 1965-1966 by Shell/Armour Star (Shellstar), and sold to DSM in 1975. The downturn in the economy in the early 1980s adversely impacted on the Severnside facility, with its then two aging units. Rather than shut them down, ICI decided to adopt its novel Leading Concept Ammonia (LCA) process, which incorporated several process improvements, including the introduction of an ICI catalyst suited to even lower pressures than the AMV process, that is, between 70 and 80 atm [4, pp. 267–268]. The LCA process was developed under the guidance of Alwyn Pinto, who also headed the AMV design team. The process was structurally integrated, with a core unit for synthesis gas preparation and ammonia synthesis. Because it included a separate utility unit, it was not fully integrated in the fashion of the modern high-capacity systems. The converter, following first-generation configurations, was tubular, based on the TVA model, with counter-current gas flow. This allowed synthetic ammonia units of up to 300–450-mtpd rated capacity to achieve the same capital performance and energy efficiency as larger units. (A TVA type low-capacity converter was also introduced by Haldor

Topsøe.) The LCA process, introduced at Severnside in 1988–1989, was among the first fully automated ammonia processes. Each of the two units installed there gave a yield of 12–15%. When the second unit was first started up, ammonia appeared after just nineteen hours, at the time a world record. In the redesign of the steam-reforming section, the primary reformer was arranged to receive heat from the process gas exiting the secondary reformer. This did away with the need for a primary reformer furnace, as was also proposed by Chiyoda of Japan in 1984.

Acknowledging the considerable startup difficulties experienced with the Kellogg process at Billingham in the 1960s, the ICI engineers were keen to point out that "The degree of change involved in the LCA design is more fundamental than the change in the 1960s [at Billingham] to energy-integrated ammonia plants. The LCA technology has been established more quickly than the progress made by ICI with its experience of the 1960s integrated designs.... The learning experience gained during the initial operation of these plants in the 1960s was the foundation of a subsequently very large number of highly reliable and efficient ammonia plants" [9]. That was already borne out by progress in ICI's ammonia and methanol plant designs. However, in many ways the LCA process, and the AMV process, came too late in the history of ICI's long involvement with ammonia to make much difference. ICI was already cutting back on fertilizer production. The late 1980s was a transformative time for the UK ammonia and fertilizer industry. At Severnside, phosphatic fertilizer production ceased in 1989.

By 1990, ICI's ammonia production had been losing money for four years. Rather than use its advanced technology to improve the ammonia units at Billingham, the corporation decided to dispose of its ammonia facilities. An offer was received from the US agrochemicals corporation Terra Industries, but the takeover was then blocked by the Monopolies and Mergers Commission. A decade later Terra Industries purchased Billingham and Severnside.[1] In 2006, Terra formed a joint venture with Kemira GrowHow Oyj, of Finland, since 1985, the owner of Ince, called GrowHow UK Ltd (Fig. 13.2).

ICI, however, licensed the LCA process and remained in the catalyst business.[2] The catalyst unit, Katalco, in 1997 purchased BASF's synthesis gas catalyst division.

From the 1960s, there were several other process innovations in the synthetic ammonia industry two of which deserve our attention: the design for a new converter, and a gas purification process introduced, respectively, by Haldor Topsøe and C.F. Braun [10]. Both were aimed at improving the performance of high-capacity units,

[1] Terra ceased manufacture of ammonia at Severnside during the winter of 2007–2008.

[2] Several novel Katalco catalysts were incorporated: Katalco 61-2 (the first low-temperature hydrodesulfurization (HDS) catalyst); Katalco 83-1 (the first application of a process gas heated reformer (GHR), isothermal shift catalyst specifically developed to resist the high operating temperature); Katalco 11-4 (a low-temperature methanation catalyst); and Katalco 74-1 (suited to the low-pressure LCA process). In addition there was the first low-temperature sulfur removal absorbent.

Towards Consolidation in the UK Ammonia Industry.
Corporate Ownership of Billingham and other Ammonia Factories

1920-1923	Brunner, Mond & Co. develops ammonia synthesis.
1924	Brunner, Mond & Co. produces ammonia at Billingham.
1926	(December 7) ICI (Imperial Chemical Industries) created from merger of Brunner, Mond with Nobel Industries, British Dyestuffs Corporation, and United Alkali.
1965	Shellstar (Shell/Armour Star) ammonia factory opens at Ince, Cheshire.
1973	Ince sold to DSM of Holland. Trades as UKF Fertilisers Ltd.
1985	Ince sold to Kemira Oyj, of Finland, whose agricultural unit becomes Kemira Agro Oy. Kemira builds an ammonia plant at Hull (1988).
1990	ICI pursues options to divest of ammonia operations.
1991	Terra Nitrogen (UK) Ltd, part of Terra Nitrogen Company, of Deerfield, Illinois, US (founded in 1991), purchases ICI Billingham and Severnside ammonia facilities.
2004	Kemira Agro Oy spun off as Kemira GrowHow Oyj.
2006	Terra and Kemira GrowHow Oyj form 50-50 joint venture, GrowHow UK Ltd.
2007	Kemira GrowHow acquired by Yara International ASA (formerly part of Norsk Hydro) of Norway. Ince modernized.
2010	Terra, and its share in GrowHow, merged with CF Industries, of Deerfield, Illinois (founded in 1946 as the Central Farmers Fertilizer Company).
2015	Yara's share in GrowHow sold to CF Industries. Yara retains plant at Hull. GrowHow rebranded as CF Fertilisers UK Ltd in November.

Fig. 13.2 The UK synthetic ammonia industry, 1920–2015

13.3 Haldor Topsøe and the Radial-Flow Ammonia Converter

Haldor Topsøe A/S was founded in 1940 by Dr Haldor Topsøe. By 1944–1945, the firm was engaged in research and development of catalysts for the synthesis of ammonia [11]. In 1962, Haldor Topsøe made a major entry into ammonia when it filed a patent for steam reforming of hydrocarbons, including naphtha, using a novel catalyst. In 1965, Haldor Topsøe's naphtha reforming process was incorporated in the new ammonia facility of Japan's Kyowa Chemicals Co. Ltd (a project undertaken jointly with Chiyoda). One year later, the same technology was used in a 300,000-mtpy ammonia unit erected by Chemical Construction (GB) Ltd, a subsidiary of Chemico, for Shellstar, at Ince [12].

The first patent for an ammonia synthesis reactor was filed in 1963. It claimed the benefits of a radial-flow configuration, particularly in multi-bed converters in which catalyst activity was compromised by a relatively large pressure drop. This pressure drop was due to the practical limitation on the diameter of converters. It had been overcome by increasing the depth of the catalyst bed, and using large-size catalyst particles, of grain size 6 to 10mm. But these catalysts were of reduced activity. The radial flow, which was suited to high-capacity units, enabled the use of smaller particles of grain size 1.5 to 3 mm, and thus of greater activity, and overcame the problem of the pressure drop.

Late in 1966, the Bechtel Corporation completed work on a 1000-mtpd single-train anhydrous ammonia unit that operated with the first radial-flow converter, based on the Haldor Topsøe design, for the Agrico Chemical Company, at Blytheville, Arkansas (Fig. 13.3) [13].[3]

In 1976, the Haldor Topsøe converter evolved into the series S-200 reactor (2-bed radial flow, with indirect cooling between the beds), of 1350-tpd capacity, as used in India, and from 1986 at Yazoo City (Mississippi Chemical). The radial-flow configuration through the two catalyst beds had the advantage of a very low-pressure drop. It vastly increased the area across which the gas flow was distributed. This allowed the particle size of the catalyst to be reduced without exceeding the pressure drop across the catalyst bed of an equivalent axial-flow reactor loaded with a conventional catalyst. Thus, the surface area of the catalyst was considerably increased, raising the attainable conversion rate and ammonia yield [14].

Around 1970, Haldor Topsøe had faced many problems, though not always of a technical nature. This included in Libya, where after Colonel Muammar Gaddafi took power in 1969, agreements with the previous government were canceled or invalidated. There were problems with payments for certain projects in Kuwait; and substantial falls in the profitability of its silicon plant in Frederikssund, near Copenhagen, and at its electronics subsidiary Univel. The losses started to pile up. The aggravated financial situation led in 1972 to a merger with Italy's Snamprogetti.

Fortunately for Haldor Topsøe, increases in agricultural production and new crop varieties in the developing economies enabled the firm to make a major contribution to the Green Revolution. This was especially the case in India.

India in the mid-1970s was racked by political strife and social unrest, the causes of widespread strikes (1974), followed by a state of emergency (1975–1977). During the second half of the decade, great efforts were made to improve living conditions, to bring about land reforms (especially in West Bengal), to enhance agricultural productivity, and to create a balanced national economy by concentrating attention on measures needed to bring about modernization and self-sufficiency, including in fertilizers. In the mid-1970s, Haldor Topsøe constructed

[3] Agrico was successor to America Agricultural Chemical Company, founded in 1899, and purchased by Continental Oil Company (Conoco) in 1963. The Agrico Chemical Company was formed in 1966, when Conoco merged two divisions. By early 1980, Kellogg had completed work on a fertilizer complex for Agrico, at Verdigris, Oklahoma, incorporating a 1150-stpd ammonia unit, a 639-stpd nitric acid unit, and a 1800-stpd urea-ammonium nitrate unit of Stamicarbon design.

Fig. 13.3 Bechtel advertisement, 1966, showing the new Agrico Chemical Company's ammonia facility at Blytheville, Arkansas. The 1000-tpd single-train unit incorporated the first radial-flow converter, designed by Haldor Topsøe. (Hydrocarbon Processing 45(December):54) (Bechtel Corporation. With permission)

four ammonia plants for the Fertilizer Corporation of India Ltd, and five for National Fertilizers Limited [15]. An important initiative was the decision of the Indian government to build a pipeline from the offshore gas field near Bombay (Mumbai), discovered in 1977, across the country, and erect a series of fertilizer plants along the way. Around 1980, Haldor Topsøe was awarded a $17m contract to build two 1350-mtpd ammonia units for the fertilizer complex of Rashtriya Chemicals and Fertilizers Limited at Thai-Vaishet, in Gujarat State. A similar order went to Kellogg's London office for Krishak Bharati Cooperative Limited, at Hazira, also in Gujarat. Haldor Topsøe and Kellogg were the leaders in the installation of high-capacity ammonia units in India [16]. This started the move away from naphtha to natural gas as feedstock [17].

Another important market for Haldor Topsøe was the People's Republic of China, where it had a presence since the 1960s, selling both catalysts and know-how. In 1973, China imported Haldor Topsøe catalyst technology suited to production of twelve varieties of catalysts [18]. Two years later, catalysts were produced at thirty-five sites in China. Russia and Brazil also gained from technology provided by Haldor Topsøe, which often succeeded in being the first presence in these growing economies.[4] From around 1990, the Haldor Topsøe Heat Exchange Reformer (HER) reduced fuel consumption by 70%, and gas consumption by 15%. These successes enabled the firm in 2007 to regain its independence. It remains a leader in heterogeneous catalysis; its catalysts are notable for their effectiveness and high selectivity.[5] Around 2012, the original S-200 converter at Yazoo City (from 2010 part of CF Industries), which had operated since 1986, was retrofitted with an S-300 (3-bed radial flow) converter basket [19].

Other firms that developed radial-flow converters were Ammonia Casale (1979–1980), Chemoprojekt (Prague), Kellogg (1979), Lummus (1978), Snamprogetti (1976), and Uhde (in 1982) ([4], pp. 267–268). Ammonia Casale's process was distinguished by its mixed axial-radial flow, with some gas entering the top of the catalyst in axial-flow, but the majority of the gas entering radially (Chap. 14).

13.4 Purification of Syngas: The C.F. Braun Process

As ammonia projects grew larger and became more expensive, purchasers demanded a higher degree of customization to suit their individual circumstances. This provided new opportunities for contractors such as C.F. Braun & Company.

[4] Though relations between the Danes and Italians were not always idyllic, especially as regards long-term strategy, the Haldor Topsøe-Snamprogetti union, as an ammonia licensee and engineering, procurement, and construction (EPC) contractor on the one hand, and a licensee of urea on the other, particularly in India, turned out to be very profitable.

[5] By around 2020, there were 130 series 200 units worldwide based on Haldor Topsøe technology, 82 fed by natural gas and 38 with other types of feedstock (coal, naphtha, coke oven gas, etc.). The largest proportion, more than 70%, are in Asia, while Europe and the Americas represent about 10% and 15%, respectively.

This concern was founded by the engineer Carl Franklin Braun in San Francisco in 1909. In 1922, Braun moved to Alhambra, California. By 1954, it had become a leading designer and manufacturer of chemical equipment, with around 5000 people on the payroll.

The Braun synthesis gas purifier process produces a high purity synthesis gas for the ammonia process. The earliest design was based on the work of B. J. Grotz, Jr. It enables the use of smaller, and less expensive, primary tubular steam reformers, by moving some of the load to the secondary reformer, into which an excess of air is introduced. Excess nitrogen is removed by cryogenic purification, when all of the methane and the excess nitrogen, as well as part of the argon, are removed in the downstream gas. The great purity of the resulting syngas, except for a trace amount of argon, reduces the required synthesis pressure and loop purge rate. This was especially useful in high-capacity ammonia units. In 1966, the first purifier was incorporated in a 750-stpd Mont Cenis single-train ammonia unit at Collier Carbon & Chemical Corporation of Brea, California. This was followed one year later in a 1000-stpd unit at Arco Chemical Co., in Fort Madison, Iowa, and a 1500-stpd unit at Esso Chemie BV, in Rotterdam [20]. In 1969, a Braun-designed ammonia synthesis unit incorporating, unusually for the time, two adiabatic reactors (converters), was started up for Arco Chemical (then a division of Atlantic Richfield, about to be merged with Sinclair Petrochemicals, of Sinclair Oil Corporation). The reactors, of an axial-flow type, were operated in parallel. Heat exchange between the two catalyst beds, each of which was held in a cylindrical container within the pressure shell of the reactor, provided heat used to warm the feed to the first bed. The reported yield of ammonia leaving the second reactor was 16.5%. The claimed advantages of the design included mechanical simplicity, no removable basket, and manway-size enclosures [4, pp. 271–272; 21]. The use of two reactors represented an alternative to the quench converters that were widely used at the time. A similar configuration, a dual-pressure system in which the catalyst is held in two converters, connected in series, was adopted by Uhde (Sect. 13.6).

By late 1971, the Braun ammonia process, incorporating the cryogenic purifier, with external gas exchange for temperature control, was promoted for capacities of up to 2000-mtpd. The purifier was then in use at the two facilities in the US, two in Holland, and one, for Asahi Chemical Industry Co., at Mizushima, Japan. Not all of the Braun ammonia synthesis units used the new Braun two-reactor design. Braun was also active in China, where its purifier was selected for use at three sites, each with a rated capacity of 1000-tpd: Hejiang and Fuling in Sichuan Province, and Jinxi in Liaoning Province [22]. In all, the Braun purifier was used in fifteen plants built between 1966 and 2001.

In 1989, Braun was merged into Brown & Root which in 1998, then a Halliburton company, was merged with M.W. Kellogg, to form Kellogg Brown & Root (KBR). Kellogg continued to market the Braun purifier with its ammonia processes [23].

13.5 Changing Organizations

A major outcome of the oil crises, particularly from 1973 and until just before the mid-1980s, was the impact on the fates of large engineering contractors following extensive reduction or postponement of investments in new plant in the petrochemical industry. Most contractors suffered considerably; the upheaval stretched through the supply chain and also to clients. Aramco, of Saudi Arabia, halted all payments to certain engineering contractors. In 1975, Chemico got into difficulties following loss of its contract with Algeria. Its operations ended in 1984. In 1979, the Islamic Revolution in Iran brought on the second global oil shock, and difficulties for contractors. This did however encourage gas production by non-OPEC producers, as had been the case during 1973–1974, notably in Norway, the UK, Alaska, and Mexico.

Though cutbacks caused loss of business for engineering contractors, they continued to work on design modifications. Kellogg, by around 1970, had designed a 1500-tpd ammonia unit, and then a two-bed horizontal intercooled ammonia converter, with a capacity of 1700-mtpd [24]. It was introduced commercially in 1972 in Japan, at Chiba, near Tokyo, and was described as "the most efficient ammonia plant in the world" [25].[6] The catalyst beds were arranged side by side in a removable cartridge. Apart from Braun's 1500-tpd unit, and the Haldor Topsøe series S-200 converter (followed in the loop with the one-bed S-50 converter), at 1350-tpd, and introduced in 1976, the large converters were until then mainly of 1000-tpd capacity [26–28]. By the end of the 1980s, converter designs of 1800-tpd capacity were available from the leading engineering contractors.

To some extent the renewed demands for ammonia plants in Asia and the Soviet Union, with finance provided by international loans, compensated for the halts to construction and canceled projects, particularly in the West, where, in addition to financing issues, and the increased cost of energy, there were growing environmental concerns. There was also some relief from projects involving exploitation of North Sea gas, while flared gas released during extraction of oil was increasingly captured for use as a feedstock in petrochemical plants. American engineering contractors continued to build up the ammonia industries in some of the countries that were actively exporting to the US. However, American fertilizer manufacturers suffered from competition in the form of cheap ammonia imported from the Soviet Union, Trinidad, Mexico, and Canada. Ammonia and methanol were the main chemicals exported from the Soviet Union, from where in 1979 Occidental Petroleum imported into the US 770,000 short tons of anhydrous ammonia, as part of the barter arrangement [29]. In the mid- to late 1980s, a debt crisis among farmers, especially in the US, cut into demand. It was not a good time to move forward with investment or expansion in the US synthetic ammonia industry.

However, by the early 1980s, Saudi Arabia was better placed to regain its move into petrochemicals. Saudi Arabian Basic Industry Company (SABIC) set

---

[6] In 1972, 40 single-train high capacity plants based on Kellogg technology were in operation.

up partnerships with leading multinational chemical engineering contractors. They included Mitsubishi Heavy Industries Ltd, which signed a contract with Saudi Methanol Co., for installation of a 600,000-tpy methanol plant at Al-Jubail, north of Dammam, scheduled to be operational by early 1983. For the Al-Jubail Fertilizer Company (SAMAD), Kellogg designed an ammonia (1000-tpd)-urea (1600-tpd) complex [30].[7] Saudi Arabia focused on commodity products, including fertilizers, which were shipped in bulk to remote export markets. Iran, by contrast, with its enormous reserves of gas, originally developed its petrochemical industry to serve the mainly domestic market. Following Saudi Arabia, the OPEC nations entered into large-scale manufacture of petrochemicals.

During the last three decades of the twentieth century, the manufacture of synthetic ammonia continued to represent an important part of the global chemical industry, with around 80% consumed by the agricultural sector [31]. Worldwide, from 1971 to 1982, the share of new ammonia licenses was 40% for Kellogg, 20% for ICI, and 8% for Haldor Topsøe. During the 1980s, Kellogg's market share in the ammonia business dropped to less than 30%. Today (Pullman-) Kellogg's successor, KBR, is a leading licensor of ammonia technology. Kellogg/KBR has since its first wartime plant was commissioned in 1943 licensed, engineered or constructed 252 grassroots ammonia units worldwide. The other leaders are Haldor Topsøe, thyssenkrupp Industrial Solutions (Uhde technology), and Casale (formerly Ammonia Casale).

Whatever the scale of operation, safety in the synthetic ammonia industry is a primary concern. Process hazards might include failures of machinery, explosions, fires, and leaks. This is highlighted by some of the examples met in earlier chapters when discussing the first decades of the industry. Under the high-pressure operating conditions, it was not unknown for the lid to be blown off the head of a converter. Primary steam reforming, in particular, requires care in operation. For all ammonia plants, routine maintenance and sensitive monitoring instruments help to minimize risks. Though ammonia itself has a low environmental impact, its toxicity, if inadvertently released to the atmosphere, poses risks to workers and local communities. Ammonia safety conferences arranged, for example by the American Institute of Chemical Engineers, have contributed greatly towards the drawing up of ammonia safety manuals and procedures.

13.6 Into the New Millennium: Still Growing

High-capacity synthetic ammonia facilities at the turn of the twenty-first century included: TogliattiAzot, Novomoskovsk, and Kemerovo, in Russia; Cosoleacaque, Veracruz, in Mexico; CF Industries, at Donaldsonville, Louisiana; Point Lisas, in Trinidad; Sluiskil, in the Netherlands; PUSRI, at Palembang, and Pupuk Kaltim (Pupak Kalimantan Timur), at Bontang, in Indonesia; Gorlovka and Cherkassy,

[7] The SAMAD complex started up in March 1983.

in Ukraine; and Umm Sais, in Quatar [32]. Iran's nitrogen industry was revived in the 1990s, following the Iran-Iraq war (1980–1988), drawing on the country's substantial reserves of natural gas [33].

The Donaldsonville plant of CF Industries, with its multiple high-capacity ammonia units, was the largest producer of nitrogen-based fertilizer in the US, prior to becoming the largest privately owned producer in the world. This happened in 2015 after the GrowHow facilities at Billingham and Ince (Cheshire) became part of CF Industries and were rebranded as CF Fertilisers UK Ltd CF Industries, with manufacturing operations in the US, Canada, and, until 2023, in the UK, is headquartered in Deerfield, Illinois.

Expansion followed in the early 2000s: P.T. Kaltim Pasifik Ammonia, in Indonesia (2000-tpd), in operation since 2004 (now part of Pupuk Kaltim); Profertil SA, at Bahia Blanca in Argentina (2050-tpd), since 2001; Oman India Fertiliser Company S.A.O.C. (OMIFCO) at Sur, in Oman (two units each of 1750-tpd), since 2005; and Qatar Fertiliser Company (QAFCO), (two new units installed, each with a rated capacity of 2300-tpd), since 2013. During 2006, at the Kota ammonia-urea plant of Shriram Fertilisers and Chemicals (SFC) in India modifications were introduced to enable the use of natural gas in addition to naphtha as feedstock. Since 2009, the plant relies on natural gas alone. In 2013, the Nangal facility, in the Punjab—which in the late 1970s had changed over from electrolysis for hydrogen to naphtha—moved over to natural gas.

A plant based on the Uhde-ICI design with a capacity of 1800-tpd was erected in 1998–1990 near Belle Plaine, Saskatchewan, Canada, for SaskFerco Products Inc. (now Yara Belle Plaine Inc.; Yara is the 2004 spinoff from Nork Hydro). This was the first application of the Uhde radial-flow converter, introduced in 1982. It incorporated two ammonia converters, with three catalyst beds, and indirect heat intercooling among the beds, placed in series; a Krupp Uhde top-fired primary reformer with a cold outlet manifold; and a secondary reformer with a circumferential vortex burner (a design introduced in 1968). Fifteen of these plants were commissioned by 1999. This evolved into the present high-capacity dual-pressure ammonia process, with synthesis in two stages at different pressures, and, generally, two converters, again placed in series. Compressed synthesis gas is fed into the first converter in a once-through stage. After separation from liquefied ammonia, the recovered synthesis gas from this first converter is then fed into the second converter, configured in a conventional ammonia loop. The process was introduced at Saudi Arabia Fertilizer Company's SAFCO IV plant, at Al-Jubail, commissioned in 2006, with a capacity of 3300-mtpd, and claimed to be the largest ammonia plant in the world. Over the following decade, similar 3300-mtpd plants were installed at Raz Az Zwor, of the Saudi Arabian Mining Company (Ma'aden), commissioned in 2011; Donaldson, Louisiana, USA (CF Industries, Ammonia 6), commissioned in 2016; and Ma'aden 2, at Raz Az Zwor, Saudi Arabia. No critical high-pressure equipment exceeding the 2000-tpd design was required. Firms in the Middle East have continued to expand, often through strategic partnerships. In 2019, the Abu Dhabi National Oil Company (ADNOC) jointly with OCI of

Holland, a manufacturer of nitrogen products established in 2013, created Fertiglobe, the largest global exporter of ammonia and urea. Fertiglobe has production facilities in the United Arab Emirates, Egypt, and Algeria.

In April 2018, Yara International, the main European producer of ammonia, and BASF opened a large ammonia manufacturing facility, with an annual rated capacity of 750,000-tpy, at Freeport, Texas. The hydrogen and nitrogen are piped in from various petrochemical complexes, some quite remote. The ammonia synthesis technology is that of KBR.

The People's Republic of China remains the world's largest producer of synthetic ammonia. In 2016, China produced ammonia at an estimated rate of 48-mtpy, followed by Russia, at 12-mtpy, and India at 11-mtpy. In 2023, estimates suggest that China's output was 43 million metric tons, and each of India, Russia and the US around 14 million metric tons. As in the past, India and, even more, China depends on high levels of synthetic nitrogen fertilizer to sustain their populations. In 2022, wheat production in China amounted to 138 million metric tons, in India 108 million metric tons, and in Russia 104 million metric tons [34]. India is moving towards self-sufficiency in production of urea.

The ammonia industry continues to remain highly sensitive to the cost of energy. In mid-September 2021, as a result of the threefold increase in the price of natural gas since the beginning of the year, CF Industries shuttered its two UK factories, at Billingham and Ince. This created a panic in the food and beverage industry, which relied for 45% of its food-grade carbon dioxide on the two facilities. For the same reason Yara International planned to reduce output at six of its factories, including in France and Italy, and in the UK at Salt End, Hull (constructed during 1988–1989). On 21 September, the UK government agreed to provide a short-term support package for the reopening of Billingham in order to maintain the supply of carbon dioxide [35].[8] Production of synthetic ammonia in the UK ceased in February 2025, with the closure of Yara's Hull facility.

13.7 Summary

This and the preceding three chapters have spanned over five decades of technology-driven changes in high-pressure and ammonia-related processes. During the early 1930s, high-pressure hydrogenation was a dominant theme. This facilitated, and was followed by, the changeover from coal to hydrocarbons as feedstock. However, the basic synthetic ammonia process, a marvel of self-contained engineering, had undergone few alterations since the 1920s. That changed in the mid-1960s with the evolution of the single-train process, enabling, with massive savings in energy and labor, and through use of centrifugal compressors, production of 1000 tons of ammonia a day. Synthetic ammonia now acted as

[8] CF Fertilisers UK Ltd supplied carbon dioxide to three UK distributors: Air Liquide, BOC (British Oxygen Company), and Nippon Gas.

handmaiden to new varieties of wheat that enabled the success of the Green Revolution. As outlined in the foregoing, from the 1960s novel proprietary ammonia technology became available from Kellogg, ICI, Braun, Haldor Topsøe, and Uhde.

By the late 1970s, however, many of the converters were aging. In the uncertain economic environment, the increasing costs of building new plants presented new risks.

Until this time, Ammonia Casale had been mainly excluded from the post-1960s synthetic ammonia endeavor. The new situation, however, would present the company with an opportunity, through novel innovations, to join the league of major licensors and engineering contractors. The main breakthrough came in 1986 with the retrofitting, using a completely new design of internals, of Kellogg converters at the Donaldsonville, Louisiana, ammonia facility of CF Industries, as described in the next chapter.

References

1. Pattabathula V (2019) Ammonia. In: Kirk-Othmer encyclopedia of chemical technology. https://doi.org/10.1002/0471238961.0113131503262116.a01.pub4
2. Nichols DE, Williamson PC (1982) The TVA ammonia-from-coal project. In: More AI (ed) Fertilizer nitrogen: facing the challenge of more expensive energy. Proceedings of the British Sulphur Corporation's Fourth International Conference on Fertilizer Technology, London, 19–21 January 1981. British Sulphur Corporation, Ltd, London, pp 53–72
3. Baddour RF, Brian PLT, Logeais BA, Eymery JP (1965) Steady-state simulation of an ammonia synthesis converter. Chem Eng Sci 20:281–292
4. Hooper CW (1991) Ammonia synthesis: commercial practice. In: Jennings JR (ed) Catalytic ammonia synthesis: fundamentals and practice. Plenum, New York, pp 253–283, on 263
5. Habermehl R (1969) Optimizing industrial catalysts. Chem Eng Prog 65(Oct 1969):55–58
6. Habermehl R (1965) Nitrogen fertilizer plants: improved catalysts reduce costs. Chem Eng Prog 61(Jan):57–61
7. Cane MCV (1985) Ammonia & protein: why use fertilizers? University of York Science Education Group, York
8. Brightling J (2018) Ammonia and the fertilizer industry: the development of ammonia at Billingham. Johns Matthey Technol Rev 62(1):32–47
9. Armitage PM, Elkins KJ, Kitchen D, Pinto A (1991) Leading concept ammonia process; first two years. AIChE Ammonia Safety Symposium, Los Angeles
10. Czuppon TA, Knez SA, Rovner JM (1992) Ammonia. Kirk-Othmer encyclopedia of chemical technology, vol 2, 4th edn. Wiley, New York, pp 638–691
11. Hansen JB, Clausen BS (2015) Haldor Frederik Aksel Topsøe: the man, the scientist and the company. J Catal 328:2–4
12. Anon (1966) Hydrocarb Process 45(Dec):54
13. Bechtel advertisement (1966) Hydrocarb Process 45(Dec):72-I
14. Appl M (1997) The Haber-Bosch heritage: the ammonia production technology. Paper presented at the 50th anniversary of the IFA technical conference, September 25–26th 1997, Sevilla, Spain, pp 8–9
15. Dybkjaer I (1982) Improved catalyst and reactor design for energy-efficient ammonia production. In: More AI (ed) Fertilizer nitrogen: facing the challenge of more expensive energy. Proceedings of the British Sulphur Corporation's Fourth International Conference on Fertilizer Technology, London, 19–21 Jan 1981. British Sulphur Corporation, Ltd, London, pp 503–525
16. Anon (1975) Construction boxscore section. Hydrocarb Process 54(Feb):37–39

17. Chandran NB (1984) Setting up and operating ammonia plants. The Indian scenario today. Econ Polit Wkly 19(50)(15 Dec):A21, A23–A24
18. Bookey JB (1986) China's fertilizer industry and the role of the Shanghai Research Institute of Chemical Industry. Chem Ind, 3 Mar, pp 168–174
19. Liddon WC, Copeland BG, Dahlstrom BJ (2013) Successful installation and startup of an S-300 Ammonia converter basket. Safety in ammonia plants & related facilities. American Institute of Chemical Engineers, New York, pp 121–131
20. Grotz BJ (1967) Syngas purifier cuts ammonia costs. Hydrocarb Process 46(April):197–202
21. Glover W, Yoars J (1973) Ammonia design proved. Hydrocarb Process 52(April):165–166
22. Anon (1989) Ammonia technology selected for plants. Hydrocarb Process 68(Feb):28
23. Ostuni R (2021) Kellogg technology. In: Cariati V (ed) Ammonia for life: the synthesis of ammonia and the birth of modern chemical industry. Casale SA, Lugano, pp 95–110, on 99–100, 106–107
24. Eschenbrenner GP, Wagner GA III (1972) A new high capacity ammonia converter. Chem Eng Pract 68(Jan):62–66
25. Kellogg advertisement (1972) Chem Eng Progr 68 (May):36
26. Quartulli OJ, Wagner GA III (1978) Why horizontal HN_3 converters? Hydrocarb Process 57(Nov):115–122
27. Zardi U (1982) Review these developments in ammonia and methanol reactors. Hydrocarb Process 61(Aug):129–133
28. Rossetti I (2020) Reactor design, modelling and process intensification for ammonia synthesis. In: Inamuddin RB, Asiri AM (eds) Sustainable ammonia production. Springer, Cham, pp 17–48
29. Anon (1980) Anhydrous ammonia from the U.S.S.R. Report to the president on investigation No. TA-406-6 under section 406 of the Trade Act of 1974. USITC [United States International Trade Commission]. Publication 1051: Washington DC
30. Guffey AA (1982). Discussion. In: More AI (ed) Fertilizer nitrogen: facing the challenge of more expensive energy. Proceedings of the British Sulphur Corporation's Fourth International Conference on Fertilizer Technology, London, 19–21 January 1981. British Sulphur Corporation, Ltd, London, pp 679–680
31. Greenberg E, Christopher T, Hill CT, Newburger DJ (1979) Regulation, market prices, and process innovation: the case of the ammonia industry. Westview Press (reprinted Routledge, London, 2019)
32. Smil V (2001) Enriching the earth: Fritz Haber, Carl Bosch, and the transformation of world food production. The MIT Press, Cambridge, Mass, p 125
33. Rose J (1991) Fertilizers: picking up the pieces. Chem Ind, 1 July, p 456
34. UN Food and Agriculture Organization, 2022 figures; https://www.fao.org/Faostat/era/#data/QCL
35. Anon (2021) Fuel, fertiliser and food: how a UK ammonia plant threatened nation with crisis. Financial Times, 22 Sept

14 Revamping, Catalyst Studies, and Green Hydrogen

Abstract

This chapter deals with three topics that are relevant to the modern synthetic ammonia industry: the revamping, or retrofitting, of inefficient converters using a novel design of internals based on the axial-radial flow of gases; the academic study and improvement of catalysts; and sustainable production of hydrogen. It was the axial-radial-flow concept that created an opportunity for the revival of Ammonia Casale. As for catalysts, academic work undertaken in surface physics provided the mechanism for the take up of nitrogen at the surface of the catalyst. There were other important developments, including in China. Finally, there is production of hydrogen, the most costly feature of the manufacture of ammonia, and which is so relevant to the rapidly developing hydrogen economy. Presently the most widely used process is steam reforming of natural gas (methane), but, as with other fossil fuels, this leads to emission of carbon dioxide. In the long term, methods for sustainable and environmentally friendly production of clean, or green, hydrogen by electrolysis of water using renewable energy sources will become necessary.

14.1 Introduction

As the basic material from which nitrogen fertilizers—urea, ammonium nitrate, diammonium phosphate (DAP), NP/NPK, calcium ammonium nitrate, and so on—are made, ammonia remains an essential commodity. Through fertilizers, it solves one of the world's most pressing challenges, by contributing to the annual global production of 800 million metric tons of wheat alone (according to the UN Food and Agricultural Organization's 2022 figures). Nitrogen fertilizers ensure food security for hundreds of millions of people.

Today, ammonia is one of the top chemicals produced globally. As a heavy chemical, it is produced in over five-hundred factories in around 80 countries.

A. S. Travis, *The Global Synthetic Ammonia Industry*, Perspectives on the History of Chemistry, https://doi.org/10.1007/978-3-031-87978-4_14

Together, these factories consume an enormous amount of carbon-based feedstocks, mainly natural gas, in the production of hydrogen. The outcome is that natural gas is used to produce 183 million tons of ammonia every year [1].

The energy and feedstock costs, and related factors, particularly the need to enhance the efficiencies of the various production steps, dominate parts of this closing chapter. Sustainability drives the most recent developments in ammonia technologies, aimed at solving another of the world's most pressing challenges, climate change, by reduction of the ammonia industry's carbon footprint. Steam reforming of fossil fuels liberates carbon dioxide; cryogenic separation of nitrogen from the air is energy intensive. The efforts to transition away from fossil fuels have been in progress for over two decades.

As we have seen, the last quarter of the twentieth century was a time of rapid expansion in the synthetic ammonia and allied industries. Important innovations in production technologies were associated with lowering production costs and increasing the capacities of converters. For the manufacturers, issues such as energy costs and over-production, and their impacts on profit margins, were, and remain, ongoing concerns. This was the case in the 1970s as fertilizer prices fell globally. As a result, there was little economic incentive for ammonia plant expansion and renewal of equipment, at least in the West. In the mid-1970s, Du Pont researchers drew attention to the fact that "[a] major part of the cost of manufacture [of ammonia] is capital cost; a plant with a capacity of 1000-tpd together with its associated facilities requires an investment approaching $100 million. Moreover in the less developed countries, these plants operate at about one-half their rated capacity" [2]. By 1980, the fertilizer industry in the West was operating at a negligible profit and often at a loss. The US, once one of the world's largest exporters, became one of its largest importers. Nitrogen fertilizer requirements were met with imports from regions and countries with low-cost natural gas: the Middle East, the Caribbean, Asia, and, especially, the Soviet Union had grabbed market share [3]. Several US ammonia facilities were barely running or had been mothballed. Some corporations collapsed or downsized. Consequently, major investments in ammonia plants ceased. When the time came to replace existing plants, as demand grew again, the high cost could not always be justified. There was a limit to how much it was worth investing in ammonia.

For those firms that survived, in the US and elsewhere, the problems of heavy usage of energy, the cost of which had grown exponentially in the 1970s, and the costs and risks of replacing worn-out plant, were compounded by the fact that older high-capacity ammonia plants, some with downstream urea facilities, including a number dating from the mid- and late 1960s, were far from optimal. This was particularly the case among emerging nations that had adopted policies based on agrarian self-sufficiency. The demand was still there, even with less than efficient plants. Global ammonia production increased dramatically in countries where economies were based on central planning. However, while this growth continues today, as does the demand for more eco-friendly production of both energy and hydrogen, geopolitical factors, and highly challenging conditions, have increasingly come to the forefront.

The disruption to typical production cycles, and profitability, has been highly significant during the current decade. Notwithstanding the demand, increases in energy costs have led to a number of closures and cutbacks. First was the impact of increased prices of natural gas experienced in Europe during 2021. And then the war between Russia and Ukraine that began on 24 February 2022. This halted the export, at an annual rate of 2.5 million metric tons, of Russian ammonia via Transammlak's Togliatti-Odessa pipeline. Pumping stopped with the outbreak of war, and in June 2023 a section in Ukraine was damaged. The price of anhydrous ammonia almost tripled, reaching $1808 per metric ton (compared with around $500 per metric ton in early 2021). In sharp contrast, and notwithstanding the European Union's economic sanctions, and embargoes, European countries allowed the import of Russian-made nitrogen fertilizers, especially urea, in order to ensure food security. As a result, Russia, drawing on its abundant natural gas as feedstock, flooded the markets with cheap nitrogen products, which helped farmers, but put pressure on manufacturers of ammonia and nitrogen products outside Russia [4]. Concerns were raised over the growing reliance on Russia, and the impact on self-sufficiency elsewhere. Significantly, the Western ammonia industry which had boomed in the months after the war started in 2022, was, within two years, as a result of the loss of Russian natural gas, forced to cut back and shelve investment plans. This was especially the case with Yara in Europe, and Canadian-owned Nutrien, where earnings fell by 71% in the first quarter of 2024 [5].

The implications of these often unanticipated disruptions affect not only the availability of fertilizer but also the availability of an increasingly important energy vector, liquid ammonia. Transporting the liquid ammonia to a site where it can be broken down into hydrogen and nitrogen is a major feature of the hydrogen economy, which includes steel making and transportation. In the twenty-first century, ammonia and hydrogen are inextricably linked.

With an emphasis mainly on technical and scientific aspects, we now review three quite different developments that have contributed towards greater efficiency in ammonia plant production and stimulated diversification: changes to the design of internals of converters; an understanding of the action of the catalyst; and research into ways for reducing the environmental impact of hydrogen production, at the same time making it more readily available. The first part of the review also fittingly provides an opportunity to bring the story of Ammonia Casale SA (now Casale SA), where we started, up to date.

14.2 Upgrading Outdated Technology: The Redesign of Converters

The leading development in the ammonia industry from the 1960s was the design and uptake of high-capacity, single-train energy-integrated units for synthesis of ammonia. The main technical advances in ammonia were achieved by M.W. Kellogg, and later by C.F. Braun, Chemico, ICI, Haldor Topsøe, and Uhde. They set in motion a complete transformation in the production of nitrogen fertilizers, or as Kellogg advertisements described it, "A New Era for Ammonia!" The novel technologies also influenced urea processes, notably those of Montecatini, Snamprogetti, Stamicarbon, and Toyo Engineering. This new wave of innovation came at a time of greatly increased need, and a deepening global food crisis, with the threat of widespread famine, particularly in Asian countries.

Ammonia Casale lacked the resources to adapt to the new challenges. Its ammonia technology, lagging far behind the competition, was mainly confined to the low-capacity production units, for which demand was low, and hardly sufficient to ensure survival. The market was dominated by the large international engineering contractors and ICI.

In an attempt to break into the new high-capacity market, Ammonia Casale designed a converter that featured modular catalyst cartridges and intercooling by gas injection, similar to the Kellogg model. This low-pressure design incorporated a thermodynamically efficient cartridge arrangement of three catalyst beds (held within the cartridge) which were cooled by the incoming gas. Such a configuration, as used in the Kellogg bottleneck converter, was incorporated into an ammonia train designed for single-train operation with a rated capacity of 1000-mtpd. Remarkably, during 1964–1965, even before Kellogg high-capacity single-train units were in operation, Foster Wheeler, the agent for Ammonia Casale, advertised a single-train low-pressure Casale ammonia process with a rated capacity of 1000-mtpd [6, 7]. Three of these new synthetic ammonia units were in operation by the late 1960s, two in France, at Carling (Ammoniac Sarro-Lorraine), and Grandpuits, near Nangis, southeast of Paris (for Société des Engrais d'Ile de France, SEIF), and one in Romania, at Piatra Neamt (for Industriali Import Romania) [8]. There were, as mentioned earlier, also four 600-mtpd units. No more were ordered.

Ammonia Casale, with its engineering operations in Terni, was not alone in facing a difficult future. The Italian economic miracle, at least for chemistry, was drawing to a close. The impact was especially felt at Ammonia Casale, whereby the early 1970s, and especially following the 1973 global increase in the price of oil, funds and technical resources were no longer available to maintain research and development into innovative process designs. Ammonia Casale became a shadow of its former self, in part because the technologies and scales of ammonia production elsewhere continued to undergo continuous and unprecedented transformations. Apart from its low-capacity ammonia units, Ammonia Casale's market share had been lost to the multinational engineering contractors. They had developed the technologies more suited to the needs of the time. Foster Wheeler no longer promoted Ammonia Casale's ammonia technology. Sales stalled, and Ammonia Casale faced a major crisis. In 1977, Ammonia Casale and its research and development unit SIRI were acquired by the plant engineering company INGECO (INGegneria e COnstruzioni), a member of the Altech Group, a consortium of contractors in Italy, including INGECO Spa, and abroad (INGECO International SA, INGECO Laing Ltd) [9]. INGECOAltech's several areas of activity included design services for petrochemical and refinery operations, mainly in the Middle East, as well as general feasibility studies.

In 1979, SIRI and its works at Terni were spun off, and the SIRI offices in Rome were transferred to a new company, Ammonia Casale SpA, with its registered office in Milan. The SIRI technical staff were absorbed by Altech's Ammonia Casale, which continued to offer services in ammonia, catalyst, and methanol technologies and claimed a methanol process that was superior to the ICI low-pressure quench-lozenge process.

However, Ammonia Casale was still not capable of bringing about the sort of innovations that were associated with the major engineering contractors. Fortunately, the solution for recovery did not involve large development budgets. The adaptive response came from an understanding of the various difficulties associated with inefficient converters, and how to overcome them.

The outcome was a new design concept suited to a more cost-effective production of ammonia. It enabled Ammonia Casale to restore the foundations of the pioneering company that five decades earlier had been the most influential force in the growth of the international synthetic ammonia business. This was achieved after 1978 when Umberto Zardi (1929–2025), an experienced and talented engineer and manager, who had led Snamprogetti's Inorganic Engineering Division, was appointed managing director of Ammonia Casale by Franco Salimbeni, former CEO at Snamprogetti, and the new CEO at INGECO's Ammonia Casale division.

Umberto Zardi's background was wide ranging. After graduating in mechanical engineering from the University of Bologna in 1954, he joined Montecatini's Inorganic Projects Division in Milan. For a decade he worked in research and technology development, eventually taking responsibility for the engineering and marketing of urea technology. In 1964, dissatisfied with cutbacks in research and development at Montecatini, Zardi joined Snamprogetti. There, he developed innovations in the synthesis of ammonia, and a new total recycle stripping technology for the urea process. He contributed to the construction of about twenty industrial plants. Snamprogetti became one of the largest licensors of urea processes in the world; its technology was employed in more than 130 facilities. With the engineer Giorgio Pagani, Zardi developed an integrated ammonia-urea process that brought about savings in energy and plant costs. This enabled a greater output of urea, at a lower cost, which further encouraged its application as a fertilizer. In 1976, Pagani invented a radial converter.

In 1978, Zardi, now charged with reviving the ailing Ammonia Casale, in an almost singular first response within the ammonia industry, planned a strategic overhaul based on the redesign of converters and revamping in order to confront the major shortcoming of existing, and especially older, ammonia converters, their inefficiencies. By this time, it was apparent that increasing equipment and production costs were presenting new challenges to the ammonia industry. The age and performance of existing ammonia plants came under scrutiny. Replacement converters, in particular, were expensive. Zardi's radical solution was to extend the lifetime of older plants through revamping (retrofitting) based on conceptual features that made converters more efficient, with enhanced yields and reduced energy costs.

The modernization and refurbishment of sophisticated chemical plant was hardly a new endeavor. After 1945, revamping was carried out on a vast scale in order to give new life to war-worn equipment. In around 1980, however, the revamping of ammonia converters was not just to get them working again. There were acute challenges of improving mechanical reliability, reducing high energy use, and the need to achieve greater overall efficiency. Zardi's innovative solution was the redesign of the internals of existing ammonia converters, which until then were almost exclusively of the axial flow type. It was an ambitious initiative.

On the basis of extensive performance calculations, and following the work of the Danish contractor Haldor Topsøe with radial flow of gas through two catalyst beds, and indirect cooling between the beds, Zardi and colleagues designed a unique converter that combined axial- and radial-flow operation in the catalyst bed. By 1979, the design of the axial-radial ammonia converter was complete. The new arrangement of converter internals promised to improve the performance of the synthesis loop. The proprietary axial-radial converter, which reduces the pressure drop, was promoted in the technical media in 1980 as part of a package for the revamping of existing ammonia converters (Fig. 14.1). The design was described in December 1980 at the Fertilizer Association of India's seminar on fertilizers, held in New Delhi [10]. In January 1981, the low-pressure drop vessel, applicable to modern ammonia synthesis conditions, was showcased by Zardi and colleagues at the British Sulphur Corporation's fourth "Fertilizer Nitrogen" conference in London. While the main theme of the three-day event was the challenge of more expensive energy, many other related aspects of state-of-the-art ammonia technology were discussed. These included the case for novel designs and innovations in revamping. The Ammonia Casale converter internals, as designed for the revamping of existing converters, offered the audience plenty of potential, not only in energy saving, but also in capital cost savings, through the upgrading of existing equipment [11]. Revamping based on Zardi's concept was promoted as a viable and economical alternative to complete plant replacement.

In the axial-radial catalyst bed, most of the synthesis gas passes through the bed in a radial direction, creating the very low-pressure drop. The rest of the synthesis gas passes down through a top layer of catalyst in an axial direction. This does away with the need for a top cover on the catalyst bed [12]. The cooling arrangement involves interbed tubular heat exchangers, which are the current standard.

In his published account, Umberto Zardi described the advantages of the axial-radial-flow converter [8]:

> The design is a vertical reactor with stacked modular catalytic beds (modular cartridge). In each catalyst bed, the gas flows through a first zone in a prevalently axial flow and in a second zone in a radial flow (beds with split gas flow). Unlike conventional radial reactor designs, sealing at the top end of each layer is unnecessary, since the axial flow zone acts as sealing pad for the gas flow. An important simplification of the gas run is thus achieved with a minimum of pressure drop[,] and a simple reactor cartridge which consists of catalyst baskets, each individual basket being individually detachable, and fitting one on top of the other with a simple sealed joint. The cylindrical bed walls are perforated along their whole

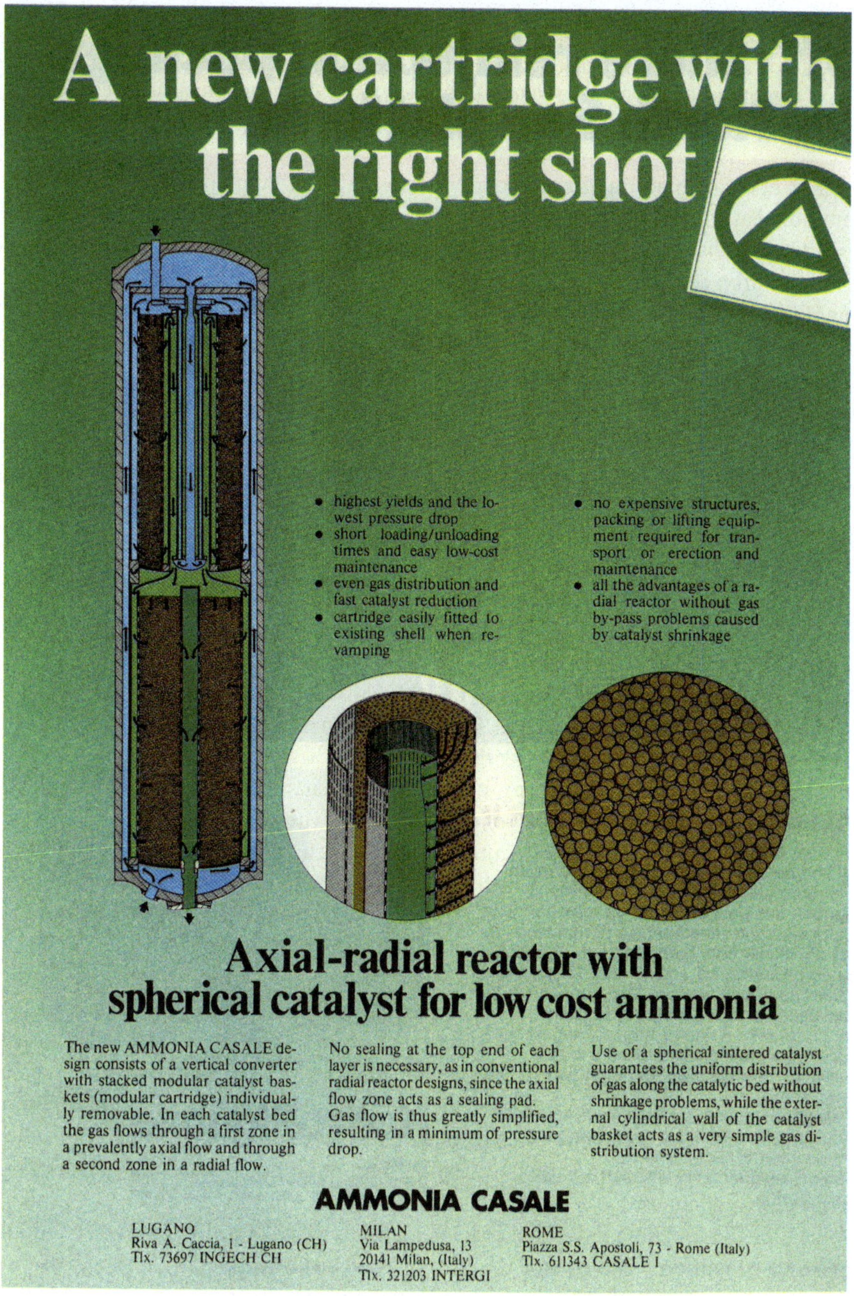

Fig. 14.1 "A new cartridge with the right shot. Axial-radial reactor with spherical catalyst for ammonia." From an Ammonia Casale 1980 advertisement emphasizing the novelty of its new axial-radial configuration and its suitability for the revamping of existing converters (Hydrocarbon Processing 58(December):164) (Casale SA. With permission)

> length with the exception of the top portion, to ensure correct distribution of the gas flow. The height of the unperforated zone in the top portion is optimized according to the type of catalyst used and to the radial thickness of the bed in order to make the best possible use of the catalyst and of reactor volume. This novel reactor design is particularly suitable for low-pressure loops involving large volumes of catalyst and very high gas flow.

For both methanol and ammonia converters, Zardi emphasized the role of the multiple gas inlets: "In each unsealed bed a split flow of gas (axial and radial flows) is achieved with optimal utilization of catalyst volume... Large column-like, low-pressure drop reactors can be designed by adopting the new bed architecture, with single trains up to 5000 metric tons/day." The new axial-radial-flow arrangement with stacked modular catalyst beds enabled the use of catalysts of considerably smaller pellet size than previously employed, and thus of increased efficiency, as discussed in Chap. 13. In 1980, Ammonia Casale claimed that its novel spherical small-size catalysts both enhanced gas distribution and contributed to a reduced pressure drop across the catalyst bed. However, the Casale spherical catalyst never went into commercial production, no doubt, as Luca Bianchi points out, because it could not compete with the catalysts marketed by ICI, BASF, and specialist manufacturers.

Ammonia Casale now offered a contract service for revamping of existing ammonia converters in order to both extend lifetimes and make them more efficient. The first industrial application of what was to become Ammonia Casale's hallmark axial-radial converter was in 1982, when a newbuild 400-tpd unit was ordered by Phosphoric Fertilizers Industry Ltd in Greece. It went on stream in 1985.

However, and notwithstanding promotional events and advertising, orders for revamping did not arrive. Part of the problem was that from late 1983 there was a downturn in the fortunes of INGECOAltech, the result of diversification away from core activities and a difficult economic climate. In a state of crisis, and to meet its financial obligations, INGECOAltech in 1984 divested its holding in Ammonia Casale to a group of private investors, including banks [13]. From Lugano, Umberto Zardi led the company under new ownership. In 1984, at Zardi's behest, his former colleague the engineer Giorgio Pagani joined Ammonia Casale from Snamprogetti. Pagani's outstanding contribution was the proposal for in situ revamping of the converter internals using the axial-radial gas flow with special reference to the Kellogg bottleneck converter, a design dating from the mid-1960s. With this untested innovation, Zardi planned a strategic overhaul that confronted the economic shortcomings of the existing ammonia industry through revamping, irrespective of the process originally worked [14].

The design of the axial-radial reactor internals was carefully tailored to suit all existing converters. The in situ retrofitting was stated to be particularly useful in those designs in which the catalyst basket could not be removed from the converter [15]. However, enlisting a customer willing to take risks based on what, to many, must have seemed little more than a guiding idea, presented demands of its own. The main challenge, and in contrast to the axial-radial converter installed in Greece, was the retrofitting of Kellogg bottleneck converters, based on Pagini's studies. How this challenge was overcome is explained by Giuseppe Guarino, another engineer who joined Zardi in 1984 and who later became CEO of Ammonia Casale [16]:

> … a great many of the largest ammonia units, with capacities ranging typically from 900- to 1200-mtpd, were licensed by M.W. Kellogg—which claimed to have sold more than one-hundred units of this type—whose converter had the notorious "bottle" shape. This shape is the result of the converter having a restricted top opening (i.e., with a diameter smaller than the vessel's inside diameter) in addition to the gas-in/gas-out exchanger (of even smaller diameter) being placed on the upper part, sitting on the restricted opening. The reason for such peculiarities in the design was the need to keep the cost down of what was an already very large (for those times) converter, which housed typically approximately 75 m^3 of catalyst and had approximately a 3000 mm internal diameter, both of which were significant figures. The restricted opening (with its smaller top flange) and the even smaller top exchanger helped to keep the weight and hence the costs down.
>
> The cartridge was one single piece, of totally welded construction, and impossible to remove after the top dished end was incorporated during the fabrication of the reactor in the workshop. This raised a major problem in the early 1980s when replacement of the cartridge was necessary (e.g., for maintenance) or desired (e.g., switching from axial to radial flow). To do so, it was necessary to replace the entire converter (the high-pressure vessel included) or, as some had dared to propose, cut the bottom girth seam, remove the cylindrical high-pressure vessel, switch the cartridge with the new one, put back in place the high-pressure vessel, and re-weld the girth seam! This was a very daunting task; fortunately none ever dared to do it.
>
> And right at this point arose the notion of in situ retrofitting. The advantages of the radial flow (i.e., reduced pressure drop and increased conversion due to the use of small-size catalyst, whose use, in turn, was made possible by the former) were already evident. If one wanted to introduce this configuration in an existing unit, it seemed that the only option, as noted above, was to replace the entire reactor: a very costly and lengthy operation. However, there was the Ammonia Casale proposal, based on the studies of Pagani, for M.W. Kellogg converters which was deceptively simple but had never been attempted before: it consisted in cutting to pieces the existing baskets (but not the cartridge), thus enabling their removal through the manways and the top opening (of course once the top exchanger was dismantled), introducing new internals, designed to permit the axial-radial flow (Ammonia Casale's hallmark design and its first [axial-radial] patent!), cut also in pieces, and assembling them inside the existing cartridge. All this without any hot work to the existing parts (the cartridge and the high-pressure vessel).

The commercial breakthrough occurred with the upgrading in 1986 of four Kellogg high-capacity (1000-tpd) ammonia units for CF Industries, at Donaldsonville, in Louisiana.

According to Giuseppe Guarino: "It all sounded easy, but nobody wanted to be the first to try it. The company who actually dared to do so was famously CF industries. Its engineers were very attracted to the idea but nevertheless took precautions: after placing an order for two sets, plus another two on condition that the former were successful, they sent an engineer to Lugano who spent months participating during the design stage and who fastidiously checked everything to his satisfaction. He was quoted as saying, 'It is so simple that it cannot but work'"

[17]. And so it was. All four units, for the Donaldsonville plant, went on stream, successfully, in 1986 [18].

Guarino recalled that "the news spread and this raised a lot of interest all over the world. I remember we spent the entire 1987 devising retrofits for virtually all sorts of converter designs, we issued lot of proposals, but none turned into orders: they started to arrive in the following year" [15].

Guarino also emphasizes that much of Ammonia Casale's success drew from an understanding of the different engineering cultures and needs of customers all over the globe. American engineers, for example, were more prepared to take risks, once they had been convinced that a technology was viable on a theoretical basis, even if it had not been tried before. This, he explained, proved especially to be the case with in situ revamping:

> Revamping is a product-in-itself, it is valid and very important, and among other things offers good margins.... For many products, you have to go through experimentation, through the pilot plant, and then you go out and sell the final product. In the case of revamping, we only ever relied on simulations, that is only on calculations, but we convinced the customer to the extent that the customer accepted the risk. Why did this happen? Because we found those with the requisite knowledge, who were able to understand without having to see what we were offering. And this, for example, went down very well with American engineers. Experienced and qualified engineers at ... CF Industries were able to understand that here was, yes, a novelty—though presenting a certain challenge, but that challenge was worthwhile even if it was not possible to physically see the product.

According to Guarino, the revamped converter with axial-radial internals worked as well as, and sometimes even better than, expected. This set the stage for the revival of Ammonia Casale as a global leader.

Simulations based on calculations did not suit every potential customer. Mere theoretical information—not backed up by working machinery—seemed hopelessly inadequate as a sophisticated sales ploy. "There were countries where if they do not see the product you want to sell, if they haven't seen it in operation for five years, they don't buy it. Here, obviously revamping was at first too much of a novelty. It couldn't have this feature of being seen, precisely because it was new. Someone had to try it but who could try it? Only those who understood the value of such an innovation, theoretically, and not by seeing it at work. From this, however, from the fact that we were able to offer novel technologies without undertaking experiments—and thus burdened with expensive research costs that would perhaps have limited sales—there came the possibility of developing these new products."

This showed itself in often different ways, because the approach of the Chinese customer was not the same as that of the Russian customer, nor the same as the American customer. "The Russian customer, for example, saw in us a continuity in assisting him, because in those early days of the revived Casale, Russia was a closed nation.... What we did was not only introduce our new technologies, but also teach the plant manager ... what could be done to improve efficiency, what it meant to have our technologies. We worked together. In fact, the great projects, the major contracts, that we obtained, arose from a particular personal closeness between us—engineer Umberto Zardi , and myself—with the owners of the plants, to whom we did not go just as salesmen but as super-consultants, advising on the best and most appropriate technology we could offer" [15].

Ammonia Casale soon moved over to complete plant revamping, including the reformers used in the production of hydrogen. The firm offered customers a backup service throughout the lifetimes of plants that it designed or revamped. The revamps improved the performance of converters to a level that was comparable to that of newly built vessels. The axial-radial-flow converter, and especially the in situ revamping concept, became a best-seller [17].

The placing of new axial-radial internals inside the pressure shell by in situ modification of the existing axial beds was soon after widely adopted for ammonia units whose capacity was limited by the original capacity of the synthesis reactor. Yields of ammonia were increased by 40%, which gave the operator the choice of significantly reducing specific energy consumption or increasing capacity. Not long after the completion of the revamping of the Kellogg converters at CF Industries, and of Kellogg converters elsewhere, Casale internals were also installed in converters of other designs, including an ICI lozenge-type converter, and a Haldor Topsøe S-100 (radial flow) type. The capacities of both the ICI converter, originally of 1200-mtpd capacity, and the Haldor Topsøe converter, originally of 1700-mtpd capacity, were increased by 20%.

Ammonia Casale also made an early entry, in 1987, into the Chinese market, which had opened up in the aftermath of President Nixon's momentous visit to the People's Republic of China in 1972 (see Chap. 12). As a direct consequence of the success of the CF Industries project in retrofitting Kellogg converters, Casale received an order for revamping five Kellogg converters in China. A sixth order for a single revamp arrived in 1988, and a seventh in 1990 [15]. The converters had originally been designed for use with natural gas as feedstock. Subsequently, and in order to reduce reliance on natural gas, they were mainly changed over to coal as feedstock. China now began a major program of replacing the hundreds of old, small, and highly inefficient ammonia converters, some dating back to the period of the Great Leap Forward (1958–1962). These were the ones that were based on local technology and relied on coal gasification. In 1990, Ammonia Casale received an order for the supply of the basic design, including the ammonia converter, of a completely new 500-mtpd coal-based synthesis loop in Heilongjiang Province. This proved to be the first of a long list of successes. The key, besides the good performance, lay in the ability to smoothly cooperate with local contractors. In addition, Casale committed to using locally produced ammonia synthesis catalysts, something that no other foreign licensor dared to do. As of 2022, the number of Casale synthesis loops in China exceeded seventy, with individual capacities of up to 2300-mtpd. Casale became the most successful foreign licensor of ammonia technology to the Chinese market.

From the late 1980s until the mid-1990s, Ammonia Casale expanded into Indonesia—where it revamped almost all the ammonia converters—Russia, India, and other countries. In addition, by the end of the 1980s, Ammonia Casale had revamped converters for Cominco American, at Borger, Texas; Pec-Rhin SA in Ottmarsheim, France; Agrimont SpA, in Priolo, Italy; DSM, in Ijmuiden, the Netherlands; Erdolchemie GmbH, in Cologne, Germany; and Ube Ammonia Industry Ltd, in Ube, Japan. In 1993, an order was awarded from India's Gujarat State Fertilizers & Chemicals (GSFC) for the first new ammonia loop of 1350-mtpd rated capacity incorporating a Casale axial-radial three-bed converter in a Linde ammonia concept (LAC) plant, designed for producing ammonia from light hydrocarbons. The single unit, commissioned in 1998, replaced Casale technology dating from the opening of GSFC in the early 1960s.

In 1995, Casale was awarded a contract for the first complete ammonia plant revamp, from Ultrafértil in Brazil, where urea and methanol units were also revamped. In the 2000s, plate-cooled converters were introduced for methanol synthesis. The adiabatic bed-type arrangement was used in the energy-saving axial-radial configuration and, especially, in the plate converters.

The useful lifetime of an ammonia converter is about ten to fifteen years. It is at this point that the converter requires a revamp. This is one of the main reasons why in the mid-1990s, the four converters of CF Industries at Donaldsonville, that had been revamped by Ammonia Casale in 1986, were provided with a second revamp, this time with Casale axial-radial internals featuring a new process layout based on three catalyst beds, with quench cooling between the first and second beds, and indirect cooling by heat exchanger between the second and third beds, further increasing the efficiency. Ammonia capacity was increased from 1270- to 1600-mtpd. In 2010 (the year in which CF Industries acquired nitric acid and fertilizer producer Terra Industries), additional retrofitting at Donaldsonville, in this case of two converters, was undertaken at what was then known as the 5th plant. CF project management collaborated with Casale in this work. One converter was changed from a 4-bed adiabatic Casale layout to Casale's latest single-bed isothermal technology. The other, a booster converter, based on Kellogg's ruthenium catalyst (see later), was changed from a 2-bed adiabatic layout to single-bed adiabatic. The converters were worked together, in series, the first isothermal, the second adiabatic, such that, per pass, the recirculation through the synthesis loop was reduced, as was the pressure drop and the load on the synthesis gas recycle compressor (with considerable energy savings). The capacity of the isothermal converter was increased from 1550-stpd (1043-mtpd) to 1800-stpd (1633-mtpd). The newly refurbished converters were commissioned in 2013 [19].

Ammonia Casale SA, from 2014 rebranded as Casale SA, by introducing energy-saving features, and novel converter designs offering even greater capacities, became a leader in the technologies of nitrogen products. In all, between 1984 and April 2021, Casale undertook over 200 ammonia converter retrofits, and 13 complete ammonia plant revamps, ranging in capacity from 500- to 2000-mtpd. In addition, during the same period Casale licensed six new 2050-mtpd grassroots plants, and 70 new synthesis loops with a capacity of up to 2000-mtpd, and installed 105 new synthesis converters, ranging in capacity from 500- to 2000-mtpd. By 2021, Casale ammonia technology, through revamping and completely new units, had been incorporated into more than 300 plants.

Retrofitting projects have been undertaken by other major engineering contractors. Thus the first application of the Kellogg Advanced Ammonia Process (KAAP), designed to produce up to 2000-mtpd of ammonia, was in a retrofitting project at Ocelot Chemical Industries in Kitimat, British Columbia, in October 1990. This was also the first application of the Kellogg ruthenium-based catalyst, developed jointly with British Petroleum, holder of the first ruthenium catalyst patent (1984). The catalyst enabled high conversion at a relatively low pressure. The KAAP low-pressure process incorporated the Kellogg Reformer Exchanging System (KRES), which thermally integrated the primary (endothermic) and secondary (exothermic) reforming steps. In addition to retrofitting, two grassroots KAAP ammonia plants each of 1850-mtpd capacity were installed in Trinidad in 1998. This was KBR/Kellogg's advanced ammonia process, offered alongside its conventional single-train process. Together they represented 190 single-stream Kellogg plants, in operation or contracted, at the end of the twentieth century.

A more recent example of revamping is that undertaken by thyssenkrupp Industrial Solutions for the Saudi Arabia Fertilizer Company's SAFCO IV plant, at the Al-Jubail petrochemical complex. The 3300-mtpd ammonia unit, based on Uhde technology, was commissioned in 2006. In 2014, thyssenkrupp Industrial Solutions received the basic engineering contract for raising the capacity to 3670-mtpd. This was the first-ever revamp of such a dual-pressure ammonia unit. The proprietary equipment, training, and commissioning advisory service were supplied to the engineering contractor ETec of South Korea [20]. Haldor Topsøe, the pioneer in radial-flow converters, is another leader in revamping technology and offers a range of converter configurations. In Russia, the engineering contractor GIAP offers a revamping service.

14.2.1 Diversification: Melamine and Nitric Acid

Building on the success of its novel converter designs, during 2013 and 2014 the Casale management, led by CEO Giuseppe Guarino, conceived of a diversification strategy that had at its core the objective of expanding the technological portfolio to sectors that would fit in smoothly with the traditional ones of ammonia, urea, and methanol.

With this in mind, the first step was the acquisition of the technology for the production of melamine from one of the largest producers and licensors in the world. Melamine was of interest because it relied on urea. The technology was available from Borealis AG, which had decided to focus on the production of plastics. Borealis was founded in 1994, as a joint venture involving Neste of Finland and Statoil of Denmark (and since 2006, is based in Vienna). Borealis acquired the melamine technology when it took over the entire business of the French company GPN (Grande Paroisse).

Melamine is an important source of stable resins for plastic products. The market for melamine, unlike the market for nitrogen fertilizers, is not subject to fluctuations in demand, thereby offering a useful route to diversification. Between

2013 and 2015, Ammonia Casale acquired various well-proven technologies from Borealis, widening its range of downstream products, first to melamine, and then nitric acid, phosphate processes, and compound fertilizers.

Giuseppe Guarino emphasized that this acquisition of a very broad portfolio of closely linked technologies was achieved through careful but deliberate negotiations, backed up by Ammonia Casale's track record [15]:

> One of the products that seemed to be interesting was melamine. For melamine, we did not have the technology, but one company, Borealis, did, and was willing to sell it, so we made it known that we were interested in acquiring it, as did many others. There were international bids, so we carefully played our cards. Why was melamine important? And why did we win in the end, beyond the commercial consideration? Because we were already present in the first stages of melamine technology, namely the urea technology. The melamine plant needs urea, which is one of Casale's products, but the melamine plant returns to the urea plant certain products. So, in order for the melamine plant and the urea plant to work, the two units must be integrated. We already had experience of this integration; we were undertaking it for those who had melamine units and so we were already present in the market. We demonstrated that we were therefore more credible as a company that could also handle the melamine technology. And, consequently, we won this bid. We succeeded in acquiring melamine because we convinced Borealis that we would be successful in replicating, and in multiplying, melamine plants around the world, because that was also one of the seller's goals. Borealis also had nitric acid technology and nitric acid is obtained from ammonia. Nitric acid is one of the most widely used chemicals in the world. It is one of the major products and so it seemed interesting—necessary I would almost say—to be able to acquire the nitric acid technology, that we did not have.

The first Borealis technology acquired, in 2013, was for the conversion of urea into melamine. The nitric acid, nitrates, and phosphate technologies followed in 2014. Nitric acid is important not only in manufacture of fertilizers but also in manufacture of explosives, and for use in the chemical industry. Casale, in 2015, completed the transfer from Borealis of the whole set of proprietary technologies, which in addition to nitric acid, include ammonium nitrate (AN), and urea-ammonium nitrate solution (UAN), and granulation technologies, for different downstream fertilizer products. With the new package of former GPN technologies, Casale acquired the entire value chain, starting with the raw materials, right up to production of the final solid fertilizers.

Giuseppe Guarino explained that there was a further important factor that convinced Borealis:

> The acquisition by Casale of a contractor whose engineers specialised in the design and construction of nitric acid plants. This was Chemprojekt Nitrogen a.s., a division of the Czech company, Chemoprojekt a.s., of Prague (established in 1950). Chemoprojekt Nitrogen a.s. was well-known to Borealis: Normally Borealis did not build the plants but had them built by others. Borealis had bought the nitric acid technology from the French company Grand Paroisse. The builder of the nitric acid plants was Chemoprojekt Nitrogen. Ammonia Casale decided that it would be useful to have a company that already had experience in nitric acid in order to convince Borealis to sell us the technology.

The acquisition was a critical component of Ammonia Casale's programs for diversification and expansion. It included a team of experienced, creative, and energetic engineers. "We badly needed manpower; we needed the capacity to develop engineering in a big way. And going through a recruitment drive would have taken too long. We had this twofold need, to significantly increase the workforce and the engineering capabilities of Casale, and to acquire skills and know-how in the field of nitric acid. As a result, we acquired Chemoprojekt Nitrogen a.s. (CHPN). Following this move we went straight back to Borealis; and this was one of the reasons, quite apart from the offer in monetary terms, that allowed us to beat all the competitors, who were as interested as we were in acquiring nitric acid. We were already well established in our ability to sell nitric acid plant. There was indeed a strategy; it was not random, and there was nothing random about it. There was a will to build and to acquire" [15]. This series of interconnecting considerations provided the background to the acquisition, in July 2014, of majority control in CHPN. Subsequently, most nitric acid plants were built by CHPN.

The CHPN acquisition, particularly the availability of its considerable expertise, represented the single most important change in Ammonia Casale's corporate structure. To underscore its full integration into the Casale group structure and its alignment with the related business strategies, the newly combined company was renamed Casale Project a.s, on 1 December 2016. Thanks to this move, Casale was now able to offer complete turnkey projects.

14.3 Academic Research: Catalysts

The BASF synthetic ammonia process based on Haber's invention is without doubt the foremost example of the early application of heterogeneous catalysis in high-pressure industrial chemistry. Catalyst research for industrial applications, most notably in the ammonia synthesis, was originally an empirical endeavor. From 1909 until 1913, BASF undertook twenty thousand laboratory experiments using 3000 compositions of iron combined with various promoters, carried out in batteries of bench-scale converters, but without greatly improving on the original low-cost iron catalyst of Alwin Mittasch. During the early 1930s, some progress in understanding the nature of adsorption of nitrogen on the catalyst was made by Paul H. Emmett and Stephen Brunauer at the US Department of Agriculture's Fixed Nitrogen Research Laboratory. Their experiments suggested that disassociation of nitrogen on the surface of the iron was the rate-limiting step.

Most studies were carried out in industrial research laboratories and included the search for catalysts that enhanced the performance of several stages, including in the shift reaction, and in steam methane reforming, where nickel catalysts encouraged the reaction between steam and hydrocarbon. Until the 1960s, manufacturers of ammonia prepared their own catalysts; a typical ammonia process required four to six catalysts. By this time, specialist firms were undertaking research and development in, and supply of, catalysts. However, certain chemical firms remained prominent in production of catalysts for sale to other manufacturers. In this area, the UK's ICI was for many years a world leader (Chap. 13). Its innovative, and profitable, catalyst division included the supply of catalysts for the manufacture of ammonia. Firms in the rare metals market also undertook research into catalysts, notably the UK's Johnson Matthey, which acquired ICI's catalyst business in 2002. Chemical engineering contractors engaged in important high-activity catalyst developments for ammonia processes included C-E Lummus, Haldor Topsøe, and Kellogg, whose ruthenium-based catalyst, while expensive, was claimed to be ten to twenty times more active than traditional iron catalysts.

Because of the industrial importance of synthetic ammonia, the role of the catalyst, as the heart of the converter, has received the greatest research effort in basic science. At a high pressure and an elevated temperature, the catalytic material is under continuous impact from the flow of mixed gases. However, the exact nature of the take up of nitrogen at the surface was not understood until the last quarter of the twentieth century when academic chemists applied new instrumental methods to follow the progress of the reaction as it happened on a laboratory scale. The lead researcher was Gerhard Ertl at the Ludwig Maximilian University, Munich, who from the 1960s was interested in gaseous reactions at metal surfaces. From the semiconductor industry, he borrowed a technique for creating high vacuums in order to study the lowering of the activation barrier presented by the triple bond—which is responsible for the large bond energy (941 kilojoules per mole)—between the two nitrogen atoms. The candidates for catalyst experiments included traditional iron-based catalysts promoted by alkali and alkaline earth metal oxides. The techniques of surface chemistry led to an understanding of the rate-limiting step, namely the dissociative chemisorption of nitrogen (the result of electron transfers from transition metals to the antibonding orbitals of the nitrogen). The exact mechanism of chemisorption of nitrogen was established. It was shown to occur in six steps. The experimental findings replicated the process taking place in an industrial converter leading to formation of ammonia. It brought about closure of decades of experimental and theoretical investigations that had started with the work of Haber and Nernst. For this important work, Gerhard Ertl in 2006 received the Nobel Prize in Chemistry [21, 22]. Ertl's work, however, did not bring about a decline of interest in the mechanism of industrial catalysis for ammonia. Far from it [23, 24].

Chinese academics have contributed to catalyst developments for the ammonia industry with a high-activity catalyst based on wüstite ($Fe_{1-x}O$), discovered in the 1980s at Zhejiang University of Technology. The technology was licensed to Süd-Chemie in Germany; production of what were called Amo-Max® catalysts commenced in 2003. This represented a major departure from the traditional iron catalysts that had been used for almost a century, as did the more recent ruthenium catalysts [25]. Jointly with the Swiss firm Clariant (which acquired Sūd-Chemie in 2011), Casale SA developed a highly active variant of the Amo-Max® catalyst, for which the two firms jointly received the 2021 Sandmeyer Award of the Swiss Chemical Society. Ruthenium and wüstite catalysts represented the main

Table 14.1 Main phases of innovation in synthetic ammonia and allied processes

Years	Innovations
1909–1930	Design of closed-loop manufacturing units, including converters, and high pressure reciprocating compressors. Promoters for iron catalysts. Hydrogen from water electrolysis, water gas, and coke oven gases. Nitric acid from ammonia. Ammonia processes adapted to methanol production
1930–1960	Hydrogen from steam reforming of hydrocarbons: primary reforming in a multiple parallel tube reactor with nickel catalyst; secondary reforming of gases, by burning with air to provide a stoichiometric mixture for ammonia synthesis. Carbon monoxide shifting and carbon dioxide removal processes. From the 1950s, methanation increasingly replaced copper solution scrubbing for removal of residual carbon oxides. Multi-bed ammonia converters from 1950s. Typically 200- to 250-tpd of ammonia at around 250–300 atmospheres
1960–1980	Energy integrated high-capacity single-train units of 1000-tpd, or more, with centrifugal compressors, and lower synthesis pressures (150 atm), from 1966. Increase in reforming pressures. Pipelines and ships for distribution. Novel ammonia-to-urea processes
1980–2015	Axial-radial converters, for revamping and grassroots plants. 1800-tpd single-train units in the 1990s. Wüstite and ruthenium catalysts. Horizontal converters
2015–present	Processes for sustainability, especially hydrogen from water electrolysis

developments in ammonia catalysts at the turn of the twenty-first century [26] (Table 14.1).

Catalyzed reactions for formation of ammonia continue to attract considerable academic interest. The following are a few examples. X-ray photoelectron spectroscopy has been used by researchers at Stockholm, in collaboration with the Deutsches Elektronen-Synchrotron (DESY) research center in Hamburg and the Montan University in Austria, to demonstrate that the rate-limiting step of the ruthenium catalyst is dissociation of the nitrogen. The ruthenium single-crystal surfaces remain metallic and are almost adsorbate-free [27]. A special focus is on catalysts that considerably lower the energy requirements of the synthesis reaction. At RIKEN CSRS in Japan, a catalyst of ultra-small molybdenum metal particles prepared from a hexanuclear molecular metal halide cluster, with removal of halide ligands by activation with hydrogen, has in the laboratory brought about a low temperature ammonia synthesis [28]. Low-cost catalysts have also attracted attention. Researchers at RMIT University in Australia claim a liquid metal copper and gallium catalyst that uses less heat and lower pressures than the conventional ammonia processes. These metals are much cheaper than ruthenium [29]. University of Chicago researchers have investigated lithium-mediated ammonia synthesis, in which nitrogen combines with a hydrogen-donating fluid such as ethanol in the presence of a charged lithium electrode. Rather than break apart nitrogen gas molecules at a high temperature and pressure, in this method the nitrogen adheres to the lithium, and then combines with hydrogen to form ammonia [30].

Finally, electrocatalytic nitrogen fixation by electrolysis, to afford ammonia, and its oxidation at the anode to ammonium nitrate fertilizer, presents itself as another intriguing possibility in the search for sustainable carbon-free processes [31].

14.4 Hydrogen: Towards Green + Sustainable

Ammonia production is cyclical, and expensive. Increases in the price of oil and natural gas—the principal sources of energy and hydrogen—impact on most ammonia operators. The synthetic ammonia industry consumes much of the 95,000 kilotons of hydrogen (2023 estimate) produced each year mainly by steam methane reforming.

As we have seen, two-stage steam reforming of gaseous hydrocarbons was developed in the 1930s mainly at ICI. In the early 1940s, ICI assisted American engineering firms to introduce steam methane reforming at six new ammonia production facilities in the US and also in Canada. For over a decade, reforming was carried out at atmospheric or low pressures. Then in 1953, Kellogg introduced high-pressure steam reforming, at first by taking advantage of the pressure of natural gas as it was released from gas wells. In Italy, with abundant supplies of natural gas, Montecatini used steam methane reforming for its Fauser process by the mid-1950s. ICI developed steam reforming of naphtha under pressure. Process improvements and the introduction in 1966 by Kellogg of the high-capacity ammonia process in which a centrifugal compressor drove all operations, transformed hydrogen production by bringing about methane steam reforming under even higher pressures, in part matching the pressure required for the ammonia synthesis. During the 1990s, single-train units of up to around 1800-tpd capacity became the norm, as promoted by Haldor Topsøe, Kellogg (KAAP), and Krupp Uhde, among others (see also Chap. 13). Early in the twenty-first century, capacities were doubled.

Global demand for hydrogen in the ammonia and oil refining industries has grown threefold since the 1970s and shows no sign of abating. Increased production of hydrogen for the synthesis of ammonia is necessary in order to avert a global food catastrophe. At the same time, the world is moving towards a low-emissions future. Climate and food are inextricably linked.

Challenges posed by the cost of energy are as relevant today as they were in the last quarter of the twentieth century. The cost of natural gas fluctuates according to geopolitical situations. Already in the 1980s rising costs, as well as cheap imports of ammonia, led to the closure of ammonia factories in the US. In 2003, Jim Harris of the Louisiana Ammonia Producers advised *Time* investigators that of nine corporate members in the producers' association in June 1998, only one, CF Industries, remained. According to Harris: "It's been devastating. The high natural gas costs have been the overwhelming reason plants have closed. It's completely depressed the whole area" [32]. This situation is particularly significant in Europe, where over the past two decades it has caused firms to slow down production and in some cases shutter manufacturing facilities. In September 2021, the CF Fertilisers UK Ltd ammonia factories at Billingham and Ince in the UK were closed. A series of sharp rises took place in the prices of nitrogen fertilizers, in particular anhydrous ammonia, ammonium nitrate, urea, and DAP. In the UK, it was only with state support, at least until January 2022, that the Billingham ammonia factory restarted production after a short stoppage, and that was in order to ensure a continuous supply of food-grade carbon dioxide of which it is, or rather was, the UK's main source. In 2022, Ince was closed, for good. CF Fertilisers UK Ltd ceased manufacture of ammonia altogether at Billingham in 2023, almost a century after manufacture began there. Yara International's Hull facility closed in February 2025. Today, Yara manufactures ammonia at Sluiskil, Holland; Brunsbüttel, Germany; Porsgrunn, Norway; Ferrara, Italy; Le Havre, France; and Tertre, Belgium.

Currently more than 90% of the ammonia industry's energy demand is fulfilled by natural gas and, to a lesser extent, other non-renewable carbon-based fuels. However, as discussed earlier, fossil-fuel-based hydrogen is associated with the release of carbon dioxide. Catalytic hydrogen production by steam methane reforming produces 9 tons of carbon dioxide for every ton of hydrogen produced. For very large ammonia units, with capacities of 4000-tpd or more, autothermal reforming, a combination of steam methane reforming and partial oxidation reforming, is available and offers greater efficiency [33]. This would suit Casale's design for a 6000-mtpd ammonia unit, and KBR's (formerly Kellogg) design for a 10,000-mtpd unit that was announced in October 2022 [34].

The need for emission-free production of hydrogen, that is, low and zero carbon hydrogen, is a pressing problem that can be tackled by carbon capture (sequestration) and storage, a technology that has seen limited, but growing, use and is an effective way to slow climate change. This is blue hydrogen which is used to produce blue ammonia. A leading firm in this area is Giammarco-Ventrocoke of Italy which produces activators that bring about the absorption of the carbon dioxide. Fertiglobe and TA'ZIZ at Ruwais in Abu Dhabi operate the largest blue ammonia facility in the Middle East and North Africa (the MENA region), with a capacity of up to one million metric tons of ammonia per year. In 2022, QatarEnergy Renewable Solutions and Qatar Fertilizer Company (QAFCO) signed an agreement valued at over $1 bn to plan and develop the blue Ammonia-7 plant in Mesaieed, Qatar. Construction of what was claimed to be the largest facility of its kind in the world, with a rated capacity of up to 1.3 million metric tons per year, began in late 2024. In July 2025, the CF industries Donaldson facility, the largest producer of ammonia in the world, with an annual output of almost 2 million metric tons, began large-scale carbon sequestration. Two million metric tons of carbon dioxide will be sent to an underground storage facility for permanent storage every year.

For the long term, methods of more sustainable production of hydrogen, and ammonia, will become necessary. This includes especially the return to hydroelectric power, and electrolysis, which was used on a large scale until the 1930s; installed production peaked around 1970. Following the development of steam reforming of natural gas, hydroelectric power became too expensive to compete with fossil-fuel sources of hydrogen. But hydroelectric power is now considered, for the future at least, the most sustainable, carbon-free and environmentally friendly source of clean, or green, hydrogen, and in turn of green ammonia. It also offers self-sufficiency. But there is a long way to go. There are, for example, concerns over the ecological impact of dams. Today, solar energy is a strong candidate for production of a sustainable source of hydrogen. The global ammonia industry, once located in Europe, based on coal and then natural gas, and in the US, based on gas, has begun to migrate to locations rich in water, solar, and wind power. There is also the potential for extracting geologically preformed abiotic hydrogen, generated deep underground by the reaction between hydrothermal water and iron-containing rocks. However, technologies that reduce carbon release require large-scale financial incentives, including government subsidies and tax benefits, to encourage investments in clean energy, and hydrogen, at least over the coming decades.

The transition to clean energy and reduction in greenhouse emissions undoubtedly represents the next phase in the development of the synthetic ammonia industry [35, 36]. However, it has also to confront the present commitment to and security in fossil fuels. Electrolysis for hydrogen has been commercialized and produces around 180 kilotons annually, just 4% of total global production. As a sustainable energy source suited to a wide range of industrial activities, hydrogen needs to be produced in a cost-efficient manner. Massive subsidies in the US, Europe, and China are encouraging research into more efficient electrolyzers, including cost-effective catalysts. The ammonia industry has also taken up the cause of green hydrogen. Casale, as a result of its long experience in the production of ammonia, has partnered with Canadian electrolyzer innovator Next Hydrogen Solutions Inc. to develop efficient hydrogen production technologies for ammonia and methanol. In addition, moves are underway for large-scale production of ammonia as an energy carrier for hydrogen. The hydrogen is released by cracking ammonia in a reformer. This system has already been developed by thyssenkrupp using the Uhde reformer. "Reversing the Haber process" offers a viable method for using ammonia as a fuel. Dissociated ammonia was used on a small scale from the mid-1940s for small-scale hydrogenation, including of cotton seed oil [37]. Other uses of disassociated ammonia included as a source of hydrogen for welding, for manufacture of metal powders, and hydrogenation of petroleum fractions.

In 2023, with backing from the German government amounting to 700 million euros, thyssenkrupp established Thyssen Krupp Nucera which specializes in the manufacture of electrolyzers. Applications include in steel manufacture at the thyssenkrupp steelmaking complex in Duisberg, and at H2 Green Steel in Sweden. More relevant here is a $100 million electrolyzer unit supplied to CF Industries at Donaldsonville for the manufacture of synthetic ammonia. CF Industries report that in 2024 the cost of making a ton of emission-free ammonia from electrolytic hydrogen is around $600, several times as much as making ammonia from gray hydrogen (without carbon capture) [38]. ExxonMobil is working on a low-carbon hydrogen production plant at its Baytown Complex, near Houston, Texas, with an annual capacity of about 900,000 metric tons of hydrogen and more than one million tons of ammonia. The project, which requires tax credits, is scheduled for completion in 2028 [39]. CF Industries is the leading producer of ammonia in the US, where it is manufactured by 17 companies at over 30 facilities, for home consumption of food and export.

Fertiglobe is among the large corporations planning the move to green hydrogen, imported from Egypt, for ammonia production based on KBR technology. Similar initiatives have been taken up globally, of which the following are a few examples. Thus the European Commission has approved a grant from the Lithuanian government that will enable Lithuania's largest ammonia manufacturer, Achema, to move over from gray to green hydrogen. In June 2023 and June 2024, the World Bank provided loans, together totaling $3 billion, to India in order to support the development of green hydrogen projects. The energy corporation Reliance Industries is leading a consortium working on a trillion rupee ($12 billion) green hydrogen and ammonia facility at the Port of Kandla, in Gujarat state. Norwegian nitrogen producer Yara has come to an agreement, through its Yara Clean Ammonia, to offtake and export green hydrogen-derived ammonia from a planned 100 MW Egyptian facility. In June 2024, Yara opened a renewable hydrogen plant based on electrolysis at the Herøya Industrial Park. The hydrogen will be used in the manufacture of ammonia and is expected to reduce annually the emission of 41,000 metric tons of carbon dioxide, from using natural gas as the feedstock.

These are part of the move towards a climate-neutral global economy. Hydrogen fuels cells in which the chemical energy of hydrogen is converted through an electrochemical reaction into electricity are being introduced for railway locomotives and railcars. However, while the use of green hydrogen as a fuel for transportation and domestic heating remains a contentious issue, this is not the case for ammonia manufacture and production of steel. Here the industrial leaders in renewable ammonia technologies are KBR, Haldor Topsøe, thyssenkrupp Uhde, Casale, and Stamicarbon.

There is also the role of transportation of ammonia to consider. Ammonia, which contains 17.6% of hydrogen by weight, requires less space and is far cheaper and safer than shipping hydrogen over long distances, including by barges [40]. It can be stored at room temperature under a pressure of ten atmospheres. The bulk transport of anhydrous ammonia by sea has been carried out for over half a century. It began in January 1964, when a tanker specifically constructed for transporting refrigerated anhydrous ammonia was launched at Verolme United Shipyards, in Rotterdam. It was the first of two supplied to Oswego Chemical Carriers Corp., for the shipping under charter from W.R. Grace & Co. of ammonia from Trinidad to US east coast terminals [41]. Today, Fertiglobe is the largest seaborne exporter of urea and ammonia, combined, produced in the UAE, Egypt, and Algeria, with access to six key ports and distribution hubs on the Mediterranean, the Red Sea, and in the Arab Gulf. At a remote location in Australia, Yara has successfully achieved ship-to-ship transfer of ammonia.

However, coal as a source of ammonia, and ultimately of hydrogen, has not been completely removed from the equation. In the US, Glenrock Energy, of Casper, and Kanata America, in Cheyenne, have proposed the construction of a coal-based synthesis gas production unit at the Kemmerer coal mine in Wyoming. The synthesis gas would be used in production of electricity, the carbon dioxide would then be removed, and the hydrogen purified for production of ammonia. The promoters emphasize that the technology would reduce the demand on imported ammonia and at the same time enhance national security in ammonia and hydrogen [42]. Ammonia has potential as a carbon-free fuel for internal combustion engines, though the exhaust gases will require treatment. For this reason it is best suited to marine engines situated in large spaces where the gases can be processed.

Finally, mention should be made of the role of instruments in the synthetic ammonia industry. The industry has since its foundation been reliant on instrumentation in process control. Great strides were made after World War II when electronic instrumentation became widely used for automatic gas analysis. Moreover, the control of the composition of gases and gas purity was of critical importance for safe operation. By around 1950, infrared instruments were used for monitoring methane, carbon monoxide and carbon dioxide in gas streams. As an example, Phillips Petroleum designed and assembled inhouse infrared instruments that were used at the Phillips Chemical ammonia installation, completed in late 1953 [43]. Within a few years the petroleum industry was a major user of the newly invented gas chromatography. Measuring the environmental impact of releases from industry requires similar analytical instruments. However, concerns over monitoring of releases into the atmosphere and contamination of water supplies from ammonia installations, almost all of which relied on fossil fuels as feedstock, did not attract much attention from the industry and regulators until around 1980. By that time, advances in plant design in order to reduce the energy input, and recovery of released gases for generation of energy, impacted not only on profits but also on the reduction of air emissions [44]. In the third decade of the twenty-first century, the situation is very different. Preventing emissions in order to combat climate change, as discussed in the foregoing, is a major driver in the moves towards sustainability [45].

14.5 Summary and Conclusion

The synthetic ammonia industry is invariably described in the literature as a mature industry based on the century-old Haber–Bosch process, and which provides us with the important nitrogen fertilizers. This has tended to mask the fact that over the past century many incremental and sometimes major innovations have been introduced, and also that the modern ammonia industry owes it origins not just to Fritz Haber and Carl Bosch but also to other pioneers. The least known, but the most significant in terms of technology and global expansion, at least until the late 1920s, is Luigi Casale. Maybe this has to do with the fact that, unlike Fauser and Claude, his process was never associated with a major corporation. Maybe it has to do with Casale's early death. Whatever the reason, this study has, in its early chapters, attempted to set the record straight.

No less important, the present historical review of synthetic ammonia production has enabled us to identify various phases, technical as well as commercial, in the growth of the industry. At first, of course, was the success of BASF in scaling up Haber's invention. Described as a great accomplishment of both science and technology, the BASF process became justified through wartime needs rather than production of nitrogen fertilizer. After 1918, by blocking rights to its ammonia process, BASF aimed to dominate the global market in ammonium sulfate fertilizer. This stimulated the parallel development of other processes, with an emphasis on novel designs of the internals of the converter, as well as novel catalysts, and encouraged not a little industrial espionage. Some of the processes faced difficulties until suitable steels for the converters became available. Some processes were confined at first to individual firms and corporations. But it was Luigi Casale who by making his process available from 1921 through licensing arrangements, procuring machinery, and providing design and startup services, became a founding father of the international synthetic ammonia industry. Casale, the engineering contractor, demonstrated that in the new era of industrial high-pressure chemistry, the combination of patents *and* knowhow could be used to sell licensing contracts. The widespread, and rapid, spread of Luigi Casale's synthetic ammonia process was also an outstanding example of technology transfer from Italy during the 1920s. Until late in the decade, the ammonia industry was confined to Europe and, through Casale, to Japan. Only from 1928 did the US take up large-scale production of ammonia for fertilizer production, in part, again, thanks to the Casale process. By 1930, at least eight rival synthetic ammonia processes were in use and the design technologies had been adapted to methanol and synthetic fuels. Within a decade the industry had expanded in both fertilizer and munitions production in the build up to World War II.

The strength of the industry relied on the supply of pure gases. Hydroelectric power enabled production of pure hydrogen but in most cases was too expensive. BASF developed the water gas process for hydrogen which was used in large factories in Europe and the US. Then from the late 1920s came coke oven hydrogen. Steam reforming and partial oxidation were available by 1940. These were taken up for natural gas (methane) and naphtha as feedstocks. Novel reforming catalysts and methods for purifying gases by removal of carbon oxides were introduced. By the 1980s, natural gas was the main source of hydrogen and has remained so.

In the 1960s, American engineering contractors made massive inroads into the design and construction of ammonia and nitrogen fertilizer plants. No doubt the most important innovation, representing the greatest transformation since the 1920s, was the development of energy-saving high-capacity single-train units, at first by M.W. Kellogg. Centrifugal compressors replaced reciprocating compressors in these large units. Kellogg innovations included in the 1940s a completely new model of converter, incorporating cartridges and quenching instead of internal tube-cooling. The Kellogg bottleneck converter of the mid-1960s enabled the shift in the scale of production of individual ammonia units from 250 to 300 tons per day to 1000 tons per day from 1966. This was followed up elsewhere. American and Japanese engineering firms, often in collaboration with European contractors and chemical firms, in particular ICI, dominated the industry. The ammonia synthesis process was now worked at a lower pressure, while steam reforming took place at a higher pressure than was the case in the 1950s. The matching of pressure requirements was a tremendous contribution towards reducing the energy requirements of the entire process. Subsequently, single-train processes of even higher capacities were introduced. New markets appeared beyond the Western hemisphere, brought about by the need to confront growing famine and political changes. The boom lasted until the energy crises of the 1970s, first the fourfold increase in the price of oil, and then the impact of the collapse of the price at the end of the decade. The resulting fluctuations in the prices of energy and costs of ammonia militated against new investments in the industry.

By the late 1970s, manufacturers of ammonia relied on aging and increasingly energy-inefficient converters to satisfy the demand for nitrogen fertilizers. This situation created an opportunity for the revival of Ammonia Casale which from around 1980 specialized in revamping (retrofitting) the internals of existing converters with a design that included the modification of axial flow ammonia converters to axial-radial flow. This can also be counted among the major innovations in the synthetic ammonia industry. No less significant are the latest KBR ammonia units. Nor is the historical record of the KBR corporation. As already mentioned in Chap. 13, since 1943, when its predecessor Kellogg undertook erection of two US government wartime ammonia plants, both based on reforming of natural gas, KBR has been involved in over 250 grassroots ammonia plants worldwide.

From the 1980s, as this chapter has shown, economic, scientific, and environmental factors determined the directions of innovative inquiry in the industry. New converter and reformer designs contributed to the lowering of the cost of production, new laboratory techniques enabled an understanding of the mechanism of take up of nitrogen on the surface of the catalyst, and new ways of making pure hydrogen, freed of the constraints of the wide variations in the costs and supply of fossils fuels, and less harmful to the environment, became a major driver of innovation.

References

1. Saygin D, Blanco H, Boshell F et al (2023). Ammonia production from clean hydrogen and the implications for global natural gas demand. Sustainability 15(2):1623. https://doi.org/10.3390/su15021623
2. Hardy RWF, Havelka UD (1975) Nitrogen fixation research: a key to world food? Science 188(9)(May):633–643
3. Anon (1980) Anhydrous ammonia from the U.S.S.R. Report to the president on investigation No. TA-406-6 under section 406 of the Trade Act of 1974. USITC [United States International Trade Commission]. Publication 1051, Washington DC
4. Savage S, Tani S (2024) "Flood" of cheap Russian fertiliser risks Europe's food security, industry says. Financial Times. https://www.ft.com/content/8d4e6543-f210-4a26-9bed-93fd0c35b6e6. Accessed 1 July 2024

5. Chapman K, Freitas G Jr (2024) Top fertilizer maker says "black swan" jolts are finally easing. Bloomberg. https://finance.yahoo.com/news/top-fertilizer-maker-says-black-103000538.html
6. Foster Wheeler advertisement (1964) Hydrocarb Process 43(Aug):251
7. Foster Wheeler advertisement (1965) Hydrocarb Process 44(July):94
8. Zardi U (1982) Review these developments in ammonia and methanol reactors. Hydrocarb Process 61(Aug):129–133
9. Cariati V (2010) INGECO: la fine dell'avventura. Impiantistica Italiana XXIII (5)(Sept-Oct):85–93
10. Zardi U, Comandini E, Passariello A (1980) Advances in catalyst and reactor design for low energy ammonia production. Fertilizer Association of India. Seminar on fertilizers in India in the eighties, New Delhi, 4–6 Dec. Unpublished paper. (Casale SA)
11. Zardi U, Comandini E, Gallazzi C (1982) A novel reactor design for ammonia and methanol synthesis. In: More AI (ed) Fertilizer nitrogen: facing the challenge of more expensive energy. Proceedings of the British Sulphur Corporation's fourth international conference on fertilizer technology, London, 19–21 Jan 1981. British Sulphur Corporation, Ltd, London, pp 173–193
12. Zardi U (1983) US patent 4372920
13. Cariati V (2009) Nata due volte: L'avventura bella Ammonia Casale. Impiantistica Italiana XXII (5)(Sept-Oct):35–43
14. Zardi U (1986) European patent 202454132
15. Shannahan N (1989) Ammonia loop modernization. Hydrocarb Process 68(January):60–63
16. Giuseppe Guarino, from an edited version of a Casale SA internal company interview, June 2022
17. Sutherland L, Wallace B (1988) Improvements to CFI's ammonia converters utilizing Casale's internals. In: Proceeding of the 38th annual meeting. Fertilizer industry round table, November 14–16, 1988, Hyatt Regency, Baltimore, Maryland, pp 59–74.
18. Anon (2008) Ammonia Casale technologies for ammonia plant revamps. Paper presented at the Nitrogen & Syngas Conference and Exhibition, Moscow, Russia, 20–23 April 2008, pp 5, 8. Paper/amm/con/finf/nit/Nitrogen paper 2008 per stampa interm. Accessed 20 July 2022
19. Dean M, Carrera D, Rembold T (2014) The revamping of CF industries #5 NH_3 plant: application of Casale's isothermal technology and Clariant's Amo-Max® Catalyst. Ammonia Technical Manual, pp 208–216
20. thyssenkrupp Uhde AG (2024) https://insights.thyssenkrupp-uhde.com/story/making-the-worlds-largest-ammonia-plant-even-larger/. Accessed 11 Dec 2024
21. Ertl G (2008) Reactions at surfaces: from atoms to complexity (Nobel lecture). Angew Chem Int Ed 47(19):3524–3535
22. Johnson B (2022) Making ammonia: Fritz Haber, Walther Nernst, and the nature of scientific discovery. Springer, Cham. https://doi.org/10.1007/978-3-030-85532-1
23. Hellman A, Baerends EJ, Biczysko M et al (2006) Predicting catalysis: understanding ammonia synthesis from first-principles calculations. J Phys Chem B 110(36):17719–17735. https://doi.org/10.1021/jp056982h
24. Ertl G (2022) My life with science. GDcH, Berlin
25. Liu H, Han W, Huo C, Cen Y (2020). Development and application of wüstite-based ammonia synthesis catalysts. Cat Today 355(15 Sept):110–127. https://doi.org/10.1016/j.cattod.2019.10.031
26. Pattabathula V (2019) Ammonia. In: Kirk-Othmer encyclopedia of chemical technology. https://doi.org/10.1002/0471238961.0113131503262116.a01.pub4
27. Goodwin CM, Lömker P, Degerman D et al (2024) Operando probing of the surface chemistry during the Haber–Bosch process. Nature. https://doi.org/10.1038/s41586-023-06844-5
28. Kamiguchi S, Kiyotaka A, Shibayama T et al (2024) Catalytic ammonia synthesis on HY-zeolite-supported angstrom-size molybdenum cluster. Chem Sci 15:2914–2922. https://doi.org/10.1039/D3SC05447K
29. Zuraiqi K, Jin Y, Parker CJ et al (2024) (2024) Unveiling metal mobility in a liquid Cu–Ga catalyst for ammonia synthesis. Nat Catal 7:1044–1052. https://doi.org/10.1038/s41929-024-01219-z

30. Kani NC, Goyal I, Gauthier JA, Shields W, Shields M, Singh MR (2024) Pathway toward scalable energy-efficient Li-mediated ammonia synthesis. ACS Appl Mater Interf 16:16203–16212. https://doi.org/10.1021/acsami.3c19499
31. Jewess M, Crabtree RH (2016) Electrocatalytic nitrogen fixation for distributed fertilizer production? ACS Sustain Chem Eng 4:5855–5858. https://doi.org/10.1021/acssuschemeng.6bO14733
32. Bartlett RL, Steele JB (2003) Why U.S. is running out of gas. Time, 13 July
33. Rouwenhorst KHR (2022) Production technology updates: from mega-scale to distributed ammonia. Ammonia Energy Association, Ashburn, VA
34. Anon (2022) KBR launches ammonia 10,000 to meet future sustainable energy and fertilizer demands. KBR Press Release. https://www.kbr.com/en/insights-news/press-release/kbr-launches-ammonia-10000-meet-future-sustainable-energy-and-fertilizer-demands
35. Rouwenhorst KHR, Krzywda PM, Benes NE, Mul G, Lefferts L (2020) Ammonia production technologies. In: Bañares-Alcántara R, Valera-Medina A (eds) Techno-Economic challenges of green ammonia as energy vector. Elsevier Science Publishing Co. Inc., pp 41–84 https://doi.org/10.1016/B978-0-12-820560-0.00004-7
36. Rouwenhorst KHR, Krzywda PM, Benes NE, Mul G, Lefferts L (2020) Ammonia, 4. Green ammonia production. In: Ullmann's encyclopedia of industrial chemistry. https://doi.org/10.1002/14356007.w02_w02
37. Anon (1950) Reversing the Haber process. Chemical Industries 67(Oct):575
38. Reed S, Eddy M (2024) A way to make steel without fossil fuels. New York Times International Edition, 3 May, p 7
39. Anon (2024) JERA, Exxon to explore development of hydrogen and ammonia production project in US. Reuters, 25 Mar. https://www.reuters.com/business/energy/jera-exxon-explore-development-hydrogen-ammonia-production-project-us-2024-03-25/. Accessed 25 Mar 2024
40. Anon (1960) Market newsletter. Chem Week 78(18)(20 Oct):104
41. Applegate FL Jr (1965) Ammonia is easy to handle. Chem Eng Prog 61(1)(Jan):66–71
42. Anon (2024) New project would make Kemmerer home to $2.5 billion coal-to-ammonia plant (2024). Cowboy State Daily. https://cowboystatedaily.com/2024/04/04/kemmerer-could-become-home-to-2-5-billion-coal-to-ammonia-plant/. Accessed 5 May 2024
43. Anon (1954) Starting up a new ammonia plant. Pet Refin 33(Dec):168–170
44. Greenberg E, Christopher T, Hill CT, Newburger DJ (1979) Regulation, market prices, and process innovation: the case of the ammonia industry. Westview Press (reprinted Routledge, London, 2019)
45. Rueda S, Miltrup P, Adekanbi A (2024) Ammonia, nitrogen, and green hydrogen production & purification. Gesellschaft für Internationale Zussamenarbeit (GIZ), Bonn and Eschborn. International-PtX-Hub_202401_Ammonia-nitrogen-and-green-hydrogen-production-and-purification. Accessed 20 Nov 2024

Index of Names

A. S. Travis, *The Global Synthetic Ammonia Industry*, Perspectives on the History of Chemistry, https://doi.org/10.1007/978-3-031-87978-4

General Index

A. S. Travis, *The Global Synthetic Ammonia Industry*, Perspectives on the History of Chemistry, https://doi.org/10.1007/978-3-031-87978-4

The manufacturer's authorised representative in the EU is Springer Nature Customer Service Centre GmbH, Europaplatz 3, 69115 Heidelberg, Germany. If you have any concerns regarding our products, please contact ProductSafety@springernature.com

Printed and bound by CPI Group (UK) Ltd, Croydon, CR0 4YY
30/09/2026
02225571-0009